# A HISTORY OF THE FRENCH NAVY

SHIP OF THE LINE: TIME OF LOUIS XIV. Rigged and with guns aboard, she is ready to slip and make sail.

E H JENKINS

# A HISTORY OF THE FRENCH NAVY

## From its Beginnings to the Present Day

MACDONALD AND JANE'S

First published 1973
ISBN 0356 04196 4




Published by Macdonald and Jane's
(Macdonald and Company (Publishers) Ltd)
St Giles House, 49/50 Poland Street, London W1A 2LG
Printed in Great Britain by Tonbridge Printers Ltd
Peach Hall Works, Tonbridge, Kent

# CONTENTS

# LIST OF PLATES



## LIST OF MAPS



## BATTLE PLANS

# PREFACE

As the Naval Service of France has been so long-lived and so important, it is strange that no history of it has hitherto been written in English. The British reader usually knows only those parts of its record that have been concerned with operations against, or in conjunction with, his own navy, from whose standpoint he sees them. In order to present the story as much from the French viewpoint as historical impartiality allows, I have worked largely from French authorities. When they have differed from British ones, or from each other, I have given the fact or figure that seemed most reliable, and all facts given will be found in the authorities listed in the Select Bibliography at the book's end. Such judgements as are made are my own. That upon operation 'Torch' may be thought to show an anti-American bias. Such is not the case: no one could more recognize how much Europe owes to the United States for its part in overthrowing the Axis; but many Americans will agree that their government and its commanders handled the African situation and landings less well than they dealt with some other problems of those difficult times.

This is a history of France's navy – of her official seagoing force. Some French maritime historians do not separate naval matters properly so-called from nautical affairs as a whole. The border between genuinely naval doings on the one hand and, on the other hand, colonization and, above all, the activities of privateers is much less clear-cut where France is concerned than it is with most other maritime powers. Indeed, her privateering and her naval policy have been at times indistinguishable, and because naval personnel and ships were so often used for privateering and colonial occupation, it has been impossible entirely to exclude these things. In the main, however, this record has kept within its proper bounds of a History of the Navy – Royal, Imperial, or Republican – of France.

Footnotes purely of a reference type, such as *Bibl. Nat.* or *AM* (*Archives Maritimes*), followed by a catalogue number, have been avoided as far as possible. Those who want footnotes which would lead to detailed research will find them in the works given in the Bibliography. Apart from a few outliers almost all the places and parts of the sea that are mentioned will be found on one or other of the four maps incorporated in the text. It is regretted that it proved impossible to provide suitable pull-out maps, so that the reader who wishes to be sure of his whereabouts may have to do a certain amount of turning back or forward. The names of French ships have been put in small capitals throughout and a figure in brackets after a ship is that of the number of guns that she carried. The names of ships of other countries are in italics.

Strictly speaking, after 1713, when the French tacitly recognized the recent Union of England and Scotland by abandoning their support of the Stuarts, one should write of Britain and British rather than of England and the English. But at least during the eighteenth century the French thought and wrote of their enemy as England and the English, as, indeed, many of their subsequent historians have done. So that terminology has mainly been used in chapters up to the Revolutionary Wars.

Many people have kindly and generously helped me. Among them I have particularly to thank M Henri le Masson, of the Académie de la Marine, who has given useful information and advice, has supplied some material and corrected some errors. For assistance in collecting material I have especially to thank Mr Norris, librarian of the Hyde Institute branch of the Barnet borough library, and the ladies of his staff, and I have also to thank Mr Leslie Fielding and my son. For assistance in dealing with correspondence I am grateful to Mr K W Carter, and for help in getting the book ready for the printer I am much indebted to Dr A N Worden, fellow of University (now Wolfson) College, Cambridge, to my old colleague, Miss J R Strongman, to Mr Neil MacKechnie, to my wife, and to Mr P C Found of my publishers. Will the several others who have aided me to a lesser extent please believe that I am grateful to them even though I do not mention them individually. For the illustrations in the plates, and for permission to reproduce them, thanks are due to the Musée de la Marine, Paris.

The work was designed to start with the naval revolution brought by the introduction of cannon soon after the close of the Middle Ages, to continue up to, and through, that caused by the appearance of steam-powered ironclads, and to end on the eve of the last naval revolution, due to nuclear power and guided missiles, which followed on the heels of the Second World War. This left the most recent twenty-five years or so unaccounted for, and they have been covered in an appendix by Mr Antony Preston, who has dealt with them more authoritatively and capably than I could have done.

This book has been a long time on the stocks, for it is long ago indeed that my old friend Dr F H Lawson, sometime Professor of Comparative Law at Oxford, suggested that I should write it. Then, and for many years to follow, I was busy with professional work which left me good time for voluminous reading and much note-taking, but gave me no clear break in which to turn to the more solid task of writing. That had to wait till I had retired. I trust that the product of so long a period of gestation will not be found to be too ill-favoured.

Ernest H Jenkins

CHAPTER I

# Introduction — Early Days

When one speaks of a navy, one thinks of a force of ships specially designed for war, manned by crews specially trained to fight them, operating from bases specially equipped to maintain them and to make good their wastages of munitions and man-power, the whole being administered and directed by a central organization or 'Admiralty'. Considered in that way the Navy of France begins its story with Richelieu, who became his country's all-powerful Chief Minister in 1624.

Certainly there was much sea warfare before that date[1], but of this only a little, and that in the later years, can be called naval operations in any modern sense. In medieval times the clumsy single masted ships – cogs and nefs – that fought along or off the Atlantic and Channel coasts were hired or conscripted trading vessels, hastily fitted out for war by the erection of structures ('castles') at bow and stern to hold the missile men while the body of the vessel held the mailed knights and men-at-arms. The sailors were mere underlings who worked the ship and were rarely expected to take any part in combat. The sea fights, therefore, even a great battle like the English destruction of the large fleet that the French had collected at Sluys on the Flanders coast in 1350, were different from land battles only in the absence of cavalry and in being limited in space to the decks of the ships that were boarded. From time to time there were a few royal ships, but they differed little if at all in design from the merchantmen; and the merchantmen on their ordinary avocations were to some extent fighting ships, for there was much piracy, and that of a very cruel type – as was all sea warfare: Froissart calls it 'felon'.

Of strategy, of the use of sea-power, there was really none. When the mercenary, Eustace the Monk, was bringing reinforcements to the Dauphin, who was operating in England in 1217, he was intercepted and defeated in the Straits of Dover by Hubert de Burgh. Such an interception was a rare thing: for the most part, when a fleet was collected, it transported troops to the enemy's shores, or raided them, without opposition at sea. The Dauphin came to England with forces to help the barons who were in revolt against King John. France sent aid to her perpetual ally, the Scots, the English sent army after army across to France before, during, and after the Hundred Years War, all with little or no attempt to meet the invader on the water. Even tactics were limited to such simple stratagems as steering to get the sun into the opponent's eyes, or timing the attack on a fleet in a roadstead so as to have the flood-tide behind one, in order to hit its ships with a

[1]See the first four volumes of De La Roncière, *Histoire de la Marine Française*.

momentarily disconcerting crash which would make grappling and boarding the easier.

Of course there were the galleys. In great contrast to the dumpy sailing ships, which were broad in the beam for their length, the galleys were long and lean, and fitted with a ram. Being propelled by banks of oars pulled, as a rule, by convicts or slaves, they were independent of the wind, and so were much dreaded by sailing vessels in a calm. They were often hired by the French from Genoa and other Italian states. For medieval France was so highly feudalized, and the big vassals so independent, that for many decades the king's rule was far more nominal than effective over most of his realm. For a long while, indeed, the French kings controlled very little of France's Atlantic seaboard, and none at all of that on the Mediterranean. Nearly all the Atlantic coast was part of the inheritance of the Angevin rulers of England until their hold on much of western France collapsed during the reign of King John in the early thirteenth century. Even then the English retained their rule in Guyenne, and Brittany remained a large separate fief, too independent to be bidden. As to the Mediterranean, it was not till the reign of Louis VIII (1223–6) that the king's direct control reached its shores; and then it held only the western part of them, that of Languedoc, which was too devoid of natural harbours to be of any but minor maritime use. It was not till nearly forty years later that the much more welcoming eastern part, the shore of Provence, was acquired, though even so it was not for a further two hundred years (1480) that the monarchy ruled it directly: till then it was under the not always biddable vassalage of the House of Anjou. Nevertheless France had obtained a fairly good hold over its capable seamen and their galleys by the late thirteenth century, and thenceforth maintained a galley fleet, which varied greatly in number and condition from time to time. These galleys could be, and were, brought round into the Atlantic when needed; but designed for the Mediterranean and to be not over far from its coasts at that, they were not really suitable for the wilder waters of Biscay and the English Channel.

In the fifteenth century the medieval period drew to its close, giving place, with increasing rapidity as the century advanced, to what is sometimes called the Age of the Renaissance. That vigorous intellectual movement, with its interest in mathematics and in the place of the Earth in the universe, profoundly affected the knowledge and practice of navigation. Rudimentary charts, and nautical instruments which were by no means all rudimentary, appeared in increasing number. The age of scientific navigation had begun.

There were two further changes which entirely altered naval affairs. One was the development of shipbuilding, in which there was a host of experiments which produced various kinds of vessel with various and intriguing names. The main direction of the experiments lay in increasing the length of the ship in relation to its beam, and adding more masts, with more numerous and differently shaped sails with, in consequence, a much more complicated rigging plan. The whole was aimed at, and achieved, greater speed and more manoeuvrability. Ships were now built specially for war. Fairly rapidly the larger of them (which were small compared with those of the seventeenth century) merged into three main types: the 'high-built' galleon, the galleasse (which combined oars and sail) and the 'round' ship. The galleasse was comparatively short-lived, and by the seventeenth century the other two types had merged into the ship (though Spain retained her galleons for some while). Below these larger vessels were numerous smaller ones of various kinds of which some were called frigates. Originally these were merely lesser sized vessels, and it is not until the late seventeenth century that one finds frigates settled into the type that remained till the end of sailing-ship days – three-masted, ship-rigged vessels, faster but smaller and more lightly built and armed than a ship-of-the-line. The galley remained, increasingly restricted to the Mediterranean and to minor activities, until the early middle of the eighteenth century.

The other great change was the development of the gun from the awkward and dangerous wood and leather bombards of the fourteenth century. Here, too, there were many experiments which produced cannon, demi-cannons, culverins, sakers, minions, and so on down to little murdering pieces. By the early seventeenth century all these, for the most part, had become simply guns or cannon, graded and named from the weight of the shot that they fired as twelve-pounders, eighteen-pounders, and so on. But in their more varied experimental forms they came into use on ships in the early years of the sixteenth century.

These two changes revolutionized naval warfare. No longer was the object to ram the enemy or to carry him by boarding, though it took the French a long time to learn that the latter was becoming *vieux jeu*, so that they continued for some while to practise it whenever possible, and even retained a hankering after it well into the eighteenth century. On the contrary, though ranges were still short, and firing a cannon a slow business, one could manoeuvre the improved ships to keep clear of one's opponent and batter him so badly by gunfire that he could no longer resist boarders or reply adequately to one's fire, wherefore there only remained to him to escape back to harbour if he could crawl away, or to strike his colours if he could not do so.

The slow improvement in tactics brought an even slower, yet positive, introduction of strategy. There had always been French schemes for the invasion of England, but they had formerly consisted of little more than collecting enough ships to convey enough troops, with no worry about interception. Why should there be? For if they met English ships on the way they could fight their hand to hand battle on sea almost as readily as on land. This was the case up to the vast expedition that Charles VI organized, but never sent to sea, in 1386. By the time of the next major expedition against England, in 1545, we are in the reign of that typical Renaissance monarch, François I, and the naval revolution in shipbuilding and gunnery has taken place. There have already been something like proper combined operations in the Mediterranean, using the galleys, and there is something like proper planning of this attack on England. A seagoing soldier, d'Annebault, was in command of superior forces to the English, and his first object was to draw the English fleet out of its protected anchorage in Spithead in order to defeat it in more open water, so that he could land his troops. This was quite a new conception of warfare by the French. As the English fleet refused to be tempted the expedition achieved nothing more than a temporary occupation of the Isle of Wight: d'Annebault did not dare to put his troops on to the mainland while the English ships could sally forth to interfere. Twelve years later, the fleet gave proper co-operation in the capture of Calais from the English.

This was in the reign of Henri II, a ruler who made a serious and largely successful attempt to introduce some regularity into his naval affairs by having ships built to just a few patterns, in 'classes', that is to say; which meant that the motley variety of guns, too, had to be grouped into grades. Fleets were still 'naval armies' and were commanded by soldiers; but some of these were beginning to show signs of being real naval officers. There had been one such even in the later years of the medieval period in the person of Jean de Vienne in the fourteenth century. This admiral fought no major battle in the open sea, but his attacks upon English towns and on fortified places on the Mediterranean coast were genuine naval operations. The first half of the sixteenth century saw another such officer in Prégent de Bidoux (known to the unlinguistic English as Perry John) even though his service was mainly in galleys.

But if Henri II got some measure of uniformity into the building and armament of his ships, he did nothing to unify the administration. France had no proper naval control of a central kind: on the contrary, she had four admirals. This term, which comes from the Arabic through Byzantium, appears first to be used in France by Saint Louis when, organizing his crusade at Aigues Mortes in 1248, he appointed three officers with seagoing experience to command divisions of his fleet. It continued in use before, during, and after the Hundred Years War, as applied to a temporary office, ie just for the duration of the command of a fleet. Hugues Quiéret, who was murdered in cold blood after his defeat and capture by the English at Sluys, held the title; and Jean de Vienne saw that the legal rights of one who held it were fixed by law in 1374. There was later an Admiral of France, the office being apparently well established by 1472. Yet he was only a regional officer, for three great regions which had for long been separate from the French Crown had established their own admiralties. Brittany, a virtually independent fief till its incorporation with the Crown by the marriage of its Duchess Anne to Charles VIII, had its own admiral by or before 1350; Guyenne, as a French province long ruled by the English, had its own admiral; and Provence, virtually independent till it fell into the hands of the French king by the will of Charles II of Anjou, had its admiral of 'the Levant'. (It should be noted that this term was retained by France for commands and for fleets in Mediterranean waters, just as the expression Ponant was used for waters off the Atlantic coast and in the Channel.) The Admiral of France endeavoured to take these admiralties into his own hands, but was stoutly and successfully resisted, though he was recognized as in command on such few occasions as he went to sea in company with ships from Brittany, Guyenne, or 'the Levant' in addition to those of his own admiralty. He was, of course, virtually always a soldier. Guise held the office, so did the Huguenot general Gaspard de Coligny; and each did much land fighting but went to sea on only one, or at most two, occasions. To the holders of the four admiralties sea service was the least important thing: the collection of dues, customs, and tolls (of which they received at least a part), and the sale of licences and of appointments, were the things that mainly mattered to them; and in these concerns of 'admiralty' they obtained something close to legal independence.

Manning the infant fleet, too, was never quite satisfactorily arranged. Indeed, to depart from chronology for a moment, it should be emphasized that, in the days of sail, manning was never an easy matter, especially in later days when the French navy had many ships. There used to be, and perhaps still is, an opinion in England that the French sailor was a poor seaman. This is just not true. He may not have been so careful in the coiling down of ropes and in other bits of spick-and-spannery dear to British tars, but the seamanship of the real sailor was excellent. The trouble was that there were not enough of him. No country with a large fleet could afford to keep all its ships fully manned in peacetime; and though countries might complete their wartime crews by shipping soldiers, convicts, and 'quota-men', as England did, it was upon taking men from their merchant ships that they chiefly relied. France was so largely a self-contained country that she had less incentive to promote overseas trade than England or Holland, and her merchant marine was much smaller. Nevertheless, though small, the French trading marine was active. Bretons were fishing the banks off Newfoundland before 1509, Frenchmen appeared in Brazil as early as 1503, and in the Indian Ocean in 1527. Above all, the Malouin, Jacques Cartier, reached the mouth of the St Lawrence in 1534 (see p. 30). Such voyages could not have been made, especially in the ships of those days, without good sailormen. Yet, again to go ahead of chronology, in the eighteenth century English ships frequently out-sailed their French equivalents. As the French ships were the better designed, this can only have happened because the English vessels were the better handled; in other words, the French crews contained too large a proportion of inferior seamen.

Though Henri II had taken the first steps towards creating a modern navy by instituting some homogeneity of ships, and something of a true naval policy in the use of them, these steps led nowhere. Henri II died in 1559, of a mortal wound dealt by accident in a tournament, and was succeeded in turn by each of his three ineffective sons. During their reigns France became riven by religious civil wars which, with intervals of uneasy truce and spiced with Spanish invasion, lasted until the Peace of Vervins in 1598. The last Valois king had then been slain nine years before, and had been succeeded by his distant relative of Navarre as Henri IV. Since the new king was the leader of the Huguenot party, he was able to end the long strife only by eventually making a politic and hypocritical change of his faith, combined with buying off the Catholic leaders. During this long period of strife which all but brought the country to ruin, everything naval had been allowed to fall by the wayside, and no royal ship able to put to sea was left. The French ambassador going to England had to sail in an English ship; and the one French vessel that provided anything of an escort suffered the humiliation (which Frenchmen took a long time to forget) of being obliged to lower its flag in salute to British supremacy at a gunshot from the Englishman. Nor, being much more concerned with the urgent questions of healing the internal wounds of France, and conducting a foreign policy on land, had Henri IV been able to take any naval measures worth mention by the time that he was murdered in 1610; and this assassination meant that for some years France suffered a weak and faction-ridden regency during which, also, nothing naval was done. A dozen dilapidated galleys were the royal naval force of France.

When Henri had brought peace to France by changing his religion he did not, indeed was not allowed to, forget the interests of his former Huguenot companions in arms, and by the Edict of Nantes he granted them, not only the right freely to observe their religion, but also the right to garrison a number of towns, largely in the south-western part of France, where Huguenotism was strong. For the moment this settlement was wise (and possibly unavoidable), for it brought much needed peace to the land. Yet, as it created a state within a state, it could not be for all time, the more so as the Huguenots were as much a political party as a religious body, and contained a large number of those nobles who were the most anxious to preserve their excessive powers and privileges against any centralization by the Crown. Of the Huguenot towns perhaps the most important, and certainly the most independent-minded, was La Rochelle. A busy seaport, it was strongly fortified to landward and had a fleet of ships which, partly to oppose pirates, and partly because of the addiction of the Rochellais to privateering, were so armed that they might well be ranked as ships of war. In 1620 the Huguenots rose in rebellion, aiming to establish a separate republic. The royal campaigns against them on land ran into much trouble before they ended in victory, and meanwhile the Rochellais were privateering unchecked. To deal with them the king collected at Blavet (Port-Louis, in Brittany) a motley fleet of such galleys as he had, brought round from the Mediterranean, a number of merchantmen supplied with some guns, a large galleon obtained from the Knights of St John in Malta, and another such belonging to the Duc de Guise, Admiral of the Levant, who was in command. A major battle was fought in the Pertuis Breton – the water to the north of the Isle of Ré, off La Rochelle – and the Rochellais had the worse of it. But as their 'admiral', Guiton, got most of his ships away, it was inconclusive. So was the general position when peace was made at Montpellier in 1622. The Huguenots had to give up their right to garrison, except in Montauban and La Rochelle, but they abated no whit of their desire for independence, especially where the Rochellais were concerned, and waited only their opportunity and, if possible, some allies to strike again.

Such was the state of affairs when Richelieu came to virtually complete power in 1624.

CHAPTER II

# Richelieu Creates a Navy

If France was to be a great naval power, geography had given her some advantages but had balanced these with some handicaps. Her Channel coast put her in a position, were she strong enough, to interrupt the bulk of the trade of the two great sea-carrier nations, the English and the Dutch, while her Atlantic and Mediterranean coasts gave her ready access to what were then the two great trading seas of the world. Against this, the separation of her Atlantic and Mediterranean coasts by the length of the Pyrenees meant that she had to maintain two fleets, and their junction was not easy, seeing that one or other would have to round the whole of the long Spanish seaboard. Nor had she many ports suitable for fleet usage, when ships had grown, as they rapidly did, to the large size that was common before the end of the seventeenth century. Ports there were in plenty that could take small ships, and even one or two large ones, but, in days before steam tugs, getting a large ship in and out of an actual harbour was a tricky and tedious business. What was needed was a roadstead in which a fleet could lie, and such a roadstead needed to provide reasonable shelter from gales, to be close to an actual harbour which could be a dockyard, and to be fortifiable in order to protect the fleet against an assault from seaward. For all this Brest was magnificently suitable, and Aix roads, behind the little island of that name and between Rochefort and the large island of Oléron, ran it a fair second; but others on the Atlantic coast were far less to the purpose, though Cherbourg could have been made so. In the Mediterranean Marseilles was left to commerce and the galleys, and Toulon with its easily fortifiable roadstead became the base for royal ships.

When Richelieu came to office he had already in mind the necessity for France to be a real sea power and the vision of her becoming so: even if he had not, the activities of the Rochellais were almost immediately to provide a sharp lesson in that matter.

The Peace of Montpellier had not been entirely honoured by the Crown, and in particular some royal fortifications near Rochelle had not been demolished, as the terms of the peace had promised. When, therefore, Richelieu purchased six ships of war and harboured them at Blavet (Port Louis), though they were in fact intended for use in the Mediterranean against the Barbary pirates, the people of La Rochelle felt it to be a threat and the last straw. Their leader was that ardent trouble maker, the duc de Soubise, and, arming eleven Rochellais ships, he made a sudden descent on Blavet in January 1625, took Richelieu's little fleet by surprise, and carried it off, though two of his captures and a few of his own ships were lost in getting out of harbour. Indeed, he might have found himself entirely trapped but for the almost traitorous slowness to act of the Admiral of Brittany. With these captured ships, having established himself on the Isle of Ré, Soubise proceeded to play the pirate on the shipping and coasts from the Gironde to the Loire, and the king had nothing with which to oppose him. Nor could La Rochelle be brought to obedience save by a siege which would be pointless as long as the city's sailors could keep it supplied by sea. To obtain a fleet Louis and Richelieu were in the humiliating position of having to beg from their allies, getting twenty ships from the Dutch and eight from England, some of which were wholly or largely French manned, others retaining their native crews. French ships were joined to these, and the Admiral of France, the duc de Montmorency, took command, the Dutch admiral being his second. Though by a neat but unwise piece of sheer treachery Soubise succeeded in destroying the Dutch flagship, when this fleet set sail from Les Sables d'Olonne, Guiton, the Rochellais 'admiral', took refuge in the tricky shallows of the island of Ré. Montmorency executed a well-planned landing of his troops, who carried the island in a couple of days. Guiton had now to get his fleet away, and his attempt to do so led to a fierce battle in the Pertuis Breton. LA VIERGE, a forty-eight-gun ship that had been the pick of those captured by Soubise at Blavet, blew up, taking with her the four king's ships that were attacking her, and ten others of the Rochelle vessels were lost. Guiton managed to escape, to La Rochelle, as did Soubise, with a few hotly pursued ships, into Falmouth.

This royal victory was by no means to the taste of the English and Dutch, who felt that they had lent ships and men to defeat their fellow Protestants. So in the next year the Dutch recalled their vessels, while the popular wrath in England was such that Razilly, the officer who returned the loaned English ships, endeavoured to appease it by saying these ships had taken no part worth mention in the battle!

England, indeed, shortly changed sides. The showy duke of Buckingham was directing affairs: he had his own somewhat personal quarrel with France, sought popularity (for he was disliked), and thought that France's utter weakness at sea would provide him with a cheap success. So, in 1627, having already seized a number of French merchantmen, he got King Charles to declare war in aid of La Rochelle, and arrived off the Isle of Ré with a fleet of ninety ships and troop transports[1]. Strategically the idea of occupying the island was sound, for while the English held it a blockade of La Rochelle by the French would be almost impossible, and as long as the Rochellais had access to the sea, the land siege would drag on as a large, and exhausting, thorn in the side of France. Buckingham landed his troops on Ré and the small French garrisons there retreated into its two fortified places. It fell to Richelieu (who had now taken charge of everything maritime) to maintain them. Fortunately for him Buckingham's blockade was singularly inefficient, and a number of small vessels manned by daring Basque seamen, which Richelieu had collected, ran the blockade by night with provisions and small reinforcements, and brought back the wounded. Time and again the garrisons were resupplied when at their last gasp and

[1]Some English authorities give a smaller number – seventy-five or seventy-nine, fifty being transports.

about to surrender. Eventually, Richelieu succeeded in getting a convoy of ships carrying troops to battle its way through by night, with amazingly little loss; and Schomberg, who was in command, swept the English off the island in some panic and with heavy casualties. Against the indomitable will of the cardinal, the somewhat happy-go-lucky Buckingham had failed, and failed disgracefully. He knew it, and took his fleet home.

But the retreat of Buckingham by no means involved the fall of La Rochelle, which could still keep itself going because of its access to the sea; though the story of its fall has a place in a naval history only as showing the plight of a fleet-less France. With the English certain to return, and with no fleet to oppose them, Richelieu embarked upon an operation as remarkable in its day as were the 'mulberry' harbours that facilitated the Normandy landings in the Second World War. He built a mole across the mouth of the waterway that led from the sea to the town. To be more exact, he built two moles, one starting from each shore, leaving a narrow ship entry between them; and as each mole progressed, batteries were placed on it. The basis of these moles consisted of old ships filled with stones and sunk. He collected all the ships that he could – none was large and many were mere light craft – and he packed some of these inside the moles, enough to hold off any attempt by the Rochellais fleet, depleted by the battle in the Pertuis Breton, to interfere with the building; the majority he packed at anchor close outside the mole. There was no question of their moving to attack the English: they were to provide a block of opposition which, with their own fire and that of the batteries, would make a breakthrough by the English highly expensive to attempt. For his great asset was that the English warships were large for their time, and close to the moles and the ship barrier there was very little breadth of water sufficiently deep for them to manoeuvre.

In May 1628 the English reappeared: fifty ships of war under Lord Denbigh with a number of victuallers, and they anchored in the Pertuis d'Antioche, the water on the south-east side of the Isle of Ré. Denbigh was nonplussed by the unusual nature of the obstacles that he had to overcome: he took a week to investigate and weigh the problem, then brought his fleet as close in as he could while preserving room to move, bombarded heavily and, finding that it had no effect, took his ships back home. In September a large fleet with a number of troop transports, under Lord Lindsey, arrived and made a more determined effort. It was equally unavailing. His heaviest bombardment did little damage to the wall of defensive shipping and occasioned only some thirty casualties. He sent in his fire-ships, and some elementary 'infernal machines', a kind of floating mine; but the large number of light craft lying in front of the ships outside the mole grappled with these and drew them safely aside. He could possibly have occupied Ré, but there was no point in doing so without relieving the city. La Rochelle was now at its last gasp of famine, and, with the English lying helpless by the Isle of Ré, it surrendered at the end of October. Lindsey sailed away and peace between England and France was signed the next year. The terms given to the defeated Huguenots were generous: their right to garrison was taken away, but the free observance of their religion was left to them, and little vengeance was taken. Indeed, Guiton, mayor and naval commander of Rochelle, later took service in the fleet that Richelieu was busy building.

For, in October 1626, Richelieu had had himself appointed 'Grand maître, chef et surintendant général de la navigation et commerce de France', in other words, absolute master of everything maritime; and the Parlement of Paris (a corps of lawyers whose registration of an edict was necessary to turn it into law) registered the title early in 1627. Internally, Richelieu's policy was that of an ardent and ruthless centralizer. As such he could control both finance and the troublesome *noblesse*. He was not a reformer in the modern sense of the word. Indeed, sprung himself from the greater nobility, he was not the least disposed to act against his class – save that its powers and privileges must not get in the way of the power to be exercised by the Crown. In foreign affairs a fervent patriot who wished to see his country 'great' for the sake of greatness, he aimed at extending its boundaries to reach its 'natural frontiers' – the Alps, the Juras, and the Rhine. This meant conflict with the House of Austria, the Hapsburgs, whose Spanish branch ruled not only Spain but also the Kingdom of Naples, the Duchy of Milan, the Spanish (or southern) part of the Netherlands, and the Franche-Comté – that large province, part of France now for three hundred years, which stretched from the Juras up to the southern frontier of Lorraine. The last two lay within the boundaries which he wished France ultimately to reach. Not only was Spain's long coastline vulnerable, but her lines of communication with these possessions lay by sea – either direct via the Channel, or to Genoa for the Milanese and the Valtelline. To cut the communications France needed a fleet, and a powerful one, since there might be interference from both English and Dutch.

The first step was to get rid of a feudal survival, the four admiralties, whose four 'admirals' were great nobles who rarely went to sea, if at all, but made much profit from the various rights and powers of their office and stood seriously in the way of effective organization. They were bought out by the end of 1627, and the title of Admiral of France was then suppressed. Orders were given for the building of ships, and ports had to be chosen and fortified as their normal bases. These were Toulon, Brest, Le Havre and Brouage. The last two were not good choices, indeed Brouage, to the south of Rochefort, became so sand-blown as to its fortifications, and so silted as to its harbour, during the troubles that came with the minority of Louis XIV after Richelieu's death, that no attempt was made to bring it back to use, and Rochefort took its place. Inadequate roadstead was the trouble at Le Havre, which therefore gradually became more of a Royal Shipbuilding Yard than a naval base.

In 1631 Richelieu caused a great survey of the whole of the Atlantic coast (including the Channel) to be made by the Sieur d'Infreville, an able official. It took two years and revealed a rather sad state of affairs, though one in which there was a hopeful feature or two. It reported that the seafaring population was declining, some sailors even leaving France for service elsewhere, that the number of merchant ships suitable for conversion to ships of war in an emergency was small, that maritime store-houses were empty and in ruins, and that many ports were in decay, with crumbling quays and silting channels. On the other hand the number of ships built or building was forty-one and there were two good gun foundries. When this survey was nearing a close, in 1633, the shorter Mediterranean coast was given a brief but comprehensive inspection. Here many merchantmen were armed, for fear of the Barbary pirates, but only five were suitable for use as fighting ships in a national war: moreover, the fortifications along this coast, so subject to piratical raids, were nearly all in a poor way, so much so that in one coastal tower the 'garrison' consisted solely of a caretaker! Steps had already been taken to remedy all this by the *règlement* of 1631 which organized the four naval bases. Each was to have a commissary-general and a chef d'escadre. The latter was to deal with ships, the former with the supplies for them and with finance. Apart from the ports, the coast was divided into eight districts each under a lieutenant-general – a civilian deputy of the 'grand-maître'[1]. Over the whole was Richelieu's immediate lieutenant – the Secretary-General of Marine with his small Marine Council. As a result of the vigorous efforts of all concerned, France, by 1636, had in the Ponant (Atlantic bases) thirty-eight ships, six fire-ships, twelve *flûtes* (naval store-ships or tenders) and some smaller vessels, while in the Levant (Mediterranean) there were thirteen ships and twelve galleys. By 1642, the year of the great cardinal's death, these numbers had risen to a total of sixty-three ships and twenty-eight galleys. The pride of the fleet was the COURONNE of 2 000 tons and seventy-two guns, with a crew of 600. She was French built in 1638; but a number of Richelieu's ships of the line were of Dutch construction.

To man the fleet the Ordonnance of 1629 (the 'Code Michaud') made provision for the compiling each year of a list of all seafaring men in their various grades, such as captains, pilots, carpenters, caulkers, gunners, deck-hands etc, and for lists also of all vessels and available guns.

This creation of a navy had been a stupendous piece of work, but it had a grave fault: the Administration was too powerful in comparison with the actual seafaring side. The jealousy and rivalry thus created between the *plume* (the pen, the administrative officers) and the *épée* (the sword, the actual fighting seamen) was to bedevil the French service for more than a hundred years. There was also another weakness, the shortage of officers. The main source for these was the galleys of the religion, that is, the ships of the Knights of St John who, from Malta, waged perpetual war against the Barbary pirates, and who, at this time, were largely a French society. But there were not enough to go round, and though it was the practice of the time that every royal officer should be of the nobility, the gaps had to be

[1]Originally administrative officials, the chefs d'escadre and the lieutenants-general soon became seagoing officers, commanding fleets and squadrons. A lieutenant-general then held a position nearly equal to that of an English vice-admiral, and the chef d'escadre held one almost equal to that of an English rear-admiral. (NB As these ranks will continually crop up in the names of those leading fleets or divisions, the abbreviations LG for lieutenant-general and CE for chef d'escadre will in future be used throughout this book.) Their places ashore were later taken by Marine Intendants, officials of such great power that they were subject only to the Minister of Marine or Surintendant.

filled by captains of merchantmen, some of whom were experienced privateers. Richelieu was willing to use their services, but not to put them in high command, for he was very much a man of his class – even in its nepotism. (Several high naval appointments went to his relatives, though he did not hesitate to dismiss those of them who proved unsatisfactory.) It was also held essential that the commander of a fleet should be a man of great social standing, probably rightly so if he was to obtain obedience from an over-proud nobility serving under him; and so the Commandeur[1] Philippe des Gouttes, whom the cardinal considered his most able officer, was used as adviser to an inexperienced chief, but was not himself put in high command. It did become possible, at not too distant a date, for a *roturier* to be given a patent of nobility to enable him to rise to flag rank, as was the case with Abraham DuQuesne, son of a Dieppois merchant captain, born in 1610 and experienced in privateer sea fighting from his youth. But such a man had to be almost more than outstanding; and this tendency to combine command with high nobility of birth was a further bedevilment of the French navy for many years. Nevertheless Richelieu had accomplished a great work, and when he was ready to make his breach with the House of Austria in 1635, he had provided his country with a fleet which, as we shall see, was able to give it splendid service.

[1]The rank comes from the Knights of St John.

CHAPTER III

# Richelieu's Navy at War: The Foundations of Empire

The last of the great religious struggles, the Thirty Years War, had been devastating Germany since 1618. To Richelieu, Prince of the Church and Chief Minister of a Catholic country, the success of the Catholic party, since that was the party of the House of Austria, was unimportant compared with the ascendancy of France, and he had made use of the situations created by the fighting to handicap the Hapsburgs at every opportunity by a diplomacy so unscrupulous as barely to stop short of a state of war. The Protestant disaster at Nordlingen in 1634 obliged him to throw aside such little pretence of neutrality as remained, and in 1635 he formally declared war on Spain. In so doing he brought the wars of religion to an end and substituted those of sheer national aggression which have lasted to our own day.

Though he unquestionably had a sense of the value of sea-power and even a vision of its future, he seems to have given little thought to the details of putting it to use: he had to learn that there was far more in getting a fleet to sea in fighting trim than just the provision of some ships, their guns and a certain minimum of officers. The Spaniards got to sea more promptly and in September 1635 they seized the two islands known as the Lérins, a mere gunshot off the French coast near Cannes, which were so feebly held that the garrison could offer only a few hours' resistance. By this astute stroke Spain had both protected her own route for troops to north Italy by way of Genoa, and had restricted the activity of the French on their own Provençal coast. The recapture of the islands was important and, since the small Mediterranean squadron and its galleys were inadequate to deal with the situation, the whole of the Atlantic squadrons were to be sent round Spain for the purpose; in 1636 they were gathered in the road-stead behind the Île de Ré. These squadrons – Brittany, Guyenne, and Normandy – had as their flagship the NAVIRE-DU-ROI of fifty-two guns, with twenty-four other ships whose armaments ranged from twenty-four to thirty-four guns, as well as a number of smaller vessels, six fire-ships, and twelve *flûtes*[1]. Since this was a full-scale expedition a great noble had to be in

[1]A *flûte* was a storeship, tender, or troop transport, designed for the purpose and usually mounting about a dozen guns. In later years the French developed the habit of sending some of their ships of the line to sea armed *en flûte*: that is to say that a number of the ships' guns – sometimes all except the upper deck battery – were either landed or lowered into the hold in order to make room for stores or for troops. If the guns were in the hold the vessel could be restored to its full strength once it had landed its soldiers or cargo, but till then it was of only partial use as a fighting unit. The expression *en flûte* is so convenient to describe this practice that it will be used throughout these pages.

command, and the appointment was given to the comte d' Harcourt, known as Cadet-la-Perle from a huge earring that he wore, and he duly hoisted his flag though his experience was entirely that of a soldier. Commanding the ships under him, and designed to take over if anything happened to d'Harcourt, was, of all people, the Archbishop of Bordeaux, Henri d'Escoubleau de Sourdis. This prelate, who in the words of one historian looked more like a musketeer than a churchman, was at one time Richelieu's personal secretary, had been firm in conducting a quarrel in his diocese with the all-powerful duc d'Épernon, had shown himself, in the operations round La Rochelle, to be a stout fighter keenly interested in sea affairs, and was clearly a personality of force and action. He led the Brittany squadron, and the leaders of the Guyenne and Normandy squadrons were his vice- and rear-admirals. But they were not his next in command: that office was to fall to his flag-captain, the Chevalier Philippe des Gouttes, an experienced fighting seaman. It seems that Richelieu hoped that des Gouttes would give the necessary advice to de Sourdis who would have the personality to force it on to d'Harcourt; but to give the actual command to his best officer when that officer was no more than a Commander of the Knights of St John was impossible in the social structure of France of that day: he would never have been obeyed. The absurdity of this chain of command was equalled in the Mediterranean squadrons where the general of the galleys (the marquis du Pont-Courlay who was Richelieu's nephew) was at loggerheads with his second in command, Forbin, who in turn was at bitter enmity over family affairs (and France could go hang) with the Baron d'Allemagne who commanded the ships. All were on poor terms with de Vitry, the Governor of Provence, as was d'Harcourt soon after he and the main fleet arrived. With everybody giving orders and nobody obeying, the campaigning season was wasted in a series of unpatriotic and unedifying rows which reached a climax at the end of the year when, at a conference called in an attempt to get at least something moving, de Vitry called de Sourdis a canting humbug and struck him. Truly the French service had some odd growing pains to overcome, and they have been described here at some length because in fact the service never did entirely rid itself of this feudal deference to, and haughtiness of, nobility of birth until the Revolution – when it overcame it excessively.

In the whole of that year 1636 the fleet did nothing to recover the Lérins or to deal the Spaniards any other sort of blow worth mention. In February of the next year de Sourdis took the fleet to sea to make a descent upon Oristano in Sardinia, then a Spanish possession. The idea seems to have been that this valuable harbour should be exchanged for the Lérins. As a fleet exercise for inexperienced officers and crews the attack on Oristano served a good purpose; but it failed otherwise, for, though the French took the place, they were quickly driven out again with some loss and humiliation. Then at last, in March 1637, d'Harcourt got his expedition to sea for its original object, the reconquest of the Lérins. The Spanish had fortified these strongly, but the troops were got ashore (after some delay and loss of equipment owing to foul weather) under cover of the fire from the ships, and the more important island, Sainte-Marguerite, was retaken after a month's hard fighting; the other island, Saint-Honorat, fell when the troops were landed after a week's bombardment of its fortifications. The Spaniards had captured the islands in a day: it had taken over a year's operations, including six weeks of hard fighting, to recover them. The bulk of the Atlantic ships then returned to their bases. A few of the remaining ships made a demonstration off Algiers, meant to scare the Barbary pirates. It did nothing of the sort.

Richelieu had learned the folly of a divided command, and the next year, 1638, saw France victorious both in the Bay and in the Mediterranean. De Sourdis, in sole command of the Atlantic squadrons, had some difficulty in getting his fleet ready for sea: indeed, but for an ordinance that had forbidden all seafaring men to leave France, he would not have been able to man it. So it was not till near the end of July that he sailed from the roads at Ré. A French army under the Prince de Condé had invaded the Basque province of Guipuzcoa and was besieging Fuentarrabia; and de Sourdis' orders were to complete the siege by blockading from the sea – despite his representations that a few ships would suffice for that, and that the fleet could carry out a further ranging and more ambitious programme. A few ships had gone ahead to take over Passages from the army, which had overrun this valuable port in its advance and had barely been prevented from destroying some valuable prizes which it had trapped in the harbour. When these ships rejoined de Sourdis off Fuentarrabia his fleet was forty-one strong plus a number of transports and smaller vessels, including six fire-ships. He had divided it into the usual three squadrons, one under the experienced Razilly which was to maintain a close blockade, one of the ships under Montigny which cruised to the westward to give warning of any approaching Spanish fleet, and the third under his own hand with his flag in the TRIOMPHE.

The invasion had taken Spain by surprise and the only naval support that she had near at hand was the Corunna squadron of twelve galleons and two Dunkirkers under Lopez de Hoces, who was given the suicidal order to sail to relieve and reinforce the besieged town. On sighting Montigny he put into Guétaria, from which he hoped later to seize an opportunity to slip along the coast. The fighting archbishop, leaving Razilly to continue the blockade, had sailed with his own squadron to join Montigny, though it took him some days to make the junction owing to contrary winds. The Spanish admiral took advantage of the respite to make his position as strong as possible, throwing up some batteries, and mooring his ships in a line in shallow water so that they could be attacked only on so narrow a front that the French could not use their full strength.

To overcome this difficulty de Sourdis adopted a plan suggested by his flag-captain, des Gouttes. Montigny was sent in with six ships, L'EUROPE, LE CYGNE, LA LICORNE, LE COQ, LA FORTUNE, and LA VIERGE. He had his flag in L'EUROPE of thirty-four guns, the other ships were thirty-gunners. (Probably it was felt to be unwise to risk the larger flagship of de Sourdis in the shallows.) They were ordered to anchor leaving gaps between each ship, and behind these gaps was a second line of vessels, firing through them as best they could: with five of the leaders went a fire-ship. The position

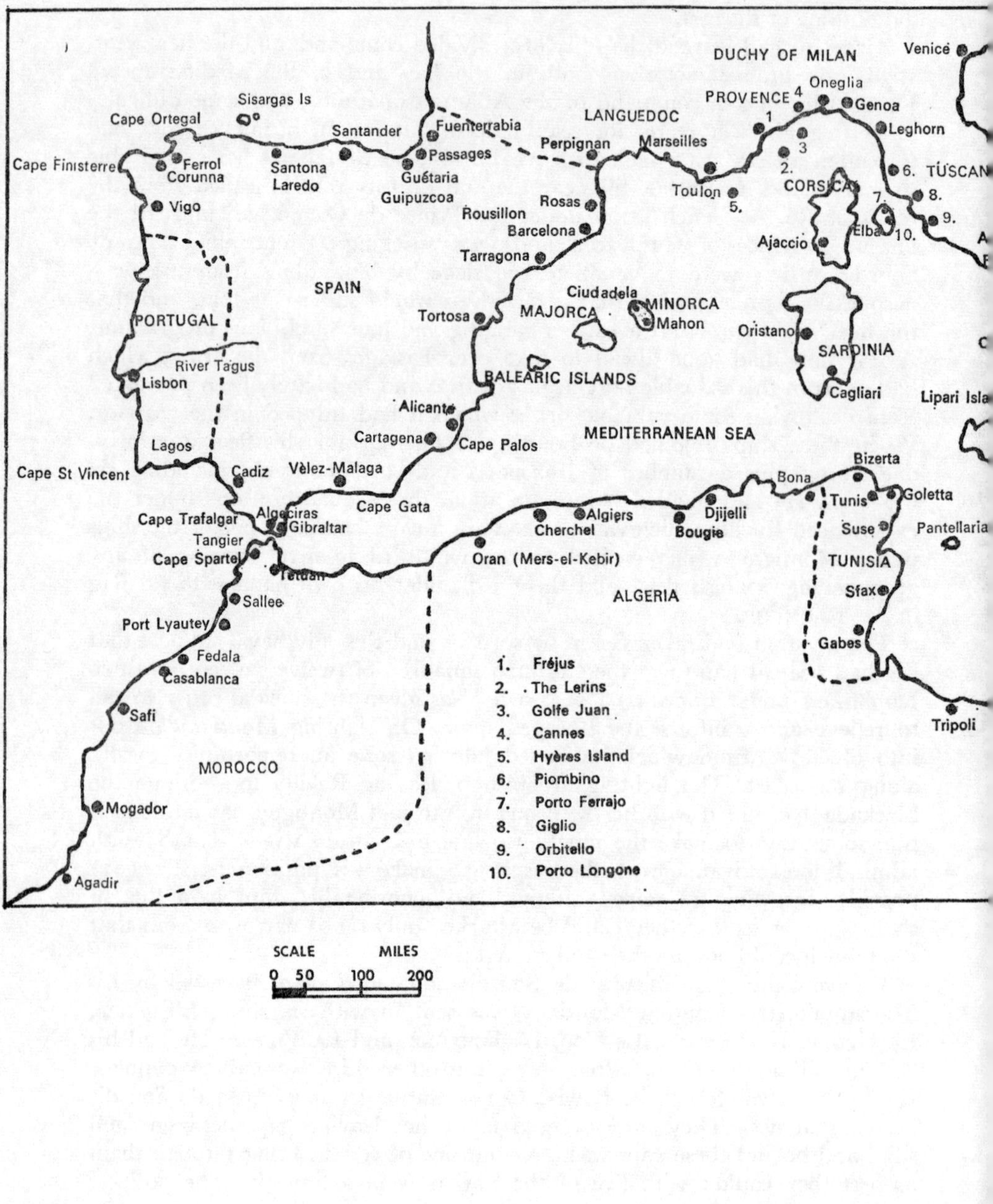
DUCHY OF MILAN
Venice
Oneglia
Genoa
PROVENCE
Leghorn
LANGUEDOC
Sisargas Is
Cape Ortegal
Santander
Fuenterrabia
Perpignan
Marseilles
Cape Finisterre
Ferrol
Corunna
Santona
Laredo
Passages
Guétaria
Guipuzcoa
Toulon
CORSICA
TUSCAN
Elba
Vigo
Rosas
Rousillon
Barcelona
Ajaccio
Tarragona
SPAIN
Ciudadela
MINORCA
MAJORCA
Mahon
PORTUGAL
Tortosa
Oristano
SARDINIA
River Tagus
Lisbon
BALEARIC ISLANDS
Cagliari
Lipari Isla
Alicante
Cartagena
Cape Palos
MEDITERRANEAN SEA
Lagos
Bizerta
Cape St Vincent
Cadiz
Vèlez-Malaga
Bona
Tunis
Goletta
Cape Gata
Djijelli
Cape Trafalgar
Algeciras
Gibraltar
Cherchel
Algiers
Bougie
Tangier
Suse
Pantellaria
Cape Spartel
Tetuan
Oran (Mers-el-Kebir)
TUNISIA
Sallee
ALGERIA
Sfax
Port Lyautey
Fedala
Gabes
Casablanca
Safi
Tripoli
MOROCCO
Mogador
Agadir
1. Fréjus
2. The Lerins
3. Golfe Juan
4. Cannes
5. Hyères Island
6. Piombino
7. Porto Ferrajo
8. Giglio
9. Orbitello
10. Porto Longone
SCALE
MILES
0 50 100 200

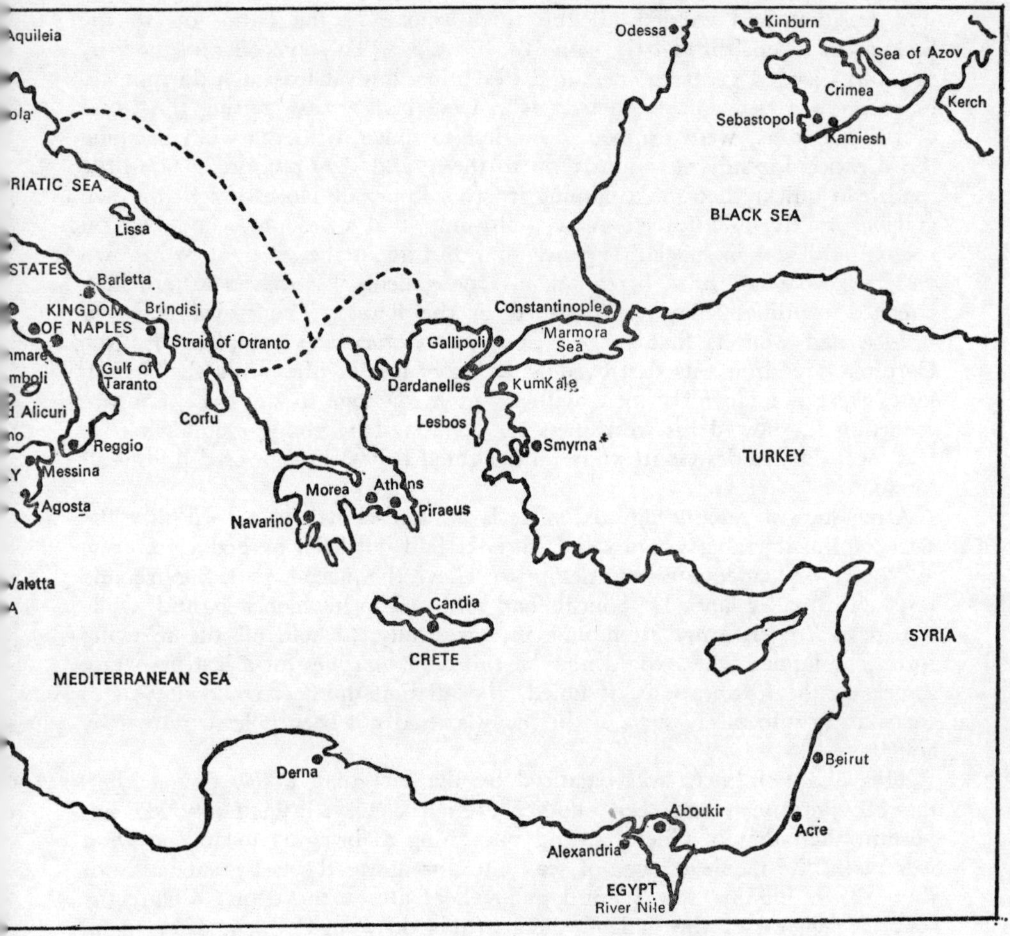
Aquileia
Odessa
Kinburn
Sea of Azov
Crimea
Kerch
Sebastopol
Kamiesh
ola
RIATIC SEA
Lissa
BLACK SEA
STATES
Barletta
KINGDOM
OF NAPLES
Brindisi
Strait of Otranto
Constantinople
Marmora
Sea
Gallipoli
mare
mboli
Gulf of
Taranto
Dardanelles
KumKale
d Alicuri
Corfu
Lesbos
no
Reggio
Smyrna
TURKEY
Y
Messina
Morea
Athens
Piraeus
Agosta
Navarino
Valetta
A
Candia
SYRIA
CRETE
MEDITERRANEAN SEA
Beirut
Derna
Aboukir
Acre
Alexandria
EGYPT
River Nile

at the opening of gunfire was therefore that six French thirty-gunners were engaging a good deal of the Spanish line, that fire-ships were sheltering under the lee of five of them, and that behind them other vessels were getting in such shots as they could without hitting their own front line, thus giving the best concentration of fire that the shallows allowed. The direction of the wind and a heavy swell made it easy for the French to get into position and carried all the thick smoke of the battle on to the Spaniards. Then, through it, came the fire-ships. To stave off an attack by fire-ships needed great coolness and discipline, and at least a little room to move one's vessel. The Spaniards, who were in any case getting the worse of the gun duel, were trapped, powerless to move, with the wind bringing the dreaded incendiaries quickly on to them, and they panicked. The little roadstead quickly became a flaming inferno. Lopez de Hoces fought his own galleon stoutly, and forced away a fire-ship that de Sourdis had kept in reserve and sent in specially against him, but at last he set light to his own vessel to save her from being taken. One galleon was driven ashore in a wrecked condition, the rest perished in the flames. The Royal Navy of France had won its first victory, and that victory was complete; for the Corunna squadron was destroyed with heavy loss of life, while the French losses were a mere forty men, with no great damage to ships. De Sourdis, reporting it, showed his worthiness for command by giving credit where it was most due, to des Gouttes; and a delighted Richelieu showered praises all round.

After such a resounding success it is an anti-climax to record that the siege of Fuentarrabia was a sad failure. It had not been pushed vigorously, and however inadequate her attempt to relieve the place by sea, Spain made no such error by land. De Sourdis had returned to his blockade, and, while the main French army attempted, unsuccessfully, to hold off the relieving forces, he landed what would now be called his marines for a last desperate attack on the fortifications. It failed, and all that the fleet could do was to carry away into safety such of the army as had not been able to retreat by land.

This disappointment was balanced by the fact that, a few days earlier, the Mediterranean had seen another French victory, little if any less important than that of Guétaria, and interesting as bringing to an end what was really the medieval type of sea-fight; for it was the last great battle of the galleys. Finding the Spanish galley fleet at sea the comte d'Harcourt sent his general of the galleys, the marquis du Pont-Courlay, after them. The two fleets met hard by Genoa. Each was fifteen galleys strong, and it happened that the 'general' of each fleet was the eighth ship in the line. The construction of these vessels allowed them to carry very few guns, and du Pont-Courlay's orders were that the French were to keep line on him, hold their fire till he fired, which would be only when close to the enemy, and then each should try to carry his opponent by boarding, coming to the aid of any Frenchman in trouble when he had done so. The result was a dreadful slogging match, with men in armour fighting with sword and buckler, the whole hand-to-hand battle interspersed with pistol and musket shots: a sheer slaughter. From the centre to the van the French had the better of it. Du Pont-Courlay carried the 'Capitana' of Sicily, after a frightful struggle. The three galleys ahead of him also captured their opponents, two others of the Spaniards in the van fled, and only the leading French galley succumbed to the Spanish attack. To the rear of the centre things went less well, for the French line did not hold firmly to its order. As a result, two French galleys were taken; but as two Spanish galleys were also captured, the victory with six captures to three losses was plainly that of the French, especially as the remaining Spaniards retreated into Genoa in disorder, with only four of their nine surviving vessels in fighting trim.

Though this was the last great galley battle, it by no means saw the disappearance of these craft. They had already been abandoned in the Atlantic, for whose heavy seas they were unsuitable, but they remained in the Mediterranean until well into the eighteenth century and for a time were much dreaded by sailing warships in flat calms, for an isolated sailing ship could be overwhelmed by numbers while lack of wind prevented its consorts from coming to its assistance. But as the gun power of sailing ships grew stronger, so the galley, unable to close before being badly shot about, became increasingly an adjunct to a fleet – one which could help to tow a damaged vessel out of action, or into position in a calm – and not a part of its fighting strength.

In the next year, 1639, de Sourdis took to sea a powerful fleet of thirty-seven ships, plus fire-ships and transports, to attack a Spanish fleet of thirty-five ships known to be in Corunna. They were in a strongly defended position, and though a complicated plan of attack was devised by the French, more cautious second thoughts prevailed and the fleet returned to Belle-Isle roads, somewhat battered by foul weather. Later in the year an attack was made on two galleons lying in the river mouth between Santoña and Laredo, fortified places some twenty miles east of Santander. Troops and sailors were landed and took both places, at which the captains of the galleons set fire to their ships: the French were in time to board one of them and extinguish the fire. It may be added in parenthesis that the reason why the Spaniards had refused battle at Corunna was that they were waiting to be joined by a second fleet from Cadiz, when the whole powerful armament sailed north to attempt to restore the situation against the Dutch in Flanders. This was a dangerous venture, and the fleet was trapped by Tromp in the Downs and virtually destroyed, Tromp disregarding the neutrality of English waters, which England made little attempt to assert.

In 1640 de Sourdis was moved to command in the Mediterranean where he did little, thanks partly to the slippery violations of neutrality by the Republic of Genoa (which was in an awkward position between the two major powers), and partly to a violent quarrel with Forbin, his rather unsatisfactory general of the galleys. The command in the Atlantic was given to Richelieu's nephew Armand de Maillé-Brézé, duc de Fronsac, an inexperienced and apparently dull young gentleman, whose seeming gaucheness concealed tremendous drive and fighting *élan*. He had (one almost writes 'of course') des Gouttes as his flag-captain and mentor. He sailed in July with twenty-one ships to intercept the Spanish convoy leaving Cadiz for the New World and successfully did so. At the approach of the French, Don Gomez de Sandoval turned his convoy back for Cadiz and formed a line between them and the enemy with twelve galleons. Maillé-Brézé, with his ships in the usual three divisions, made an all-out attack. There was nothing wrong with the dash and vigour of his assault save that he lost control of his fleet, whose ships got in each other's line of fire, steered out of their proper divisions, and generally produced a clumsy action in which only one galleon was taken, though the Spaniards, as a whole, suffered the heavier casualties and greater damage[1]. Nor did Maillé-Brézé get his orders for the night round his entangled fleet sufficiently quickly, with the result that at daybreak, with the enemy entering Cadiz, he had only six of his ships with him, the others having failed to take the right course of pursuit. The young fire-eater would have pursued his foe into Cadiz harbour itself with these reduced forces but was rightly dissuaded by des Gouttes.

Spain was now clearly under great pressure and at the end of the year Portugal, which had passed into Spanish rule when the main line of its royal house died out, rose in revolt and proclaimed the duke of Bragança as its king. Early in the next year, too, Catalonia, always separatist, revolted. As both had an extensive coastline with some good harbours France's naval task was distinctly facilitated. Richelieu took advantage of this by sending an army to besiege Tarragona and ordered de Sourdis, who was destroying the shipping in Rosas and neighbouring ports, to take the fleet off Tarragona to complete the blockade. The fighting archbishop thought it silly to besiege a port in Catalonia without first conquering the province of Rousillon, that lay between Catalonia and France, the more so as there were insufficient troops for a proper investment and no port of shelter for his ships, and he said so; but he had to obey orders. He had a mixed fleet of sailers and galleys, and repelled the first attempt to relieve the place by sea. This was made by the galley fleet of the duke of Fernandina: it was driven away, and the lesser French ships and one or two galleys made a dashing attack on such Spanish galleys and storeships as had slipped into the harbour during the battle, and destroyed almost all of them before their cargoes could be unloaded. So far so good, but the Spaniards were collecting a large fleet at Cartagena for a further attempt. (They were helped in this by the failure of Maillé-Brézé, in the Atlantic, to block the straits of Gibraltar – a failure due to his waiting for the tardy arrival of his Dutch and Portuguese allies.) De Sourdis wished to attack this fleet in harbour, while it was forming, à la Guétaria; but a council of war having opposed the idea, he did not overrule it. As a result he found himself attacked by a force of more than twice his own strength, which was escorting a large convoy. The Spaniards had the advantage of the wind, and, beating up stoutly against it to fight the galleons, the French were attacked on their quarter by the Spanish galleys, which had been escorting the convoy. They held their own well against great odds, but could not prevent the convoy from entering Tarragona and completely relieving the place; nor could they prevent them-

[1]Authorities differ greatly as to the Spanish losses, some giving them as five or six galleons. But it seems that only one 'warship' was lost (to fire-ships), and any other vessels taken must have been merchantmen which lagged astern of the convoy during its flight.

selves from being carried to leeward of both port and enemy. De Sourdis would have renewed the battle, but his galleys were out of water, some of his ships were much damaged, and a council of war advised, as they were to do only too often, a timorous course – the return to Toulon. The Spaniards, who had achieved their object, the relief of the besieged town, let them go without further fighting.

De Sourdis found himself in disgrace. He was to leave his command to go to live in Carpentras – a sort of exile. Forgotten were Guétaria and the fulsome praise for it and his staunch work since then: the spiteful former general of galleys, Forbin, and other officers whom he had upset, had got at the cardinal behind his back, and he had, too, been so indiscreet as to criticize some of Richelieu's plans and be friendly with people whom Richelieu did not like. Therefore he was removed with contumely, despite a lengthy protest formally written and presented by fourteen of his captains, headed by the Chevalier de Cangé, who had proved himself a stout fighter from Guétaria onwards.

So Henri d'Escoubleau de Sourdis, Archbishop of Bordeaux, passes out of the pages of naval history – but not without leaving some mark. He never, of course, became a real seaman, and in such matters as weather-wisdom and the best use of the vagaries of wind and tide he had to rely on the advice of those of his officers who had true nautical experience; but he had an admirable grasp of naval strategy, as it could be applied at that time, and would have shown it even more clearly had he not been handicapped at councils of war by those who were less naturally fighters than he was, for he was essentially a man of action. He was also a man of personality, and though that personality was obviously contentious, the contentiousness seems to have been kept mainly for those of powerful position whose belief in their exalted birth made them certain of their right to give (or disobey) orders, and unwilling to receive them. As a leader he failed to assuage the harmful rivalry between the shipmen and the galleys; but his ship captains believed in him, and showed it by their protest at his dismissal – a protest that, being made in support of one out of favour with the ruthless cardinal, was courageous, and made even more courageous by those of them who went openly to visit him in his exile. He was the first admiral, in a modern sense, that France had, and, for his day, he was not too bad a one.

His place in the Mediterranean command in the next year (1642) was taken by Maillé-Brézé who brought the Atlantic fleet round while de Cangé brought that of the Mediterranean out from Toulon to join him, and the Spaniards failed to intercept their junction. The struggle for Rousillon and Catalonia was continuing and Maillé-Brézé was off Barcelona when the Spaniards hove in sight at the end of June. They had only thirty-six ships against the French forty-four, and less galleys; but they had the advantage of the wind. Maillé-Brézé beat out against it, his fleet in three divisions under himself, de Montigny and de Cangé; and a clumsy two-day action followed at the end of which the Spaniards drew off to Majorca having had two galleons destroyed by fire-ships and a third taken, with heavy casualties and damage on some others. The French lost one ship – the GALLION-DE-GUISE – which inadequate handling of the fleet as a whole had allowed to be cut off from the rest. The loss was the more serious in that her captain, the chef d'escadre le Chevalier de Cangé, who was wounded, chose to go down with his ship. This Breton was not only one of the most seamanlike officers in the early French navy but at Guétaria, the taking of Laredo, and the battle off Tarragona had shown himself to be a fine fighting leader. When Perpignan fell in September Richelieu considered the campaigning season to be ended, and though the Spaniards were known to be at sea again he recalled the fleet to Toulon.

It was the last major order that the great cardinal was to give, for he died at the end of the year. Starting from nothing, he had laid the foundations and built a good deal of the edifice of a fleet worthy of his country. Circumstances, as we shall see, were to destroy the edifice; but the foundations remained and later were to be built upon, and with notable success, by one who, as a naval administrator, was even greater than the cardinal: Colbert. Nor was it for the naval service alone that he had laid foundations: from his work comes the future French Empire, though as its foundation was not at all the concern of the royal navy (even if its preservation was very much to be so in the future) we can here write of it only very briefly.

The great voyages of the early explorers had begun some hundred and fifty years before Richelieu came to power, but France had been later in the field and less vigorous in her oversea activities than any other country, except Denmark, that had an Atlantic seaboard. Like the Dutch and the English she could afford to ignore the arbitrary papal division which allotted all new found territories east or west of a given longitude to Portugal or Spain and follow the line of her king, François I, when he said that he would like to see the clause in Adam's will that sanctioned such a decision. But her late start meant that her claims to new territories and trading areas would bring her into conflict with Spain and Portugal in Central America, the Antilles, Brazil, and Africa, with the Dutch in the East Indies, and even, to some extent, with the English in the less sought after North American lands. Nor, before Richelieu's time, was she ever quite certain whether she wished to colonize or whether she would prefer to use her seamen's undoubted genius for playing the corsair by privateer raids on the trade of the other countries that had already established settlements. She tried, in fact, to do both, and nearly fell between two stools.

It is not known how early the Bretons started to fish the Newfoundland banks. They certainly were doing so soon after 1500. It was a Breton from St Malo, Jacques Cartier, who discovered the mouth of the St Lawrence. He made several voyages up that great river between 1534 and 1542, getting to Strathcona (now Quebec) and Hochelaga (Montreal); but his attempts to colonize this 'New France' failed, partly through inexperience in preparation, partly from the severity of the winter, and partly from the hostility of some native tribes. Meanwhile a master director of privateers, Ango, had sent ships to North America, the East Indies, and Brazil. Serious attempts were made to colonize Brazil and the southern coasts of North America but were brought to nothing by the hostility of Portugal and Spain. French privateers attacked the trade in the Spanish Main, and one voyage was even made to Archangel. None of these activities was carried out by royal ships. They were the work of private companies of merchants from towns on or near the coast, such as Rouen and Caen, companies formed sometimes to take the profits of a single voyage, sometimes with more permanent objects. It was not until Henri IV had brought peace to his country at the end of the religious wars that they received even encouragement from the Crown.

The real founder of Nouvelle France was a certain Samuel Champlain. Born of a poor fisherman's family in Brouage about 1567, he became a soldier in the religious wars and did so well as to be given a patent of nobility in 1592 or 1593. The wars ended, he turned for a living to the sea that was in his blood, and after a voyage to America on which he wrote a report he was made a court geographer by Henri. His first voyage to Canada was in 1603, but it was only with great difficulty that he could get any people to man his new settlement of Quebec. A settlement was also made, in 1605, at Port-Royal (the present Annapolis) in Acadie (Nova Scotia); but it failed because the two companies that made it fought among themselves, and so left it defenceless against the attack of a British ship from Virginia under a Captain Argall. With Richelieu in power the struggling and somewhat transitory companies were put on a firmer basis. He created the Company for St Christophe, in the West Indies, in 1626, and next year gave his support to the hitherto somewhat random Company of a Hundred Members for New France. The astute cardinal got the financial support of its merchant members by giving them patents of nobility, and obtained his colonists mainly from Normandy where there was a large *petite noblesse* which was so poor and struggling that it was glad to accept seigneuries in the new lands. Of such came Pierre Belain, sieur d'Esnambuc, who, more than anyone else, was responsible for the French foothold in the West Indies. Born in 1585 of a noble house that was destitute, he had voyaged as a corsair to Brazil and the Antilles; and when the Company of the American Isles was created in 1635 he led the expedition of 4 000 colonists that not only made effective the French occupation of St Christopher but also established France in Martinique, and then in Guadeloupe. When, in 1635, Champlain died in Quebec, Canada was well established, the Senegambia and Guinea companies had been formed in the previous year, and in 1642 Richelieu formed the Compagnie des Indes with a twenty years' monopoly of trade, and it was planning to make a settlement in Madagascar as a half-way house to the Orient. The foundations of Empire had indeed been laid: it was for the navy, which had had no part in laying them, to defend them.

CHAPTER IV

# Lean Years and Collapse

Richelieu's successor was the man he had himself designated, the Italian Cardinal Mazarin, whom he had taken into the French service on noting his skill as a papal negotiator. As Louis XIII died six months later, leaving his queen as regent for an infant son, all power was really in Mazarin's hands. A subtle diplomat, he had not Richelieu's understanding of sea-power, nor his vision of its future. He did not take into his own hands Richelieu's office of Surintendant de la Marine, but handed that to the great cardinal's nephew, the dashing young Maillé-Brézé. Willing enough to maintain the fleet as it was, and to use it to further his policy, he never gave it a high priority. So when internal troubles came to disorganize France, which they were soon to do, it was one of the first things to be allowed to go to the wall.

For the moment the French effort in the Mediterranean was concentrated in supporting the rising in Catalonia, where the Spaniards still held Rosas, Tarragona and Tortosa, and so exercised an awkward pressure on Barcelona. Based on that city, Maillé-Brézé, who had retained the active command of the fleet, gave support by blockading the coast. In August 1643 some Spanish vessels were sighted in the distance. They were in fact three small warships and three transports carrying supplies and reinforcements to Rosas. Maillé-Brézé sent his galleys and some of his ships to attack them. For the ships it was a frustrating sort of chase, for, owing to a lack of wind, they had to be towed by the galleys and made no gain on the enemy. So the galleys, led by an officer named Baillibaud, cast off the tow and attacked on their own, successfully capturing all the enemy vessels. Next, hearing that the Spaniards were collecting a considerable fleet in Cartagena, Maillé-Brézé, with the inevitable des Gouttes to advise him, and leaving his galleys to maintain the Catalonian blockade, sailed south with his ships to attack it. Despite the very strong defences of the port, he would have done so had not a gale sprung up and driven him to the south of Cape Gata. Here his smaller vessels, sailing ahead as a light squadron, signalled the approach of twenty-five sail of Spaniards coming from the Straits of Gibraltar. This was a fleet under Martin de Mencos and the Fleming, Peeters, which was awaiting some galleys, and then was to proceed with them to join the ships already in Cartagena to create a force of overpowering strength. Maillé-Brézé had twenty-four ships and some fire-ships. As the Spaniards had three large galleons, one of 1 500 tons, and the Dunkirk ships of Peeters were also formidable, they were, perhaps, rather the stronger. The young French admiral had already instructed his captains that in any fleet action he intended to obtain the wind and then to make an all-out attack on the nearest

'LA COURONNE' The pride of Richelieu's Fleet

'SOLEIL ROYAL' The pride of Louis XIV's Fleet. (Note the extensive decoration of stern and poop)

PLATE II

## SEVENTEENTH-CENTURY ADMIRALS

HENRI D'ESCOUBLEAU
DE SOURDIS

ARMAND de MAILLÉ-BRÉZÉ
Duc de Fronsac

ABRAHAM DUQUESNE
Lieutenant-General
(*by Edelinck*)

of the enemy vessels, and he spent the night of September 3–4 in manoeuvring for this purpose. The heavier Spanish ships were indifferent sailers, so when day came their fleet was somewhat too scattered properly to receive the French attack. Nevertheless, there was a hard-fought action lasting more than three hours in the course of which the 'Admiral of Naples' was destroyed by a fire-ship, the 'Vice-admiral of Spain' was carried by boarding, and a third galleon was taken. De Mencos and Peeters got the rest of their rather battered vessels away into Cartagena – some going ashore.

The Spanish loss in men had been heavier than that of the French, and their fleet was unfit for further action. Maillé-Brézé sent his more severely damaged vessels to Toulon for refit, and brought his fleet there at the end of the campaigning season to receive a letter from Mazarin which combined the warmest of congratulations with an earnest plea to use the strictest economy of material in restoring the fleet to order! This victory, known as the battle of Cartagena, but more properly named that of Cape Gata, had made the French undoubted masters of the western Mediterranean. In the next year a blockade of Tarragona failed because of the defeat of the land forces, but in 1645, with the fleet blockading it, Rosas fell to the French, giving them a suitable port for future Catalonian operations.

Mazarin, himself an Italian, now turned to attack the Spanish domination in Italy, where it was far from popular. In a maritime sense it was assailable in two areas, the long coastline of the Kingdom of Naples and the Tuscan Presidencies. These were a series of small ports along the coast of the Grand Duchy of Tuscany, cut off from their hinterland by the malarial strip of the Maremma, but much used by the Spaniards to rest the crews on their galley route from Sicily and Naples to Genoa – the port through which they supplied and reinforced their troops in the Duchy of Milan. The general scheme was to take Orbetello, which seemed to be the key position among the Presidencies, and later to attack Naples, which was ripe for revolt, in order to put on its throne Prince Thomas of Savoy. In return he would hand Gaeta to France; and, with that port, the Presidencies, her own Provençal harbours and Rosas, she would hold a semicircle of bases that would ensure her permanent control of all the western Mediterranean except its southern waters, where ranged the Barbary pirates.

A full scale expedition was therefore fitted out. The fleet of sixteen sailing warships, twenty galleys, and eight fire-ships escorted a large number of troop transports, which were mostly small. The soldiers were landed and Prince Thomas began the siege of Orbetello, where he met a stout defence, and showed little vigour or capacity. Moreover, because Mazarin was essentially close-fisted and never understood that a fleet needed arsenals well supplied, the expedition, instead of sailing in February 1646 as had been intended, had not left until the end of April. This, combined with the slowness of the siege, allowed Don Francisco Pimienta to collect twenty-five ships and thirty galleys and bring them to the relief of Orbetello. Maillé-Brézé made his usual headlong attack on them off Giglio Island. The Spaniards were pursuing a deliberate tactic of orderly retreat, hoping to get the French fleet scattered; but Maillé-Brézé, with the tow by a galley to help, was pressing them hard when he was cut in two by a cannon-ball. At the news of the death of the young admiral, his second, the comte du Daugnon, who commanded at Brouage, broke off the action. This may have been wise, for, though French casualties were light, the odds were against them, and their ships had been as hard hit as those of Spain; but du Daugnon's next step was sheerly disgraceful. Instead of falling back to protect his transports he took the fleet back to France, when he himself scuttled off to his own command base at Brouage. Mazarin apparently thought him too powerful a noble to be brought to book – as he should have been. The siege of Orbetello was raised and, as the Spaniards had duly got among and cut up the transports, not all the troops could be got away. What should have been an easy conquest had turned out a shameful disaster.

Maillé-Brézé was only twenty-seven when he was killed, and it is interesting to speculate what would have become of the French navy had this young surintendant lived. He was undoubtedly a fighting leader of dash who had gained sea experience, but otherwise seems to have been a rather dull man, hardly suited to direct major strategy or to put the drive into the national naval administration that he could put into a fleet. His death showed the handicap under which the service laboured because of the pretensions of the greater nobility, for the Prince de Condé claimed the position of Surintendant de la Marine on the grounds that Richelieu had intended it to remain in his family, into which Condé had married. He had no knowledge of maritime affairs nor any interest in them, but he had great interest in the profits that could accrue from the office. Mazarin stalled for a while, then asked the Queen Regent to take the office into her own hands: this she did until 1650, when it was put into the hands of the duc de Vendôme.

Mazarin at least took prompt steps to make good the Orbetello disgrace. The fleet from the Atlantic had come round to Toulon, and thirty ships under the Marshall de la Meilleraye, who was purely a soldier, were sent off to the Tuscan coast. They took two of the ports, Piombino and Porto Longone, and left garrisons there before going home for the winter. (Des Gouttes was, at last, raised to the well-earned rank of Lieutenant-general, for dodging four ships through the Spanish blockade with supplies for them.)

With Anne of Austria as surintendant the command arrangements of the navy were as follows: The general of the galleys was the young duc de Richelieu. (He had the appointment as a child with des Gouttes to advise – as usual.) His other mentor and the real commander was the experienced galley captain Vincheguerre. Young Richelieu, now (1647) aged eighteen, was also admiral in the Mediterranean, with the strict proviso that the sailing vessels were to be commanded by des Gouttes (LG)[1]. There were now six CEs[1], one having been added for Catalonia, and one for Dunkirk, temporarily in French hands. (This was DuQuesne, who had departed for a while to the Swedish service, but had now brought home four Swedish ships which France had bought.) These senior, or flag, officers were no longer just port administrators, but took their fleets or squadrons to sea. Such an arrangement, with a young nobleman in chief, yet only nominal, command, meant

[1]See footnote page 19.

confusion of leadership, with frequent councils of war. That is probably why the campaign of 1647 was a failure.

Naples had revolted under Masaniello, and the duc de Guise, then in Rome, being of the house of Anjou which had once ruled in Naples, claimed the throne and had popular approval. France, which would have preferred someone more suitable, decided to back him, and the duc de Richelieu sailed in des Gouttes' flagship with twenty-nine ships and some fire-ships[1]. Everything was in his favour: the revolt was widespread and the Spanish fleet was scattered; but he proceeded to do virtually nothing. He despatched five ships under du Mé (CE) to deal with three Spanish ships of war and two transports sighted off Castellamare: four were set on fire or sunk, and the fifth, a grain ship, was taken. He fought off an attack by a somewhat superior force that had been collected by Don Juan of Austria; then he sailed for home. In the next year the rising collapsed, because it lacked any support from France. For France was in the grip of that complicated movement, or set of revolts, known as the Fronde.

Briefly, this had its origin partly in hatred of Cardinal Mazarin, a foreigner of notable avarice who, in directing the affairs of France, was clearly feathering his own nest amply, partly in distrust of the Queen Regent, another foreigner who was plainly a mere mouthpiece for the cardinal, and largely in the desire of the turbulent greater nobles to overthrow the centralizing bureaucracy that Richelieu had imposed on them and thereby to recover their local privileges and the power to abuse them. The result was a confused period of four years of riot, uneasy and brief pacifications, and violent civil war, during which Mazarin had sometimes to negotiate for France from a place of exile, and the Court had to flee from Paris to the provinces, until, in 1652, with the better disposed of the nobility rallying to the Crown, the royal authority was reimposed, and Mazarin could resume the direction of affairs, until his death nine years later. There had been a small amount of internecine sea-fighting, none of it of major importance.

During such a period of confusion, and of great strain on official finances, the navy, in which none of the leaders of factions had any great interest, went to the wall. The garrisons at Piombino and Porto Longone could be neither supplied nor withdrawn and had to yield, and, indeed, Spain was able largely to recover both the territories and the maritime position that she had lost. For though Mazarin had been able to end the Thirty Years War with the successful batch of treaties known as the Peace of Westphalia, the struggle between France and Spain continued. There was nothing with which to oppose Spain's recovery of her lost position; for the supply of finance to the navy virtually ceased, and the Fronde had been in its baleful progress only a year when the Intendant at Toulon wrote that he had been obliged to abandon all attempt to maintain the ships, and would soon have to take the same course with the galleys, whose crews (largely convicts) were facing starvation. The one successful attempt to provide supplies to the doomed garrisons in the Presidencies was made by a naval captain, the

[1]The French had a high reputation among the revolted Neapolitans because the chevalier Paul, reconnoitring the Neapolitan coast with six ships earlier in the year, had successfully beaten off several attacks by the Spanish galleys.

chevalier Paul, who fitted out a royal ship at his own expense and battled his way through to Piombino and Porto Longone against odds.

Incidentally, this officer had a career which shows that, while the higher command went normally to the greater nobility, it was just possible for a commoner to rise to it. He was the son of a Marseilles washerwoman, and was born in a boat during a storm while his mother was taking the laundry to the Château d'If in 1597. The governor of the Château was naturally interested in this unusual accouchement and made himself responsible for the boy's upbringing. The circumstances of his birth had apparently put both storminess and the sea into his blood. After a voyage as a stowaway he was entered in the service of the Knights of Malta, but was dismissed it for duelling with a superior, and joined a privateer. His personality and capacity were such that, when its captain was killed, Paul was elected to the command by the crew, and, basing his ship on Lesbos, carried out such a successful piratical war against the Turks as to get considerable booty, a considerable reputation, and readmission to the ranks of the Knights of St John in 1637. Richelieu promptly gathered him into the Royal Navy as one of his captains *entretenus*. These were commoners from the merchant (usually privateer) shipping and were the equivalent of the 'tarpaulins' in the British service. Paul had shown his seamanship and fighting capacity at Guétaria, in the Mediterranean campaigns of de Sourdis and Maillé-Brézé, and in leading a forceful reconnaissance of the Neapolitan coast in 1647. Now, as a reward for helping to prolong the resistance of the Presidency garrisons, he was given a patent of nobility by Mazarin and promoted to chef d'escadre.

In 1650, during the Fronde, the Queen Regent handed over the office of Surintendant de la Marine to César, duc de Vendôme, a bastard of Henri IV, though there was little for him to command. In July 1652, after much delay and difficulty, he sailed from Brest with thirteen ships, mostly of low fire-power, and some small vessels. His principal task was to try to relieve Dunkirk, which was hard-pressed; but first he had to drive away the Spaniards who were blockading La Rochelle and the mouth of the Gironde. His enemy had eighteen vessels, but they were a very mixed bag, and included four under du Daugnon, who, like the duc de Richelieu and Vincheguerre, was a Frondeur. These were unwilling to fight their own countrymen alongside the Spaniards, and in a battle in the Pertuis d'Antioche victory went to Vendôme. Advised by the commandeur de Nuchèze, for he had no sea experience, he sent in his fire-ships under cover of his broadsides. One Spanish ship was burned, another taken, and the remainder fled. Vendôme could not follow up his success by proceeding with his main task for, though his ships had been but a short while at sea, they had exhausted their supplies, and no more were to be had. So Dunkirk fell. A few ships sent in a last desperate attempt to relieve it had been seized by Blake as a punishment for the shelter France had given to the piratical English royalist ships of Prince Rupert.

The Fronde ended at this time. But the former fine French fleet was reduced to about thirty vessels, of which only three were of sixty or more guns, and the depots and arsenals were empty. In short, the navy, as a force, was all but ruined. Nor was there anyone greatly concerned to restore it. A small force was kept going on a cheeseparing supply of funds; and that was all.

In 1655 Vendôme, with the Chevalier Paul, now LG of the Mediterranean ships, as his second in command and adviser, took a squadron round to the Catalonian coast to support French operations. With his nine ships he fought a battle with a Spanish fleet of twelve ships off Barcelona and drove the Spaniards back into port, but he was unable to establish any sort of blockade. In 1659 the war was brought to an end by the Peace of the Pyrenees, by which France gained Rousillon and Artois, but the valuable Dunkirk went to the English, who had joined in the attack on Spain, and had played a large part in capturing it. In the short remains of Mazarin's life the only naval event of the least note was that the Chevalier Paul was sent with fifteen ships to negotiate with the Barbary pirates for the return of enslaved Frenchmen. So stormy a person was hardly likely to be a good negotiator, but he did buy back some wretched captives from Tripoli and make a face-saving treaty of no value with its pasha. At Goletta and Algiers he could do nothing before shortages of supplies compelled him to return. He did, however, put in a comprehensive report on how these places should be attacked, and urged that the crews of attacking vessels should be from the Atlantic coast, as the Provençals were so used to being terrorized by the 'Barbaries' that they had no belief in victory when they fought them! Mazarin died in March 1661 before the report reached him; in any case he had maintained no force with which to implement it.

CHAPTER V

# Colbert

Though Louis XIV had been officially declared of age when he was a mere boy, he had been kept very much in leading strings by his mother and her beloved cardinal, who had signed everything in his name. He was twenty-two when Mazarin died, and, to the surprise of the Court, he promptly took charge of everything, instructing his ministers to send out no important order, and to take no important decision, without first obtaining his approval. This began fifty-four years of complete autocracy, during which he identified the State with himself, worked tirelessly to increase his, and therefore its, prestige and greatness, and did far more harm than good. For the Roi Soleil, as his subjects called him, was at bottom a would-be Caesar in Europe, like Napoleon or Hitler, and though preferable in personality and in his acts to either of them, he ended, like them, by leaving his country exhausted both in finance and in energy. The dying Mazarin had 'bequeathed' him an able official of his personal staff, who proved to be one of the great administrators of all time. His name was Colbert, and his work for the navy was so stupendous that we shall do well to depart from chronology and consider it as a whole.

Jean-Baptiste Colbert was born in Rheims in 1619, the son of a clothier. He obtained a clerical post under François Surlet des Noyers, who was one of Richelieu's right-hand men in the new naval administration, and Mazarin, noting his ability and thoroughness, took him into his own service, making him his personal intendant. Colbert had to handle the cardinal's enormous fortune, huge collections, and complicated personal affairs. It was an admirable training. Louis appointed him as an intendant of finance. His surintendant was Nicholas Fouquet, the most prominent of the Ministers, who was plundering the public purse almost recklessly and seems to have been plotting to make himself a virtually independent ruler of a part of France. Colbert exposed his defalcations and schemes to the king, who tyrannically overruled the judges who tried Fouquet, and imprisoned him for life. Colbert took his place: Intendant of Finance with special reference to marine affairs. The title of 'Intendant de la Marine' was added in 1663, and in 1669 he was made Secretary of State, with responsibility not only for finance and all maritime affairs but also for Arts, Manufactures, the Royal Household, Commerce, and Oversea possessions. That he should hold all these offices was no disadvantage from the naval standpoint, for if the king, who was very much his own chief minister, should reduce the desired appropriation for the navy, Colbert could borrow the necessary extra funds from the Treasury, or the Arts appropriation, and include them for repayment in his estimates for the next year. It was, perhaps, also convenient that the holder of Richelieu's old office of surintendant, who was the duc de Beaufort, son of the duc de Vendôme, should be killed in Crete that year. It was probably at Colbert's suggestion that the king then allowed the office to lapse and, in lieu, revived the old office of Admiral of France, and gave it to a royal bastard aged two(!). Thereafter it was always treated as a royal appanage. At least this absurdity left Colbert free from interference by any superior other than the king himself.

Colbert had not waited till then to get on with things where the navy and trade were in question, though at first, after the overthrow of Fouquet, finance was his main concern. He handled this so successfully that in a single year a succession of budget deficits was turned into a substantial surplus. Thanks largely to this France was able to buy Dunkirk from the English in 1662. From the English viewpoint it was a ridiculous sale to make, for it surrendered what would have been complete command of the Straits of Dover. Colbert did not delay in arranging for the port to be extended and for its channel of exit through the shallows to be deepened and widened. His vision of sea-power was that of Richelieu. (The king used to jest that Colbert was always saying to him: 'But, Sire, the great cardinal would have done such-and-such'.) Perhaps, indeed, it was wider than that of Richelieu, for the cardinal's eye had been mainly on power, with trade as a secondary consideration: with Colbert they ran together. Trade would increase the finances of the country, and, by passing into French hands, would weaken the finances of certain other countries: the navy would protect that trade and draw a reserve of crews from the merchantmen. In fact, the direction of his thinking, of the taxes and ordinances that he persuaded the king to approve, was closely akin to the Mercantile System beloved of, and operated by, eighteenth-century English Whiggism.

He had to start almost entirely afresh. The once fine fleet had now only twenty ships, of which some were useless or beyond repair, and six decrepit galleys. Richelieu's naval budget of several million livres had been reduced to a mere 300 000. Crews were scattered: many seamen were serving abroad, and officers had become permanent landsmen, or had been without seagoing employment. (DuQuesne, the ablest, and about the most senior of them, was ashore for twelve years!) Port installations were in disrepair, arsenals and storehouses were empty, harbours were becoming silted. The Fronde, and Mazarin's failure to re-establish the Service after it, had thrown conditions back almost to what they had been when Richelieu took over; but his ideas, and something of his organization and work, remained. In rebuilding on these foundations Colbert had no great or influential support: his surintendant till 1669, the duc de Beaufort, son of César de Vendôme, was too inexperienced to be a help, and too haughty a noble to take advice; he did, however, largely refrain from interfering. The king was not greatly interested. It is wrong to say, as some historians have done, that Louis cared nothing for his fleet: since other powers had navies, he certainly must have one, even if as a mere matter of prestige. But its early growing pains, its failure to achieve notable successes, disappointed him: the constant complaints of DuQuesne about the unseamanlike ignorance of his officers and his crews irritated him; and in any case he did not really understand what was involved. His armies he could visit without inconvenience: the panoply of land war with its drums and flags and marching men, and especially that of sieges carried out in due form, delighted him. The incomprehensible ships in remote ports, sailing and fighting somewhere out of sight on principles that were hard to grasp, took second place. Colbert went to much trouble and expense to create a miniature fleet on the so-called 'Grand Canal' (the lake at Versailles), and the king greatly admired its displays, but more as a pageant than as showing him one of his services in action. He visited his armies often, but it was not till 1680 that he saw one of his ships. Then, when he visited Dunkirk, a vessel cleared for action for his benefit, and he and his Court condescended to be amazed at what the sailors had to do, and could do, so rapidly and efficiently. Certainly he must have a fleet and it should contribute to his glory; but always, if the purse strings had to tighten, or questions of strategy clashed, it was the army and its ideas that came first. Colbert intended the fleet to protect commerce: Louis wanted strategic success in Europe, and for this the fleet became a luxury. Yet when Colbert died in 1683 he had wrought so well that he left a navy of 117 ships of the line (though some were not really strong enough to be so rated) with thirty galleys and a number of small vessels. To man them there were 1 200 officers and 53 200 men, they operated from harbours that were properly fortified, dredged and buoyed, and the port installations and the supplies in the arsenals were such as could properly equip and maintain fleets of such a size.

To build the ships he imported foreigners – Deane from England, a Dutchman or two, and, above all, a Neapolitan, Biagio Pangallo, who became known to the French as Maître Blaise. He it was who produced the finest designs and by under-cutting the hull a little forward, and the sheer a little aft, got some extra speed, and taught the French to be the best designers of warships in the world. The ships were also splendiferously decorated, by leading sculptors, notably by Puget. Beautiful finery, but a great waste of money, especially as, in one or two cases, it handicapped the proper working of the ship. For the ports, or at least for their buildings and fortifications, Colbert relied upon Vauban, the greatest military engineer of his day. Brouage had become so sand-blown and silted that it was abandoned, and its place, thanks to the representations of some powerful and interested nobles, was taken by Rochefort, which, though useful, was never entirely satisfactory. Le Havre would have shared the fate of Brouage but for the urgent, and mistaken, representations of DuQuesne. Vauban would have substituted Cherbourg, and later, after the battle of La Hogue, France was deeply to regret that this had not been done. The already admirable Brest was improved and strengthened, and Toulon, which had begun to silt, was developed out of all knowledge, the galleys being relegated to their old home of Marseilles. By the late 1670s Dunkirk was made impregnably suitable for ships up to a little more than medium size. Vauban wished to add Port Vendres, but his proposals for this and Cherbourg had to be dropped, for even Colbert's national financial pocket was not bottomless.

To be sure of obtaining crews Colbert introduced the great *Inscription*

*Maritime*. It was built up as follows: In 1668 there was a census with compulsory registration of seamen. Next year the names obtained were divided by their bishoprics into crews of 300. This was followed in 1670 by the 'classes', a system of call-up by rotation with a retaining fee for those not called, and a ban on merchantmen against shipping any man of a class that was being summoned to serve. This *Inscription* met with a mixed reception (eg it went down well in Brittany and badly in Marseilles), but it soon settled down and it worked well. For the children of those serving were educated free, and pay was made to a man's home, so that sailors returning after months at sea in the hard life and cramped quarters of shipboard did not spend their all in one drunken and dissolute orgy when they set foot on shore. There was, too, a *Caisse des Invalides*, created in 1673 to care for the wounded. Gunners were in short supply but a corps of marine artillery was formed. Also, two regiments of Marines were established, but Louvois, the jealous Minister of War, who had the king's ear, soon appropriated those for the army. For officers it was odd that, himself a commoner and with the example of the rankers DuQuesne and the chevalier Paul (who did not die till 1669) before him, Colbert would draw only on the nobility, holding that only a man of 'good family' could set an example and command obedience. This was a grave mistake, for they were only too often proud and quarrelsome, and it meant passing over some able men, while getting soldiers to try, often ineffectively, to convert themselves into seamen. Fortunately there were still some of the old officers left, and there were younger men who had seen service in the Galleys of the Religion to draw upon. (For the future a body of cadets – Gardes de la Marine – was set up.) The Galleys of the Religion – those of the Knights of St John at Malta – were the great source of supply of French naval officers in those days, and continued to be a valuable one for much of the next century. The Order drew pronouncedly more upon France than upon any other nation: its perpetual campaigns against the Barbary corsairs made it a first-class training school; and a large number of French naval officers, including some of the most prominent, began their careers in its service.

In his early days in office Colbert and his port Intendants had difficulty in stopping sheer waste of materials, and in weaning his captains from the habit of making 'perks' out of the service. Nevertheless he managed to instil into his naval officers a realization that promotion was by merit and not by influence or mere seniority, and he was tireless in his exhortations to officers to live and act 'for the good of the Service'. Nor did he limit himself to exhortations. Rousselot, marquis de Château-Renault, a Musketeer converted into a capable, though 'difficult', senior officer, was told in 1675, when he questioned an order, that if there were another such questioning the king would have him put under arrest on his own quarter-deck; and a captain who was absent from his ship when she was wrecked in a roadstead was beheaded. Colbert almost achieved the spirit of absolute obedience that he required; and that it did not last (the more the pity) under his weaker successors was due not to their admitting a few commoners to the wardroom but to a haughty pride of birth among noble officers that needed a Revolution to bring it entirely to heel. Promotion was from lieutenant, to fire-ship captain, to frigate captain, to (in English) 'post-captain', to chef-d'escadre, to lieutenant-general. There were originally to be only six CEs and two LGs, but the numbers soon increased with the growth of the fleet. Among them was the vice-admiral of the Ponant (the Atlantic). The general of the galleys was at first in chief command in the Levant (Mediterranean), but there was soon a second vice-admiral – of the Levant. The Admiral of France, as we have seen, was usually a royal *fainéant*, and at a later date ensigns (sub-lieutenants) were introduced. Colbert's first appointment of vice-admiral (Ponant) was d'Estrées, a brave soldier and a great noble, which was what mattered: it was, on the whole, a pretty disastrous choice.

Colbert saw to it, too, that there were laws to cover maritime affairs: an ordinance controlling the cutting of timber (to ensure supplies for ship-building), a Code Noir on Negro placing in the colonies, and finally, in 1681 after a commission had sat for ten years, an almost complete code covering seafaring matters.

Side by side with his development of French sea-power went his encouragement of trade and colonization. His handling of custom duties and taxation to build up his 'system' is not the business of this history; but at his death France had more than three times as many deep sea merchantmen as when he took office. To develop this trade and to increase the prosperity of the colonies he relied at first on monopoly companies, such as that of the East Indies (for which the creation of a special port – L'Orient, on the south coast of Brittany – was started in 1666). These companies broke down, largely from lack of general financial support because the wars that Louis waged were so costly. But the colonies remained, and Colbert boosted colonization. So it was in sort a rounding off of his efforts that, in the year of his death, Robert Cavelier de La Salle voyaged down the length of the Mississippi to its exit in the Gulf of Mexico, to 'annex' Louisiana in the next year. At one time Colbert had built up for his king such a position that, had Louis cared to remain at peace in Europe and devote his country's wealth and energies to oversea expansion, he could have become the commercial master of the world. But Louis did not like commerce, was not interested in 'distant ventures', and preferred to try to become the military master of the Continent. Some years before Colbert died he had lost much of his influence with the king, who was busy carrying out that policy of territorial acquisition which has been described as 'one long lawsuit conducted sword in hand'.

The age of Colbert was, too, a time of great change in seafaring matters. He spent much on the study of navigation. It had not been too difficult for the early navigators to devise instruments with which to get latitude: but to sail till one reaches the latitude of a place and then to run down it till one gets there is hardly good navigation. Longitude presented a much more difficult problem; but when it was realized that it depended on good chronometers Colbert footed the bill for achieving them. He also provided the means for good hydrography by creating the French surveying service, though it was not till 1693, ten years after his death, that the first Neptune Français appeared – an atlas of first-class charting from Norway to Gibraltar. It was eagerly sought by English and Dutch and showed that, in scientific navigation, France could lead the world.

Tactics, too, altered. The first and second sea wars between English and Dutch, one during the Commonwealth and the other in 1665–7, had produced battles that were mêlées, as those fought by Maillé-Brézé had been, and had shown that a mere unlucky shift of wind, or momentary error by a helmsman or of an officer giving orders in handling sail, might lead to one or two ships being cut off from the rest and surrounded by enemies who could batter them almost to bits before their friends could beat up to their rescue. From this came the tactical idea that was to dominate naval warfare almost pedantically for well over a hundred years, that of fighting in line with the two fleets presenting their broadsides to each other and each ship doing its best to preserve station on its next ahead, so that no ship might be caught between two foes. Ships strong enough to stand the enemy's broadsides in such a method of fighting became known as ships of the line. As the big ships increased in size till they carried a hundred or more guns, so they perforce became the heavier and slower sailers; and the smaller vessels, lighter in build and in weight of artillery, became standardized according to their rig as frigates, corvettes, or brigs. With these changes went a certain amount of elementary signalling.

The ships were 'rated' in imitation of the English practice; and in the year of Colbert's death, 1683, the ratings were as follows:

First rates were three deckers, carrying from 76 to 110 guns, with crews varying according to the number of guns from 520 in a 76 to 1 050 in a 110-gunner. The guns in the heavy battery were 36-pounders (with a few of 48 pounds); in the second battery they were 18-pounders, and in the third battery 12- or 8-pounders.

The second rates were two-deckers (apart from a few three-decked 74s) with guns counting from 64 to 74, 24-pounders in the main battery, 18-pounders in the second battery, and lighter guns on the upper deck, with crews of 380 in a 64, and 480 in a 74.

The third rates were two-deckers ranging from 50 to 60 guns, usually 24-pounders and 12-pounders.

The fourth rates were two-deckers with 40 to 46 guns, 12-pounders or lighter, and crews of 260.

The fifth rates had 30 or so 8- and 6-pounders and crews of a 150 to 180.

All vessels with fewer guns were then still called frigates, but they soon merged with the fifth rates to form new classes of vessel (frigates, brigs etc) according to their rigs. Otherwise these ratings remained through the meagre naval first half of the eighteenth century, though it was fairly soon recognized that only the first three rates at most were fit to take their place in the line[1].

Yet for all his breadth of vision and the thorough nature of his work,

[1]There was a change of rating in 1758, and with the revival that followed the disasters of the Seven Years War and throughout the Revolutionary and Napoleonic period changes in the rating of ships were somewhat frequent. It will be noted that the French main gun, the 36-pounder, was heavier than its British 32-pound equivalent: French crews, too, were commonly larger than those of the their British opposite numbers. For these reasons, plus their superior sailing qualities and heavier scantling, it has sometimes been held that a French 84 was the equal of a British 98- or 100-gunner. This is doubtful, though it was certainly a formidable vessel.

there was one thing that Colbert notably failed to provide. There was no Naval Staff or, in the modern sense of the term, any Naval Intelligence. Though he consulted Tourville, Colbert kept the administrative and sea-going branches of the service apart; and his successors followed the same line with rigidity. As a result France never had a Strategic Plan. Only when war was imminent, or had actually been declared, was there any planning of a strategic nature; and then such planning was mainly in the hands of landsmen, and lacked the necessary long-term preparation. Allowing William to land his army in Ireland in 1690 (see page 75) is an obvious example of this: a number of others could be given. It was not until the nineteenth century that this lacuna was made good.

As we have said, some six years or so before he died Colbert had lost much of his influence with the king, and at his death he was almost execrated because of the heavy taxes that he had had to impose to pay for his master's wars, but he was able to pass on his office to his son, the marquis de Seignelay. Introduced into the administration when little more than a youth, Seignelay had been given a thorough training by his sire, and much valuable admonition. 'The quick-witted intelligence of a Frenchman makes him apt to treat serious matters lightly: youth increases this tendency. You must resist it. The will to do the work makes it pleasant, and that pleasure enables one to give one's mind to it.' But though he knew the Service thoroughly, Seignelay was cast in a different mould from his father. It was not the realities of sea-power but the prestige side of naval matters that he put first, and in any case, being a fast-living person, he survived his great parent by only seven years. He was followed by a succession of ministers of whom only an occasional one understood both the Service and its objects: France was to have no second Colbert, and one must turn to the activities of the fleets that he created.

CHAPTER VI

# Dutch Wars and Barbary Pirates[1]

The first activities of the reviving French navy were directed against those pests to all Mediterranean countries, the Barbary corsairs. They can hardly be said to have been successful. In 1663 DuQuesne and the chevalier Paul operated off the Barbary Coast with no great result. Indeed Paul, caught out of sight of other French ships by a number of corsairs, was hard put to it to get away by a ruse. Having fought them off in daylight, he launched the ship's long-boat by night with sails set to run before the wind and a stern light showing, while he took the opposite course. Because of this failure a major operation was decided upon for the next year: a permanent base was to be conquered and established on the North African shore. The duc de Beaufort set sail with a full scale expedition: fourteen of the line, eight galleys, a number of smaller vessels, and transports carrying 5 000 troops; and seven Malta galleys joined him. The command was somewhat discordant, for the chevalier Paul was his adviser, but Louis had written privately to Vivonne, general of the galleys, to keep an eye on Beaufort to prevent him from doing anything rash, and the commander of the troops would take no naval orders with regard to them. Paul advised the capture of Bougie, with its fine harbour and tiny garrison, but Beaufort had made up his mind that Djijelli was the best objective. One cannot tell why, for the harbour was not especially good and the town was in a quadrant of hills which commanded it in a military sense. The troops were landed, and took the place, but when they tried to make good a hold on the commanding positions inland they were heavily defeated, and had to be re-embarked leaving their guns and their wounded behind them. The fine expedition returned to France with its tail very much between its legs and, to crown the disaster, the ill-found 48-gun LUNE, used as a troop transport, sank in sight of land with the loss of nearly all on board. It had been a shameful business, and an angry Louis ordered that nothing should be said of it, but that vengeance should be taken next year. So in March 1665 Beaufort appeared off Goletta with five ships, his flag-captain being Mathurin Gabaret, one of a family that provided France with a number of naval officers, and in a stiff fight in the harbour they destroyed three Algerians. Later in the year, with his flag in the magnificent new ST PHILLIPE (74) and Paul under him in the ROYALE, Beaufort carried out a further destruction of corsairs that had taken refuge in Cherchel, and, ranging the coast, made other prizes. A number of Algerian teeth were certainly drawn, but that was hardly adequate compensation for the Djijelli fiasco.

In 1665 Louis found himself faced with an awkward decision. France had

[1]For map of Mediterranean see p 24, and for map of North Sea see p 170.

an offensive and defensive alliance with Holland: war had broken out between England and the Dutch; and the latter claimed the assistance from Louis to which the alliance entitled them. Louis disliked the alliance intensely. He felt that it committed him to act against his interests, and as an autocrat he detested the Dutch for a lot of republican traders. He therefore hedged until the next year, but then announced that he would abide by his word and send his fleet to join the Dutch in the Channel. So in April the duc de Beaufort, with Paul again as his adviser, left Toulon with twenty-two ships of forty guns and upwards, several smaller vessels, and ten fire-ships. But his method of uniting with the Dutch was peculiar. He cruised for some time off the coast of Portugal, to cover DuQuesne who was escorting a French princess to be the bride of the Portuguese king; and it was not till August that he got as far north even as La Rochelle. Joined here by DuQuesne's squadron he made at last for the Channel. On reaching Dieppe he found that the Dutch, after fighting two heavy battles with the English, had returned to the Texel. So Beaufort put back to Brest[1]. Of course there were 'explanations', but the real truth was that, though Colbert wanted a battle to show the king that his fleet was worth while, Louis did not want to risk his ships, especially to aid the Hollanders. So the only fighting was in the West Indies.

Here all went at first in favour of the French and Dutch, who overran several islands. Then LaBarre, commanding all fighting forces in the Antilles, made for the English fleet at Nevis. He had nineteen ships, mostly French, of which seven were of thirty or more guns, though some were converted merchantmen, and he had also some buccaneers, who for all their bloodthirsty ferocity were never reliable in a straightforward action. LaBarre had hoped to surprise the ten rather stronger English ships at anchor, but found them under weigh and ready for him. He was let down by several of his captains who refused to obey his orders to come to close quarters and board, and his own ship was saved only by the stout and competent fighting of his Dutch allies. So the battle was indecisive, save that LaBarre had to put back to Martinique. Here he moored his ships in a defensive position in front of St Pierre to await the attack of the British, who had been reinforced by six ships under Sir John Harman. This officer handled his attack skilfully. Coming in close to bombard on three different days, he tempted the French to a great expenditure of powder and shot. Then he made a fourth and earnest attack when their ammunition was low. Many of the French panicked and made for the shore: so LaBarre ordered his ships to scuttle themselves, and left the English dominant. The treaty of Breda (1667), ending the war, restored all conquests on either side.

In 1667–8 occurred what is known as the War of Devolution. On the death of Philip IV of Spain, Louis evoked an obscure and outdated law of Brabant to claim that province for his wife! The fleet put out to intercept any possible Spanish reinforcement of the Netherlands by sea; but there was no naval fighting, for Louis, finding a startled Europe arming against him, made peace on receiving some of the fortresses along the southern frontier

[1] A few of his ships, separated from the fleet in bad weather, were pursued by the English, and the Rubis (50) was taken after a notably stout defence.

of the Spanish Netherlands. The war is important in the shift that it gave to European policies; for the Dutch must now make every effort to preserve the southern Netherlands in the hands of an obviously weakening Spain as a barrier between them and the aggrandizing France.

For the moment, however, French interests were shifting otherwhere. The Ottoman Empire, with the Kiuprili as Grand Viziers, was in the last of its several periods of expansion – one that was to end with its defeat outside Vienna by John Sobieski in 1683. In the course of this it had attacked Crete, then a possession of the Venetian Republic, and had now conquered all the island except the city of Candia, splendidly defended by Morosini, but nearing its last gasp. Venice, backed by the Pope, put out a desperate appeal for aid, and Louis was willing to engage in a sort of last crusade. (In fact the French had been assisting Venice for some time with volunteers who went to Crete, and more by vigorous and successful privateering against Turkish shipping.) As Louis wanted no breach in the valuable French Levant trade the strange arrangement was made that the French should serve as nominal mercenaries under the Papal banner, and the ridiculous position was reached that Frenchmen were trading amicably in Constantinople while a powerful fleet was bearing troops to fight the Turks in Crete. Beaufort was in command of fifteen ships of forty guns or more, with some smaller vessels; there were over 7 000 troops, and he was to be joined by thirteen galleys under their general Louis Victor de Rochechouart, comte de Vivonne, duc de Mortemart, and brother of the reigning royal mistress, Madame de Montespan.

To uphold the dignity of Louis, Beaufort and Vivonne were to have second and third voices in council to the Papal commander, Fra Vicenzo, and in fact were to retain a virtually separate command. Beaufort had his flag in the MONARQUE(94), his second in command was the marquis de Martel (LG), and among his captains were the marquis de Grancey and the Chevalier de Tourville. The career of the former shows the odd background of some French captains at the time, for de Grancey had started as an abbé, had turned soldier, and now was in a naval command. The latter was to become the outstanding admiral of his day. The third son of M de Costantin, baron de Tourville, he had been born in 1643 and christened Anne-Hilarion. He had been enrolled among the Knights of Malta when only four years old, and was at sea in their galleys under a M d'Hocquincourt before he was nineteen. When this gentleman built and manned his own 36-gun privateer Tourville went with him as first officer, and quickly earned a name in successful actions against Turkish vessels and barbary corsairs. So in December 1666, at the age of only twenty-three, he was called home to be commissioned in the Royal Navy as captain of the 40-gun COURTISAN. 'If only other officers were like him,' wrote the Intendant of Toulon with regard to his handling of ships at sea and their refitting in harbour. In the Crete expedition he was commanding the CROISSANT(44).

Beaufort reached Candia in June 1669, before the arrival of the Papal galleys and Vivonne. The troops were landed, and against Morosini's advice it was decided to move against the Turkish besiegers at once, and to do so by a night attack. As a landing party of seamen was to join the troops,

Beaufort put himself at the head of them. The attack was going well when the explosion of a powder magazine started a panic: there was a shameful rout, and during it Beaufort was killed.

With the arrival of the galleys in July it was decided that the combined fleet should move along the coast to enfilade the Turkish fortified position by bombardment. Against Vivonne's advice the Papal commander ordered a somewhat confused arrangement of vessels, and the galleys towed the ships into position. The bombardment lasted for four hours, and for most of that time the ships were firing blindly through a bank of smoke that lay static in the dead calm. When a breeze arose to clear the air it could be seen that the attack had done no real damage. But the ships had suffered quite heavy casualties, and the THÉRÈSE(70) had blown up, thanks to crass incompetence in handling her own ammunition. The crews were now much reduced by battle losses and by sickness, and only about a third of the troops remained. They were re-embarked: the fleets sailed away; and Candia surrendered. High hopes had ended in a bad failure.

Nor were renewed operations against the Barbaries in the next year much more successful. De Martel blockaded Goletta for a month, while Tourville burned some shipping in Suse. M d'Estrées had been made vice-admiral of the Ponant. He had no sea experience at all, but he was a great noble who had shown dash as a soldier. He failed to do much in an attack upon Sallee. Though de Martel got the usual worthless treaty, and Château-Renault, whom d'Estrées had left behind, destroyed some shipping, none of this effort was worth a great deal. France's new navy was now to be given a much more severe test.

Though the treaty of Aix la Chapelle, which ended the war of Devolution, had been in some sense a slight check to Louis' aggression, he had no intention that the check should be anything but temporary. His wrath was particularly directed against the Dutch, who were concerned to maintain the integrity of the Spanish Netherlands, and whom he both hated and despised. He first proceeded to isolate them by adept diplomacy. In the course of this, in 1670, he made the notorious secret treaty of Dover with Charles II of England. From the English standpoint it was no bad bargain for, apart from the undesirable, but large, subsidy that Louis would pay, victory was to leave in English hands Walcheren Island, the key position to control all seaborne trade from the Baltic to Calais – an arrangement that shows clearly how little Louis understood the value of both sea-power and commerce. But the treaty, when it came to light, was far from popular with the English. They had fought two wars with their Dutch rivals in seaborne trade and by the Treaty of Breda had arrived at something like an equable settlement; and now, though they did not fully understand what was involved, they had rightly an innate suspicion of Louis XIV, and saw no need for a third Dutch War.

However, war was declared in 1672. Louis began an invasion of the Netherlands which the Dutch, deeply split between the Republicans, in power under the brothers de Witt, and the Orange Party, were in no condition to meet. At the same time d'Estrées sailed to join the English with the following fleet (with some fire-ships, smaller vessels, and storeships):

| Ship | Guns | Captain |
|---|---|---|
| ILLUSTRE | 70 | marquis de Grancey |
| TÉMÉRAIRE | 50 | de Larson |
| ADMIRABLE | 68 | de Beaulieu |
| TERRIBLE | 70 | DuQuesne (LG) |
| CONQUÉRANT | 70 | de Thivas |
| PRINCE | 50 | marquis d'Amfreville |
| BOURBON | 50 | de Kerven |
| HAZARDEUX | 38 | de Lavigerie-Treillebois |
| ALCYON | 46 | Bitaut de Bléor |
| VAILLANT | 50 | chevalier de Nesmond |
| FOUDROYANT | 70 | Louis Gabaret |
| BRAVE | 54 | chevalier de Valbelle |
| AQUILON | 50 | chevalier d'Hailly |
| TONNANT | 58 | des Ardents |
| SAINT-PHILIPPE | 78 | comte d'Estrées (V-adm) |
| GRAND | 70 | Gombaut |
| DUC | 50 | chevalier de Sèbeville |
| ÉOLE | 38 | chevalier de Cogolin |
| ORIFLAMME | 50 | de Kerjean |
| EXCELLENT | 50 | de Magnou |
| ARROGANT | 38 | de Villeneuve-Ferrières |
| FORT | 60 | comte de Blénac |
| RUBIS | 46 | de St-Aubin d'Infreville |
| GALANT | 46 | chevalier de Flacourt |
| SANS-PAREIL | 66 | de la Clochetterie |
| SUPERBE | 70 | de Rabesnières-Treillebois (CE) |
| SAGE | 50 | chevalier de Tourville |
| HARDI | 38 | de Laroque Garseval |
| HEUREUX | 50 | Panetié |
| INVINCIBLE | 70 | commandeur de Verdille |

The royal orders to d'Estrées are much concerned with matters of prestige, who shall salute whom, and so on, and there may just possibly have been a royal hint to him that the more the English and Dutch knocked each other about, without much being done to the French, the better; but he was certainly ordered to 'put up a good show' before the English. The question of being under English command raised no problem, for d'Estrées was a vice-admiral, whereas England was to be led by her Lord High Admiral, the Duke of York, who was a royal prince to boot. The combined fleet was organized, in the custom of the time, into the three divisions of van, centre (or main 'battle'), and rearguard, each of these divisions being similarly subdivided in its turn. The French were to form the van, or white squadron, of the allied fleet, the English providing the centre and rear; and in the French squadron d'Estrées was in the centre with DuQuesne commanding the van, and Rabesnières the rear. The new tactic of fighting in line was to be used, so was the new practice of signalling, according to a signal book recently got out by the Duke of York.

The Dutch admiral, de Ruyter, leading an inferior fleet on which, nevertheless, so much depended, pursued the wise tactics of withdrawing behind his difficult shoals and banks when a westerly wind favoured an allied attack, to come out with an easterly wind which might enable him to catch the allies napping yet would allow him to break off any engagement should he wish to do so. He got to sea too late to prevent the junction of the French and English fleets, and when they came after him he fell back inside his coastal shallows. The allies, being short of water, made back to England and anchored in Sole Bay (Southwold), which was then more of a true bay than it now appears owing to spits of land which have since eroded.

Here, on June 7th 1672, they were caught all but unawares by Ruyter, who came down upon them with a wind from ENE. They had watering parties ashore, and when the ÉOLE, which was on outpost duty, came flying in firing signal guns to denote 'enemy in sight', they had to cut cables and beat out to get sea-room hurriedly and therefore clumsily. The Duke stood out on the starboard tack, ie northwards; but either his sailing orders to d'Estrées were not understood, or the French admiral deliberately broke them, for he sailed on the port tack, ie southerly. It was certainly the easier course on which to get some sea-room, but it had the effect of leaving his own van to leeward, so that it was some time before DuQuesne's ships could beat up into line, and, worse, of making a complete separation of the French or van part of the combined fleet from its English centre and rear. This meant that, unless d'Estrées could beat back, the allied advantage of 107 of the line to the Dutch 75 was largely lost. In fact the battle was really two battles, for Bankert, commanding the Zeeland squadron, peeled off from the rest of the Dutch fleet and, hauling his wind into line opposite the French, engaged them at a distance, while de Ruyter threw the bulk of his force against the somewhat disordered English line. Here the fighting was heavy and grim. The duke had twice to change his flag to other ships; and the powerful *Royal James*, flagship of his vice-admiral, the Earl of Sandwich (Pepys' 'my lord'), was destroyed by fire-ships and the admiral was killed. No less than nine English ships were so damaged that they had to put back to harbour during or after the battle. The Dutch lost three ships, one taken, one sunk in the action and another in the night that followed, while two others had to make for home. This is not to say that there was no fighting where the French were concerned, for they and the Zeelanders fired at each other heartily; but Bankert continually paid off to prevent the French from getting to close quarters, and, with DuQuesne unable to bear up into line until midday, the weight of such fighting as there was fell upon the French 'rearguard', whose CE, Rabesnières, was mortally wounded.

The casualties on both sides were heavy, particularly among the English and the main body of the Dutch, and towards evening Ruyter broke off the action. During the night the wind shifted to north-west, putting the Dutch to leeward, and in the morning the Duke of York ordered a pursuit. Such of the allied ships as could do so sailed in proper order, the French leading, and at one time DuQuesne in the French van overlapped the rear of the retreating Dutch. Seeing this, the Duke hoisted the flag for him to engage; but DuQuesne did not do so, and the Duke called off the pursuit.

DuQuesne's behaviour is inexplicable. Some French historians claim that, as a strict disciplinarian, he was waiting for the 'executive' (or completion of the signal) which was the firing of two guns. The Duke did not fire them, so DuQuesne did not attack. This explanation will not do. Besides being the most experienced of the French officers, DuQuesne was both a fighter and an able tactician who knew well enough when he held an advantageous position: he was also a man of strong individual personality. Goodness alone knows why he held his hand.

Such was the battle of Solebay, and it led to a fine old row all round. When the report of it reached the English ashore they loudly proclaimed that their ships had been deliberately left in the lurch by the French; and the French, while rejecting this, had a furious dispute among themselves. The dying Rabesnières roundly accused some of his captains of having failed to do their duty, and d'Estrées, admitting disappointment that the French had not done more, threw the blame on to one or two captains and particularly on to DuQuesne. DuQuesne, for his part, fiercely accused his chief of having 'scandalously abandoned the English'. The French ambassador, hurrying from London to Chatham, found the French captains divided into two fiercely contending parties, one for d'Estrées, the other for DuQuesne, and, having held an inquiry, reported to Colbert that, frankly, he could not determine which was in the right. Two things, however, were clear. One was that the English had been given a distrust of the French that was likely to be permanent. The other was that the bitterness between DuQuesne and d'Estrées was so great that they could not serve together again. To disavow the noble vice-admiral was difficult, so DuQuesne was ordered to come ashore, and for a while France lost the services of her best senior officer. The real trouble was that the comparatively new French navy did not yet know how to handle a fleet action, either in the major sphere of tactics or in the lesser one of how best to fight their ships. (Indeed, Grancey wrote that he regretted his waste of powder and shot, which he had blazed off because others were doing so.) They would have been wiser to put their heads together to find out what had gone wrong than to let their fiery and factious tempers get the better of them.

Meanwhile, in Holland, the French invasion was carrying all before it until the Dutch saved the situation for the moment by cutting the dykes. At the same time the de Witt brothers were overthrown and murdered, and power passed into the hands of William of Orange, that cold-blooded fish of a man whose enmity to Louis was implacable, and who was to spend the rest of his life in forming coalitions against him. With the French army cut off by the inundations from its attack on the great cities of the province of Holland, which were the Dutch strength, and with a coalition forming behind it, the obvious course for the allies was to land troops on the coast of the province. It was Ruyter's business to prevent this, and to get the way clear for the home-coming trade by which the cities lived. He succeeded in doing both things.

At the opening of the campaigning season of 1673 Ruyter sailed to blockade the English in the Thames. Finding them already at sea he fell back to the Schooneveldt passage, the shallows north-east of Zeebrugge at the southern point of the Dutch coast. Here two battles were fought in June, a week apart. The allies had effected their junction in the Channel. D'Estrées, with his flag in LA REINE(104), had twenty-seven of the line, with ten fire-ships, his other flag officers being Grancey(CE) in the ORGUEILLEUX (70) and des Ardents(CE) in the TERRIBLE(70). The English had fifty-four of the line, giving an allied total of eighty-one to Ruyter's fifty-five. As the Duke of York, being a Roman Catholic, had had to resign his office when the Test Act was passed, the command was in the hands of Prince Rupert. His order of sailing was peculiar, for the French, though they were the white, or van, squadron, were in fact placed in the centre, between Rupert in the lead and Spragge in the rear. This was a plain, and almost insulting, precaution to prevent them from acting as at Solebay. Arriving off Ruyter's anchorage, Rupert sent in a light squadron, including nine Frenchmen, to tempt him out. The riposte was quicker than he expected, for the Dutch weighed with surprising speed and were assailing the light squadron before it had properly rejoined the main body. The allied line was therefore formed in some disorder, and both fleets stood out in a northerly direction with their officially named rear squadrons, under Spragge and Cornelis van Tromp, in the lead. The actual rear squadrons of Rupert and Bankert, less hotly engaged than the others, began to drop astern, leaving something of a gap ahead of them. Grancey, seeing a chance, led his section of the French fleet into this to cut off Bankert. But the art of fighting in line was as yet ill understood and his change of course left Ruyter free to act elsewhere. Since he wanted primarily to preserve his fleet for future defence the Dutchman did not spend time on the hard task of overwhelming the French centre, but led some of his ships to disengage Bankert; then, going about, he disengaged Tromp, who had been having rather the worse of things, and had had to shift his flag. The action then terminated indecisively as a mist came down. The nearest to any positive success was when Gabaret in the FOUDROYANT boarded the *Deventer*, and though he was repulsed that ship later sank. Ruyter had handled an inferior fleet with the greatest skill, but this time no one could complain that the French had not done their share. The allies remained off the coast, and a week later, having a favourable wind, Ruyter sallied forth and fought a second action. It proved to be nothing more than a heavy cannonade with no decisive result. Ruyter returned to his anchorage among the banks, and the allies, being short of ammunition, returned to the mouth of the Thames. The great Dutchman had staved off the threat of invasion, for the nonce, and the way was open for the home-coming trade.

Both fleets had suffered heavy casualties and a good deal of damage, and it seems that in those days there could be no French sea-fight without a subsequent squabble. In this case Valbelle, whose family had provided several generations of galley captains, and who was as vocal and pushful for promotion as he was courageous, charged Grancey with rashness and Jean Gabaret with cowardice. The latter asked for a court of inquiry which snubbed Valbelle. It also says much (or little) for the discipline of French officers that even Tourville, on reaching England, absented himself from his ship, the SANS-PAREIL(66), without the leave of his admiral, though an apology plus his fine record, especially in the recent battles, saved him from anything but the mildest reproach.

As the allies had failed to clear the way for an invasion, they decided to make a second attempt, this time taking the troops on board with them. The marquis de Martel (LG) had arrived to take command of d'Estrées' van. He was a little piqued at having to give up his chief command in the Mediterranean to act as second-in-command in the North Sea, and was not made any the more pleased by the cold hauteur with which his chief received him. But d'Estrées was always more concerned with his own *superbia* than with anything else. The allies were now in their proper sailing order, with the French in the van, Rupert in the centre, and Spragge in the rear. They arrived off the Texel in August, to find the Dutch hugging the coast as usual; but on August 21st, the wind having gone round to the ESE, the the Dutch sallied forth, Bankert in the van, Ruyter in the centre, and Cornelis van Tromp commanding the rear. The battle, known as the Texel but fought down the coast roughly as far as Kamperduin, took an extraordinary shape. Spragge had come to consider that he had a personal feud with Tromp, vowed to the king that he would get him or lose his own life and, finding the Dutch rear lagging a little astern, hove to the whole of the English rear to wait for them! This meant that for a long time two separate battles were being fought, one between Spragge (who was killed) and Tromp, and the other between the remaining two divisions of both fleets. Here Ruyter, who in all had only seventy ships to the allied ninety, sent Bankert ahead with a mere ten ships to engage the French so that he could throw a superior weight on Rupert in the centre. When it was seen that the French were much more numerous than the division opposing them, the signal was made to Martel in their van to stretch ahead of the Dutch and then to go about so as to catch them between two fires. This he proceeded to do, but as he was about to come into action on the port side of Bankert's ships, engaged already to starboard by d'Estrées' line of his centre and rear, Bankert's vessels put up their helms and broke clean through the French line. This showed fine fighting seamanship on their part, and a sad lack of it on the part of the French. Moreover, having broken through, Bankert led his ships on to attack Rupert, who therefore became heavily engaged in a mêlée with superior numbers.

The battle had started at eight in the morning and it was well on in the afternoon before Rupert fought his way clear and then bore down northward to join his own distant and heavily embattled rear division, followed of course by Ruyter and Bankert. All this while the French had done nothing, later pleading the weakness of the wind for preventing them from coming to the aid of the hard-pressed allied centre. (But it had been sufficient to bring Bankert on to Rupert.) Only now, in the evening, did they sail into action again. The firing in the centre had ceased from about four till six while both sides were sailing to the support of their separately engaged rear divisions, and as the French came up Rupert signalled to them to close in on the Dutch rear. D'Estrées quite simply disobeyed this signal, so that the French came formally into line, as it were, against nobody. Ruyter, on their arrival and having the wind, disengaged and stood back for his anchorage.

The battle had been, in one sense, indecisive, but in the broader sense it was Ruyter's victory; for the allies were so damaged that, with the Dutch fleet still unbeaten, invasion was out of the question and, as they must put back home, the trade could come in and out with safety.

The row that had followed the battle of Solebay was a mere breeze compared with the hurricane of recriminations that followed the battle of Texel. Rupert rightly blamed the dead Spragge for having hove to so as to make the rear division fight a separate battle out of Rupert's control, and he rightly and roundly accused the French of having betrayed him by fighting feebly against much inferior opponents, by failing to get into action later, and by disobeying his signals. D'Estrées claimed that his failure to get into action was due to the lightness of the wind and to the heaviness of the swell, which would not have allowed him to open his gun ports to use his lower batteries. These feeble excuses were so well argued by the captains who represented him at an inquiry held by the order of Charles II that Rupert was for the moment discredited. But now a bombshell was released in the form of a pamphlet by Martel (LG), charging d'Estrées with having dishonoured France by failing to fight. Martel was promptly clapped into the Bastille for several months; but the damage was done. The pamphlet so agreed with Rupert's charges as completely to rehabilitate him and, indeed, to find acceptance among so many Frenchmen that Colbert wrote sadly that the French were becoming known everywhere for their dissensions. The angry Louis ordered a French inquiry to be held by de Seuil, the intendant at Brest, and d'Estrées admitted that the king had wished to 'husband his ships', since he no longer trusted the English owing to their increasing popular opposition to the war. De Seuil's report was a diplomatic whitewashing of all concerned, but gave the game away by saying that Ruyter seemed to have sent a mere ten ships of the Zeeland squadron against the French fleet just to keep it occupied (*'pour l'amuser'*). Trust between the allies had now completely vanished on the English side, and, moreover, the coalition that formed against Louis at the end of 1673 included Spain, who could attack England's Mediterranean trade. So England made a separate peace with Holland early in 1674.

The war had seen hardly any colonial fighting, just a little in India. Louis had sent out a small 'Persian Squadron', to impress orientals, under the command of another soldier-seaman, de La Haye, in the BRETON(56). Though his ships lessened through wastage of crews and nautical damage till only his flagship was left, and even that eventually sank in a storm, this officer put up a stout enough opposition to superior forces for two years, without much actual sea fighting, till he was obliged to surrender to numbers in St Thomé with the honours of war. In the war as a whole the French navy would appear to have failed; but in fact it had been undergoing its growing pains, and had rid itself of many of them. The captains who were good had gained in experience and had come to the front; the more junior officers had learned what to do and what not to do and were becoming fit for command: crews had come to know their business. Now that France had to go on the defensive against a powerful coalition, something like a really fine Service had developed.

Nevertheless, their English allies withdrawn, Louis and Colbert were shy of sending their fleets to sea against the full power of the Dutch. The United Provinces, too, freed from the fear of invasion by the coalition that William of Orange had formed against France, could now take the offensive. So during 1674 the French ships kept mainly in harbour while a large Dutch fleet, with a number of troop transports, came down the Channel. Off Start Point it divided, Tromp sailing with part of it to attack the west coast of France, while Ruyter led the rest to make a complete conquest of the French West Indies. Tromp was able to achieve nothing. He occupied some of the islands off the Biscay shore and took what little plunder they provided, but his attempts to land on the mainland were repulsed. He carried on southward, where Vivonne, ordered to take the Toulon squadron to attack Cadiz, turned back on hearing that Tromp was at sea. The Dutch followed Vivonne into the Mediterranean, but with no chance of overtaking him, so Tromp returned empty-handed. Ruyter had fared even worse. He opened his campaign by attacking Fort Royal in Martinique. Though every effort had been made to keep his destination secret, it had become known, and a fast vessel from France had apprised the French governor, de Baas. His forces were much inferior to the Dutch, and he had only one dilapidated royal ship, the JEUX(40), under the marquis d'Amblimont. But he had prepared a strong defensive position and d'Amblimont seconded him admirably. The Dutch troops were landed but were hotly received, and by good luck or good marksmanship all their senior officers were shot down. They got out of hand, lost heavily, and became utterly discouraged. So Ruyter called off the attack, and took the whole dispirited expedition home.

The Spaniards' government of their kingdom of Naples or the Two Sicilies was neither good nor popular; and in the summer of 1674 the people of Messina rose in revolt and invited Louis to take them under his protection as a start to his conquest of the whole of Sicily. Though Louis had no intention of sparing the troops necessary for such a conquest, he was only too willing to plant an unpleasant thorn in the side of his Spanish enemy: so he accepted, and Valbelle (CE) sailed from Toulon in September with six of the line carrying a few troops and a small quantity of food. He helped the Messinese to drive the Spaniards from any positions that commanded the town and then, seeing that a long siege would follow and that starvation would compel surrender, he hurried back to France for supplies. A Spanish fleet now appeared in force to complete the blockade, and when Valbelle reappeared at the end of the year, with his flag in the POMPEUX(72), with five others of the line and a convoy, he found twenty-two Spanish ships and nineteen galleys barring the way into the port. He had the wind and, after lying off for a day and finding that the Spaniards made no move against him, he resolved to slip in by night – and did so, with little firing and few casualties. It was a bold and seamanlike achievement, though he was, of course, promptly blockaded.

To please the Sicilians Louis had appointed a viceroy, none other than our old acquaintance Louis-Victor de Rochechouart, duc de Mortemart and de Vivonne, who was now both vice-admiral of the Mediterranean and general of the galleys. He sailed for his 'viceroyalty' (ie the city of Messina) in February 1675, with nine of the line, some lesser vessels, and a large convoy. His ships were in three divisions, each of three ships. He commanded the centre with his flag in the SCEPTRE(80), DuQuesne (LG) led the van in the SAINT-ESPRIT(70) and the marquis de Preuilly d'Humières (CE), a sound officer who had commanded a ship in the battles of the Schoonevelt and the Texel, led the rear in the SAINT-MICHEL(60). Approaching the Lipari Islands to the north of the Messina Strait, Vivonne found himself faced by the Spanish fleet of twenty sail under Don Melchior de la Cueva with, behind them, nineteen galleys. Despite the heavy odds against him Vivonne, leaving de Preuilly's division to guard the convoy, bore up bravely against the Spaniards, who had the wind and were bringing DuQuesne's little division under heavy fire. As the action continued the hard-pressed Vivonne, seeing that the Spaniards were making no move against the convoy, whose course would now bring it under his own lee before the Spaniards could reach it, signalled to de Preuilly to join him, and that officer, of his own initiative, worked up with his three ships to windward of the Spaniards, coming into action on their port quarter. It was now about midday, and gunfire was heard coming from the south to windward of the battle. This was Valbelle with four of his ships, able to get out of Messina now the blockading fleet had withdrawn to fight Vivonne, making for the sound of the guns like a good fighter, and brushing out of his way the Spanish galleys who tried to oppose him. At his appearance the Spanish line broke in a disorderly flight, and, of his own initiative, DuQuesne led his ships and some of Valbelle's in pursuit, taking the *Nuestra Señora del Pueblo*. Towards evening Vivonne signalled a recall, for Cueva seemed to be re-forming his line, the road was clear to Messina, and the main object of the battle had been to get the convoy through to that city. This battle of Stromboli[1], besides revealing the ineptitude of the Spaniards, shows the French service in a favourable light, not merely because of its courage in tackling such odds, but in the insight shown, and in the right way, by the three subordinate flag-officers, DuQuesne, Preuilly and Valbelle.

Vivonne, who was always more of an epicure and a courtier than a man of action, sat placidly in his little 'viceroyalty' throughout the summer, apart from taking the fleet out to attack Naples, and then changing his mind and bringing it back to port. He received reinforcements in the shape of some galleys, and a few ships under Alméras (LG)[2]. One of these ships was the SIRÈNE(44) with Tourville as captain, and he was sent with Léry in the TÉMÉRAIRE(50) and the frigate GRACIEUSE into the Adriatic to intercept some troops that were being brought to Barletta. They arrived too late for this but attacked three warships in the harbour. Under cover of their fire lieutenant Coëtlogon, who was later to reach flag rank, cut out a ship of fifty guns and a frigate and fired the third vessel. On their return, near Messina, they fell into a flat calm. The French galleys came out and towed in the two ships and their prizes but the GRACIEUSE drifted helplessly into

[1]There is some chance of confusion, for one or two writers give this name to the next action that we shall describe – an action known to the French as Alicuri.

[2]Or d'Almeiras: both spellings are found.

Reggio and was taken. Tourville and Léry thereupon returned to that port and under cover of their guns sent in a fire-ship which destroyed her.

Being quite unable to cope with French activity in the Mediterranean, Spain called on the United Provinces for aid. The Dutch were now very strained in resources by the long wars, the subsidies they had had to pay their allies, and the depredations of French privateers on their commerce, so that they could not fit out a large fleet. Ruyter was sent, but he had only eighteen of the line with a few smaller vessels, and the ships were not so well found or well manned as formerly. Their admiral knew his fleet to be too weak for its task, and that the Spaniards joining him would be worthless, so gravely had Spain declined. Nevertheless the news that he was sailing occasioned alarm in France and stirred Vivonne into activity. First, to prevent Ruyter having a good harbour near Messina, he sent Alméras and Tourville to take Agosta (Augusta). The former led the troops while the latter, with six ships, bombarded the harbour and silenced the batteries. Then he sent DuQuesne back to France with twenty ships to collect the provisions for the necessary revictualling of Messina. This was to take a big risk, for he had sent the larger ships, retaining Alméras with only the lighter ones. However, though the Spaniards appeared in force, they turned tail when Alméras came out of harbour. Indeed, with the misgovernment of the Court cabals which acted in the name of poor Charles the Sufferer, everything concerned with Spain was now at its lowest ebb. Ruyter found this when he arrived expecting supplies for his ships and a junction with the Spanish fleet, which was not ready. After waiting for some time – a delay which possibly saved Messina – he ceased to expect the fulfilment of courteous but worthless promises, and sailed for Sicily at the end of the year.

Unable to reach and attack Messina because of contrary winds, he was cruising off the Lipari Islands in early January 1676 when DuQuesne and his convoy came in sight. That officer's twenty ships were in the usual three divisions. He commanded in the centre with his flag in the SAINT-ESPRIT (70) with Valbelle (CE) and Tourville (CE) as his next ahead and astern. His van was commanded by Preuilly d'Humières (CE) and his rear by Louis Gabaret (CE). Ruyter had eighteen Dutchmen and a Spaniard, as well as some Spanish galleys. Though there was no great disparity of numbers between the fleets, the French was much the more powerful in weight of gunfire. It was perhaps for this reason that Ruyter forbore to attack, though he had the wind. He may have resolved on a purely defensive action. During the night, however, the wind veered to the advantage of the French, and DuQuesne came in to fight. Both fleets were on the starboard tack, sailing southerly, when the French put over their helms and came down obliquely on to the enemy. Using this tactic, the assailants are exposed to the enemy's broadside while being able to make only a partial reply, and are liable to suffer damage, though this can be partly avoided by good station-keeping. For this some of the French captains were not competent enough as seamen. In the van, the leading ship, the PRUDENT, did not live up to its name, but, failing to come up into the wind in time, came under heavy fire and had to pull out of line. This exposed the next astern, the PARFAIT, which also had to give way, and, as the others of Preuilly's ships overlapped and so masked

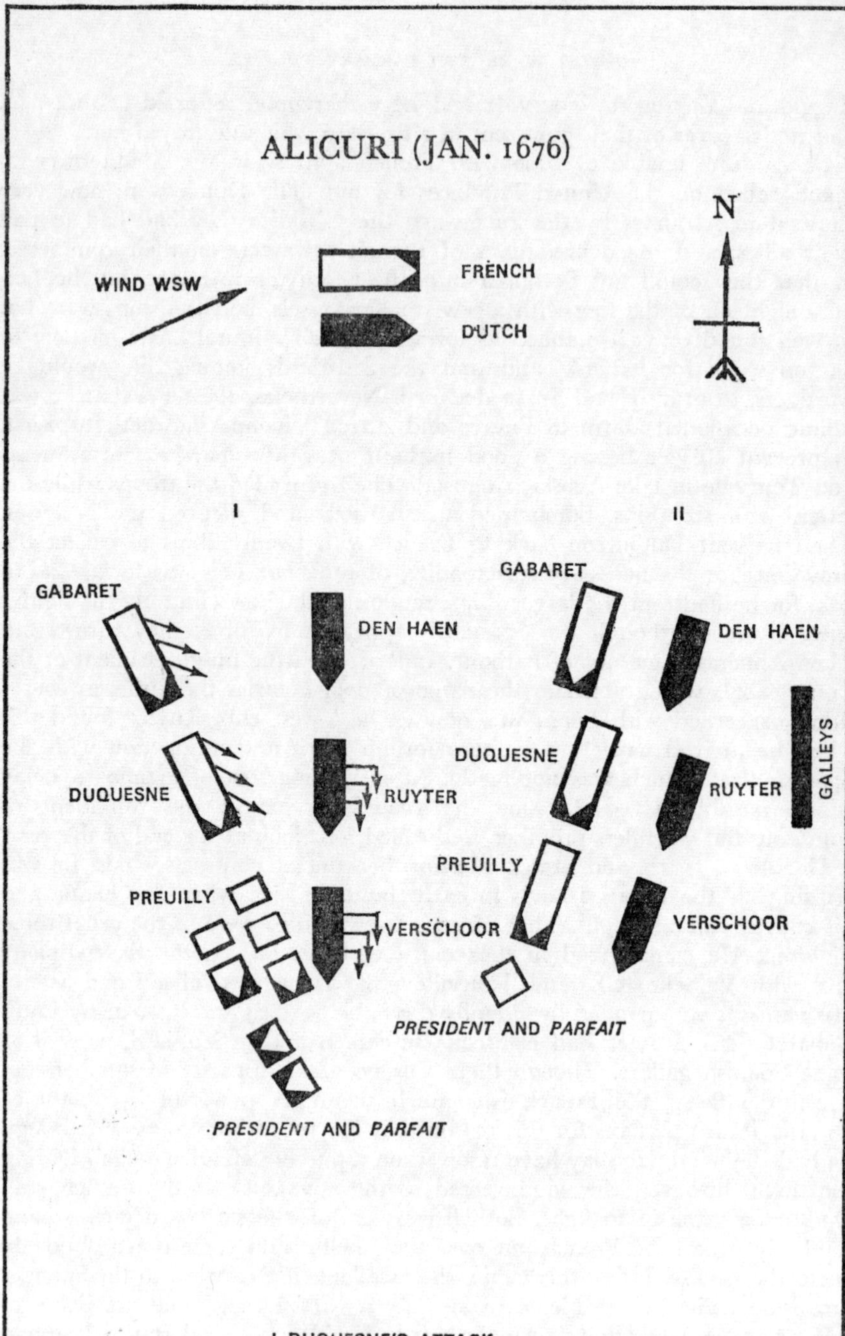
ALICURI (JAN. 1676)
WIND WSW
FRENCH
DUTCH
N
I
GABARET
DEN HAEN
DUQUESNE
RUYTER
PREUILLY
VERSCHOOR
PRESIDENT AND PARFAIT
II
GABARET
DEN HAEN
GALLEYS
DUQUESNE
RUYTER
PREUILLY
VERSCHOOR
PRESIDENT AND PARFAIT
I DUQUESNE'S ATTACK
II POSITION AT BREAK-OFF OF ACTION

each other, for a while Preuilly had to take the fire of several Dutch vessels. DuQuesne's leading ship in the centre also ran into trouble, while in Gabaret's rear squadron some captains handled their vessels so poorly as to run aboard each other. However, the action became general, and very fierce in the French van and centre, though less so in the rear. It is a good illustration of the slow pace of fighting in those days that DuQuesne sent an officer in a boat to Gabaret to order closer action. Ruyter consistently paid off a little and his leading squadron kept station on him by doing the same. Den Haen with the rearmost ships did not do this, and so a dangerous gap opened between the Dutch rear and the centre. The galleys tried to close it, but were driven away by a couple of broadsides from Tourville. Yet the wind was too light for the French to take advantage of it. After more than six hours of fighting, firing ceased in van and centre, though it continued for some time among the rear ships. The galleys came up and towed the worst damaged Dutchmen off to Milazzo, where Ruyter fell back with his other ships.

Such was the battle of Alicuri (Alicudi). Both sides had suffered much damage, and a Dutch ship sank during the night; but the French casualties were the heavier. Though DuQuesne was then joined by Alméras (LG) with the ships from Messina, Ruyter was at last joined by the Spaniards, and DuQuesne thought it better to take his convoy right round Sicily rather than to risk an action in the straits. He then wrote complaining of the poor execution of signals by some of his captains through lack of seamanship experience.

The combined fleet, twenty-seven strong, now made to attack Messina; but to do this it was necessary to get the French fleet to sea and to defeat it. As the French would not be tempted out of harbour the allies sailed on to attack Agosta. The Spanish admiral, Don Francisco de la Cerda, was in command, and rejected Ruyter's proposal that the Dutch and Spanish ships should be intermingled. So Ruyter had the van, la Cerda the Spanish centre, and den Haen the mainly Dutch rear. Vivonne, urged by the king to do something in person, had intended to take command of the fleet, but was restrained by the Messinese, who begged their viceroy to stay ashore. So DuQuesne commanded the French, twenty-nine strong, with Tourville (CE) and Preuilly (CE) as his next ahead and astern: Alméras, with Valbelle (CE) next astern, led the van and Gabaret the rear. The battle of Agosta consisted mainly of a furious fight between the two van divisions; for la Cerda kept the Spanish centre back at long range and almost out of action, and den Haen was obliged to keep station more or less on him, though his sternmost ships got into action against the last French. La Cerda had some excuse in that his government had sent him to sea distinctly short of powder, so that it was only late in the day that he came into action to disengage his Dutch van. This had driven the Lys out of line, with Alméras killed, but otherwise had been heavily handled; for the shyness of the Spaniards had allowed the leading ships of the French centre to join in the attack on it. Three Dutch vessels were towed out of line in a desperate state by galleys and Ruyter had been mortally wounded. The allies withdrew to Palermo and the victory was plainly to the French. It is therefore interesting to find

DuQuesne writing to complain of disaffection among some of his officers. This was, indeed, the persistent quarrelsomeness of the French gentry, and Valbelle was mainly responsible, for, though a good seaman and an almost over-daring fighter, he was boastful, self-centred, and a perpetual malicious intriguer. A 'character', and not a pleasant one.

Meanwhile Vivonne, longing for the fleshpots of Paris, had been asking for recall, and this had been granted. But he knew from the tone of the royal correspondence that the king was displeased with his personal inactivity, so, wishing to return with some *éclat*, he led his twenty-nine ships to attack the allied fleet in Palermo. La Cerda had been dismissed from his command (though he remained as a volunteer), and Don Diego Harra had replaced him. The bay of Palermo had on one side a mole which projected half a mile into the sea and then turned at a right angle for a quarter of a mile. It was fortified, and opposite it on the other side of the bay was a powerful battery: there were other batteries along the shore. Harra had moored the fleet in a curved line stretching from the mole to the battery opposite, and had filled any gaps with galleys: a rigid defensive position. Why Den Haen, now commanding the Dutch, had accepted this instead of insisting on having sea-room to fight it is hard to understand. But the Spaniards were of poor quality anyway, and the spirit seems to have gone out of the Dutch with Ruyter's death.

Vivonne sent Tourville and Gabaret in a felucca to reconnoitre the position and they advised an attack on the ships at the end furthest from the mole, with a bombardment of the rest of the line by the remaining French vessels, followed by sending in the fire-ships under cover of their guns and through the smoke. As Vivonne held fire-ships to be 'the law and the prophets' he readily accepted the proposal. De Preuilly d'Humières attacked the further end of the line with nine vessels, the rest came in and opened fire, and the fire-ships were released with the right wind at the right moment. The allied fleet largely panicked. Cables were cut and ships drifted, some to go ashore at the bottom of the bay, and others to huddle anyhow behind the mole. The Dutch *Steenberg* was fired and drifted against two other Dutchmen, to fire them also. Both the Spanish admiral and Den Haen were killed, and the Spanish admiral's flagship with three others of his ships and two galleys also were destroyed by fire. The victory could have been completed by a well-planned entry into the bay to attack the discouraged and disordered allies, for after all, only seven out of twenty-seven had been destroyed. But the fat and lazy Vivonne felt that enough had been done to secure his personal reputation, and took the fleet back to Messina, whence he later returned to France. The duc de la Feuillade was sent out as a replacement 'viceroy', but with instructions to evacuate the troops and bring them and the fleet home; for the Sicilian venture had never been more than a side-show in the mind of Louis. This he did early the next year, and the revolted Messinese were dishonourably left to the tender mercies of their Spanish rulers. The fine work of the French Mediterranean fleet had served only the indirect purpose of distracting one of Louis' enemies.

The battle of Palermo had been a fire-ship victory. Yet with it comes the end of the fire-ship age, although these craft were not discarded for some time. We have seen them successful and dreaded from Guétaria onwards; but it had become increasingly clear that they were deadly only to ships that were in harbour or, like the *Royal James* at Solebay, too hemmed in to have any sea-room to avoid them. They had failed in the other North Sea battles, and in the Mediterranean Ruyter, though heavily engaged, had had little difficulty in avoiding them. Moreover, the gunnery of ships soon improved so much that it became doubtful if they would succeed even against those in harbour. Their day, like that of the galleys, was coming to a close.

While DuQuesne had been gathering laurels and encomiums in the Mediterranean, his superior officer, vice-admiral le comte d'Estrées, whom he had criticized so gravely after Solebay, had been left on the beach. He now proposed to Louis that he should take a squadron to the West Indies to destroy the Dutch trade and take Dutch possessions there. He was willing to pay part of the expenses of fitting out the squadron in return for a proportionate share of the prize money. So he sailed from Brest in the autumn of 1676 with his flag in the GLORIEUX(60), four fifty-gunners, two lesser of the line, and two frigates. The Dutch had anticipated him by sending, earlier, a small squadron under vice-admiral Binckes; and by the time d'Estrées arrived, Binckes had taken Cayenne and Marie Galante, had raided San Domingo, and established his headquarters in Tobago. The few French ships in the islands had done nothing, for their commander, Grancey, had refused to obey de Baas, the governor, who had sent him home. D'Estrées first made for Cayenne and, personally leading the landing party, took it readily enough at the end of the year. Then he went to Martinique to pick up the three ships there and some troops, and sailed in February 1677 with ten of the lighter sort of ship of line and some smaller vessels to attack Tobago.

The roadstead of this island was adequately roomy, but hard to assault, for the entry channel was uncharted, rocky, and commanded by a fort. Moreover, if one did succeed in entering but failed in one's attack, it would be highly difficult to get out against the head wind that had brought one in. So when, approaching the entry, the INTRÉPIDE grazed a rock, the fleet came promptly to anchor. The troops tried to rush the fort and failed; but when a local pilot was captured who undertook to see the fleet through the channel in return for his freedom, a combined land and sea assault was determined upon. Binckes was distinctly inferior to the French. He had his flag in the *Bescherming*(50), but only this ship and the *Huis te Kruiningen* (56) and the *Zeelandia*(44) were at all equal to the French vessels. In all, his ten ships carried more than eighty guns less than their enemy, and they were short-handed. He had not expected the French to risk the entrance channel, but nevertheless had drawn up his ships on an incurving arc opposite the entry. Between them and the little town of the island were some merchantmen, packed with civilians, sheltering there because the assault was expected to be made by land.

The French came in under heavy fire, impressively and bravely in two lines ahead, one led by d'Estrées in the GLORIEUX, the other by the senior captain, Louis Gabaret, in the INTRÉPIDE. They then engaged the Dutch at the closest quarters with intent to board. D'Estrées made for the *Kruiningen*, probably thinking that, as she was the biggest, she was the flagship. So the honour of engaging Binckes fell to the Précieux, which was so roughly handled that eventually, largely dismasted, she drifted down the roadstead and went ashore – at which the unwounded among her crew left her. The Intrépide also suffered severely; Louis Gabaret was killed, his lieutenant fled the ship and refused to return (for which a court martial sentenced him to death), and with only a handful of men still able to fight she struck, and then perished in the fire[1]. For, from being a scene of battle, the roadstead had quite quickly become an inferno. The *Leyden* caught ablaze and set fire to the Marquis, which blew up. D'Estrées had succeeded in carrying the *Kruiningen* by boarding, only to find that she too was furiously alight, and before he could get the Glorieux clear the Dutch ship also blew up, firing the Glorieux, which d'Estrées and his officers then abandoned in a boat, promising to send help to the crew. They never did so, for the boat was sunk and the admiral had a job to struggle ashore, while the Glorieux blazed till she vanished. The debris of the exploded ships set fire to the rest of the Dutch, only Binckes succeeding in towing the *Beschermíng* clear, and to the merchantmen full of civilians, while the remaining French ships struggled to get away. The final tableau was that the Dutch still had the *Beschermíng,* and the grounded Précieux, which they took: two of their ships were ashore hopelessly wrecked, and all the rest had burned. The French had lost the Glorieux, the Marquis, the Intrépide and Précieux, while the Galant and Fendant had lost masts and took three days to be towed clear of the channel. It was a horrible business and for the French, in one aspect, a shameful one. For the officers of some ships had left their crews to their fate. Indeed, the captain of the grounded Précieux, told by d'Estrées to fire his ship, would have done so, committing his many wounded to a dreadful death, had not the least hurt hurt among them prevented him. But then it was an age when 'class' mattered though it apparently had no responsibility for its inferiors. The spirit of the Service, which Colbert had begun to get into the Mediterranean fleet, had not yet reached the officers of the Atlantic. The land attack on Tobago fort had failed, with heavy loss.

D'Estrées came back to France but, as Tobago still remained a base for Dutch privateers, he was ordered to return in the autumn with another fleet, and to take it, mopping up the Dutch establishments on the West African coast on the way. This he did, and arrived off Tobago in early December, with eleven of the line and the usual smaller vessels, his flag being in the Terrible, with de Grancey (CE) as second in command in the Tonnant. Though Binckes had now only the *Beschermíng* and the *Précieux,* d'Estrées was not going to try the roadstead again, but landed his troops on the other side of the island, giving them a long and trying march to attack the fort. The French had some mortars, and only the third shell fired from

[1]De la Roncière states that she surrendered the next day; but as nothing more was heard of her, and Binckes would probably have refitted her, as he did the *Précieux* (unless she was too hopelessly damaged), the account I have given seems the more likely – especially as Troude states that she took fire.

these landed in the Dutch powder magazine – the explosion killing Binckes and over 300 men. The fort was taken without a further shot; and the town, and the ships trapped in harbour, could only surrender.

D'Estrées then sailed to take the Dutch possession, Curaçao, and some French buccaneer vessels sailed with him. As we have already seen d'Estrées was a very haughty person, who thought that to show fitness for command one should be cuttingly offensive to one's immediate juniors. His officers dreaded approaching him and his flag captain considered sailing with him to be a particularly harsh kind of purgatory. He was also one of those men who will not admit their ignorance of any matter. As he had never learned to be a seaman this was particularly unfortunate; for, having kept his immediate juniors at a distance, he could hardly seek or accept their advice, and for his navigation he relied on his own ignorance and the faulty opinions of his private household. His approach to Curaçao was by running down a latitude and altering course southward when he judged that he had sailed far enough. So, on the night of May 11th 1678, he ran his fleet on to the Aves Islands. These are islands only in the technical sense, for they consist of a wide range of reefs of which some just rise above the surface. He had ordered the ships to take in topsails for the night, so that when a buccaneer sailing ahead fired an alarm gun, it proved impossible to get the ships quickly into the wind, and most of them ran on to the rocks. Daybreak revealed an appalling scene. Some vessels that had been well astern or had grounded only lightly were hove to clear of the reefs, but twelve others, including seven of the line, with both flagships, were well on the rocks amid the breakers, and the wind that had taken them there was driving them on further to break them up. There was nothing to do but to use the boats to get away as many of the crews as possible into the ships that were still afloat. Though most were saved, the loss of life was heavy; so was the loss of seven of the line. This running aground of a whole fleet through faulty navigation is unique in naval history and d'Estrées should have been dismissed the service. Nothing was done to him, perhaps because his family influence was too powerful, and he was even raised to the dignity of a Marshal of France in 1681. Colbert was aghast at the disaster, but did literally retrieve something from the wreckage. For he sent out an experienced old captain named Forant with some ships, and, by great skill and hard work for two months, he managed to recover 364 of the guns that had been lost together with some 3 000 shot.

In the Mediterranean, after their losses at Palermo, the allied fleet had retired to Naples; but as the French never intended a conquest of Sicily, their fleet had nothing more to do than bombard a fort or so and burn some shipping till the evacuation of the island. There was, however, more activity on the Atlantic coast. Most of this was privateering and the men of Dunkirk alone, led by the increasingly famous Jean Bart, claimed to have taken or sunk over 300 ships, mostly Dutch. The king joined in this by hiring out royal ships of the lower rates, and allowing his officers to command them. Among all this privateering there were two naval actions. Both were fought by François-Louis de Rousselet, marquis de Château-Renault, another soldier who had turned sailor, more skilfully than most, and with influence to support his capacity had had rather rapid promotion to CE. In 1677, cruising off Ushant with three 50-gun ships, a 40-gunner and some frigates, he made skilful use of having the wind to attack the head of the escort of a convoy, while his frigates cut off four of the merchantmen. In the next year he was sent south to intercept a reinforcement which the Dutch were sending under Evertzen to join and bring home their battered Mediterranean fleet. Château-Renault came upon the Dutch at the entry to the Straits of Gibraltar. Evertzen had eleven of the line, while Château-Renault, with his flag in the COURTISAN(70), had four 68s and a 40-gunner. The Dutch admiral apparently had not expected to meet any enemy so soon, for he was sailing in no order, and the French attack on the leading ships was made so swiftly that the Dutch line was never properly formed. The action was long and hard fought and, though the INVINCIBLE was badly mauled, the Dutch suffered the heavier casualties and much the greater damage to their ships, so that they had to retreat into Cadiz for prolonged repairs. Rather oddly, the king and Colbert reproached Château-Renault for not having gained a victory, as they expected when any of His Majesty's ships fought the Dutch! Apparently to compel a superior force to abandon its voyage and seek refuge in harbour for its damages was not enough.

This was the last naval action of any importance in the war, for both sides were nearing exhaustion. Holland was willing enough to accept the *status quo* to allow her grievously damaged commerce to recover; Louis was willing to accept Franche Comté from Spain, while he took a breather before making his next assault on the peace of Europe. So the treaties known as the Peace of Nimwegen were signed in the late summer of 1678.

With the Continental war off their shoulders Louis and Colbert could turn to deal with those perpetual pests, the 'Barbaries', and these were much hunted by single French vessels for two years. Meanwhile, in 1679, there occurred an event which showed that all was not yet perfectly well in the French service. Tourville was bringing four 50-gunners round from Toulon to Rochefort when they ran into a full-force Bay of Biscay gale, and were separated. Tourville's ship, the SANS PAREIL, lost her bowsprit. This brought down the foremast, she opened up forward, and was plainly doomed. Coëtloglon closed her in the ARC-EN-CIEL, and Tourville got his boats away with nearly eighty men; but their discipline was not yet good enough to stand up to trial, and all boarded the ARC-EN-CIEL, letting the boats drift instead of bringing them back. Coëtlogon got a boat away, which closed the SANS PAREIL, and Tourville, advising his officers to follow him, stripped to shirt and drawers, dived overboard, and swam to her. No one followed, and the ship slowly sank. The CONTENT managed to reach Belle-Isle and beach herself while barely afloat. The fourth ship, the CONQUÉRANT, in a sinking state, was closed by an English vessel, at which her captain, Chabert, gave a *sauve qui peut* to his men (of whom the Englishman picked up twenty or so), but declared that he would go down with his ship, and did so. One feels that he acted more nobly than Tourville, though no one reproached the latter. Colbert's wrath for this disaster fell upon the intendant at Toulon, who had refitted the ships, and who was told that the king had no further use for his services 'from this moment'. This was largely due to

## THE GREAT ADMINISTRATORS

RICHELIEU (*by Philippe de Champaigne*)

COLBERT (*by F de Poilly*)

CHOISEUL

PLATE IV

TOURVILLE

Colbert's son and understudy, Seignelay, being anxious promptly to find a scapegoat, and was somewhat unjust; for DuQuesne, holding an inquiry, pointed out that the ARC-EN-CIEL, a fairly new ship built at Toulon, had stood the storm readily enough, and that the other ships, built under contract at Brest, were old and should not have been sent to sea if there was any chance of rough weather. This was an early example of what was to become an old story in the French service – the unwisdom of having a navy administered by landsmen with no great experience of seafaring: the rivalry and enmity between the *épée* and the *plume*.

In 1681, as the single ship hunting of corsairs was not over-successful, DuQuesne was sent to sea after them with seven of the line. He chased a group of them into the harbour of Scio, a Turkish possession, and when its governor refused to expel them, he first bombarded the place and then established so tight a blockade that 'only the fish could get in and out'. This violation of neutrality was not to Louis' taste, as he did not want a war with Turkey on his hands; but by the French merchants who traded in Constantinople paying 80 000 crowns the wrath of the Porte was appeased.

Clearly a mere patrolling of the Mediterranean was not going to stop the depredations. The pirates must be attacked in their bases, of which Algiers was the most important. It was not easy to do this effectively, unless troops in force could be landed, and Louis, who always had his eye on war in Europe, was not willing to weaken his available armies by sending any considerable number of men oversea. The actual harbour at Algiers was made by a mole built between a strongly fortified nearby island and the shore; and the rest of the bay on which the town stood was also made formidable by a number of batteries. Ships were bound to suffer from the fire of these, while their own guns, 34-pounders at most, could effect not a great deal more than retaliatory damage.

But now France was going to try a new weapon, the bomb galiots. These were small vessels, sailing platforms, designed by a naval constructor, Renau d'Eliçagaray, and besides a few guns each carried two bomb-throwing mortars on a pivoting mount, with a range of about one and a half miles. So in 1682 DuQuesne sailed for Algiers with eleven of the line, his flag being in the SAINT-ESPRIT(76), some galleys to tow the ships into place, and five of the new bomb galiots. Though bad weather prevented him from getting his ships into position for a month, he did so by the end of August, and the galiots, covered by the fire of the ships, did much damage. Their bombs were a new and alarming form of attack, much as were those from the air in the twentieth century. The Dey sent the French consul to offer terms, but they were not sufficient and DuQuesne refused them. Then, with bad weather approaching, he sailed for Toulon. Nothing had been settled, but the Algerines had been left apprehensive.

These apprehensions were realized the next year, when DuQuesne returned with thirteen ships (Tourville, LG, being second in command as in the former year) and seven bomb galiots. Lessons had been learned from the previous attempt. The galiots were now moored on a long hawser so that they could lie off by day and be hauled in close by night; and it was by night that the bombing was done. The damage was great and the Dey capitulated, promising to free slaves and give hostages. About 600 slaves arrived on board the French ships, but when it came to hostages one of them, Mezzomorto, promoted a revolt, slew the Dey, and proclaimed himself in his place. At this DuQuesne renewed his bombardment, and the Algerines blew the unfortunate French consul, Father Levacher, from the muzzle of a cannon. With the supply of bombs exhausted DuQuesne withdrew; but the expedition had been a success. A large number of slaves had been freed, and with nearly all their shipping and much of the town destroyed the Algerines had had a severe lesson. Indeed, early in 1684 they signed a peace for a hundred years!

Colbert had died in 1683, and had been succeeded by his son, the marquis de Seignelay. He had been trained in naval administration from earliest manhood by his father, who had designated him as his successor. But he did not have his father's earnest thoroughness, nor his understanding of the proper value and use of sea-power. A child of his time, Seignelay was a flamboyant person to whom the prestige aspect of naval affairs was more important than their true values. The Republic of Genoa had always been unfriendly to France and had allowed Spain, under a treaty, the use of its harbour for landing troops *en route* for the Spanish Duchy of Milan. On the frivolous excuse of supporting the claims for loss of property of the Fieschi, who had been expelled from Genoa more than half a century before, it was decided to give the Genoese a lesson. So DuQuesne, with his flag in the ARDENT(74) and Tourville as his second in the FERME(64), with twelve other ships and ten bomb galiots, was about to sail, when Seignelay arrived to direct the expedition. DuQuesne objected: Seignelay was certainly his chief, but was a civilian, and his place was ashore. Seignelay insisted, so DuQuesne, like Achilles retiring to his tent, went into his cabin, and remained there for the whole of the operation. Though their republic was in decline, the Genoese had their pride, refused the French demands, and opened fire. The French landed troops who took and sacked a suburb, and their bombs did tremendous damage so that the Genoese had to submit. But the French had quite heavy casualties and injuries to their ships, the indemnity paid went in part to the Fieschi and in part to rebuilding churches destroyed by the bombs, and the French merchants in Genoa were ruined. A thoroughly pointless bit of prestigery.

In 1685 the vice-admiral d'Estrées put to sea again, this time to deal with Tripoli and Tunis; and thanks to the alarm caused by the new bomb-vessels he was successful in getting freedom for a number of slaves. He was less so three years later against the Algerines, who had quickly forgotten their 'hundred years of peace'. He had ten ships and ten bomb galiots, but he found wind and weather against him, the powder supplied to him was of poor quality, and, worst of all, his bomb galiots had become strained by the shock of their own artillery and had to be withdrawn piecemeal from the action. Yet though his bombardment was a failure Algiers was willing to make a peace, and it proved more lasting than any former one. This failure was not d'Estrées' fault, but it was almost his last piece of active service, though he lived until 1707. He was a brave man, but he had been a bad leader, and a poor appointment[1].

In that year of 1688, too, there occurred an unusual action, which is yet illustrative of the time. Louis had laid down that his officers should demand a salute to the French flag from ships of all nations except England. When Tourville, sailing with three ships to reinforce d'Estrées at Algiers, met the Spanish vice-admiral, Papachino, with two vessels, he demanded this salute, and, on a refusal, entered into a hot engagement. The Spaniards eventually struck, when they agreed to fire the salute, which was ceremoniously returned. All ships had causalties and were badly damaged. It was to modern minds a strange age in which ships, the object of a voyage (Tourville had to put back to harbour), and the lives of men upon whom families depended should all be risked upon a mere point of punctilio. But then Louis was more pleased with the destruction of the captured little GRACIEUSE (see above, page 56), though the matter was utterly unimportant, than he would have been at a considerable tactical success, such as landing or re-embarking troops safely in difficult circumstances. If a ship of his proud majesty did happen to fall into the hands of the enemy, it must not remain there!

In 1688, too, died Abraham DuQuesne. He does not, perhaps, quite come into the ranks of the great admirals. His judgment was sometimes at fault, as in the retention of Le Havre as a base, and in his failure to act during the pursuit that followed the battle of Solebay. He could be, and often was, contentious and quarrelsome. He growled at his inexperienced officers instead of helping them, and so perpetually complained of their incapacity and of things wrong with his ships that Colbert had to tell him that his continual grumbles handicapped rather than aided the Service, as they irritated the king. But, having been a Dieppe privateer in his youth, he was a fine seaman, a stout fighter, and, as he showed in his engagements in the Mediterranean, a considerable tactician. He was a disciplinarian, and could be imperious – even over-hastily so, as at Scio. Except for the short time that he was in the Swedish service and the years on the beach because of the dispute that followed Solebay, he had fought and handled his ship with distinction in nearly all French actions from Guétaria to the bombardments of Algiers, and was not afraid to take an independent line and defy his superiors, as with Seignelay in the Genoa affair, which was DuQuesne's last piece of service. Under him, and largely because of him and his precepts, there had grown up in his last few years a school of officers of whom no complaint could be made, for they were seamen and fighters too – a very

[1]The activities in the Mediterranean, and especially the victories over the redoubtable Dutch and their famous admiral, for the first time gave the inland Frenchman some interest in this new armed force of his king – this navy, which spoke a new and incomprehensible language. Nautical terms had been listed for the education of the little comte de Toulouse, the royal bastard who had been made Admiral of France (and who, unlike other holders of that office, was not to be a mere *fainéant*), and this list by M Desroches was published in Paris in 1687. This, as stated in its *Préface au Lecteur,* was for the benefit of those who want to speak properly of naval matters. Of course our Mr Pepys obtained a copy: this is in his library at Magdalene College.

different corps from the often makeshift officers of his earlier years. (Though, alas, one much given to intrigue for promotion; even Tourville was guilty of trying to undermine DuQuesne by letters to Seignelay: they were in disagreement, owing perhaps to a difference in age of over thirty years.)

In 1685 Louis had most unwisely revoked the Edict of Nantes, so that the Protestants were no longer allowed freedom to practise their religion: a persecution of them followed, wherefore some thousands of seamen and a number of good officers left France and are later to be found in the service of Louis' enemies. But there was no persecution of DuQuesne. Though a staunch Huguenot, DuQuesne was left peacefully in possession of the estate the king had helped him to buy, and of the patent of nobility that the king had given him. Starting as captain of a small ship in Richelieu's navy, he had seen that navy perish in the quicksands of the Fronde, and had helped Colbert to re-create it; and though, as a Protestant, he was never given the baton of a Marshal of France, he had ended as the leading naval officer of his day, with the possible exception of his last second in command, Tourville, who was to succeed him and transcend him.

CHAPTER VII

# English Succession — Spanish Succession — Guerre de Course

By now war clouds were rolling up ominously. After the peace of Nimwegen Louis had used some dubious legal decisions and an apt application of force to make annexations towards the Rhine, including much of Alsace. It had therefore been easy for William III to re-form a coalition against him, the League of Augsbourg. Events in England supplied the actual *casus belli.* A parliamentary revolution in December 1688 expelled King James II for autocracy used in favour of Roman Catholics, and gave the throne to his daughter, Mary, and her husband, William III. James fled to France, and Louis, himself both an autocrat and a keen Catholic, could not stand aside while his arch-enemy became ruler of wealthy England with her powerful fleet. He promised, and promptly gave, active support to James.

So in March 1689 Louis Gabaret (CE) took him and a few troops to Ireland, to be followed shortly by a proper expeditionary corps, escorted by Château-Renault (LG) with twenty-four of the line. The English admiral, Herbert, having failed to intercept Château-Renault, had made for Kinsale. The Frenchman passed to the south of him and entered Bantry Bay. Here he was disembarking the troops and unloading the transports when the English appeared. Herbert had twenty-two of the line; Château-Renault, with his flag in the ARDENT(66), had twenty-four, Louis Gabaret (CE) leading his van in the SAINT-MICHEL(56), and Forant the rear in the COURAGEUX(56). The French, who had the wind, came out of the deep bay with each division in line ahead, and made a muddle of getting into line of battle in its eight-mile-wide mouth. Thereby Château-Renault's division, carrying more sail than Gabaret, found themselves masking the fire of their own van, and the need shortly to go about increased the confusion. The action was therefore fought with the centre division leading the line, and the van falling into a place between it and the rear. The cannonading was heavy, and the DIAMANT suffered badly aft because her captain, Coëtlogon, though an experienced officer, had got up too much reserve powder for his stern guns, and a lucky shot exploded it. Herbert kept giving way till the fleets were some twenty miles out to sea; then Château-Renault, thinking that Herbert was drawing him away from shore so as to out-manoeuvre him and get at his transports and the troops, broke off the action and stood back to Bantry to protect them.

A good deal of damage had been done to both fleets, and both claimed a victory. The battle was in fact indecisive, but in favour of the French. Their damage and casualties were less than those of the English, and the object of

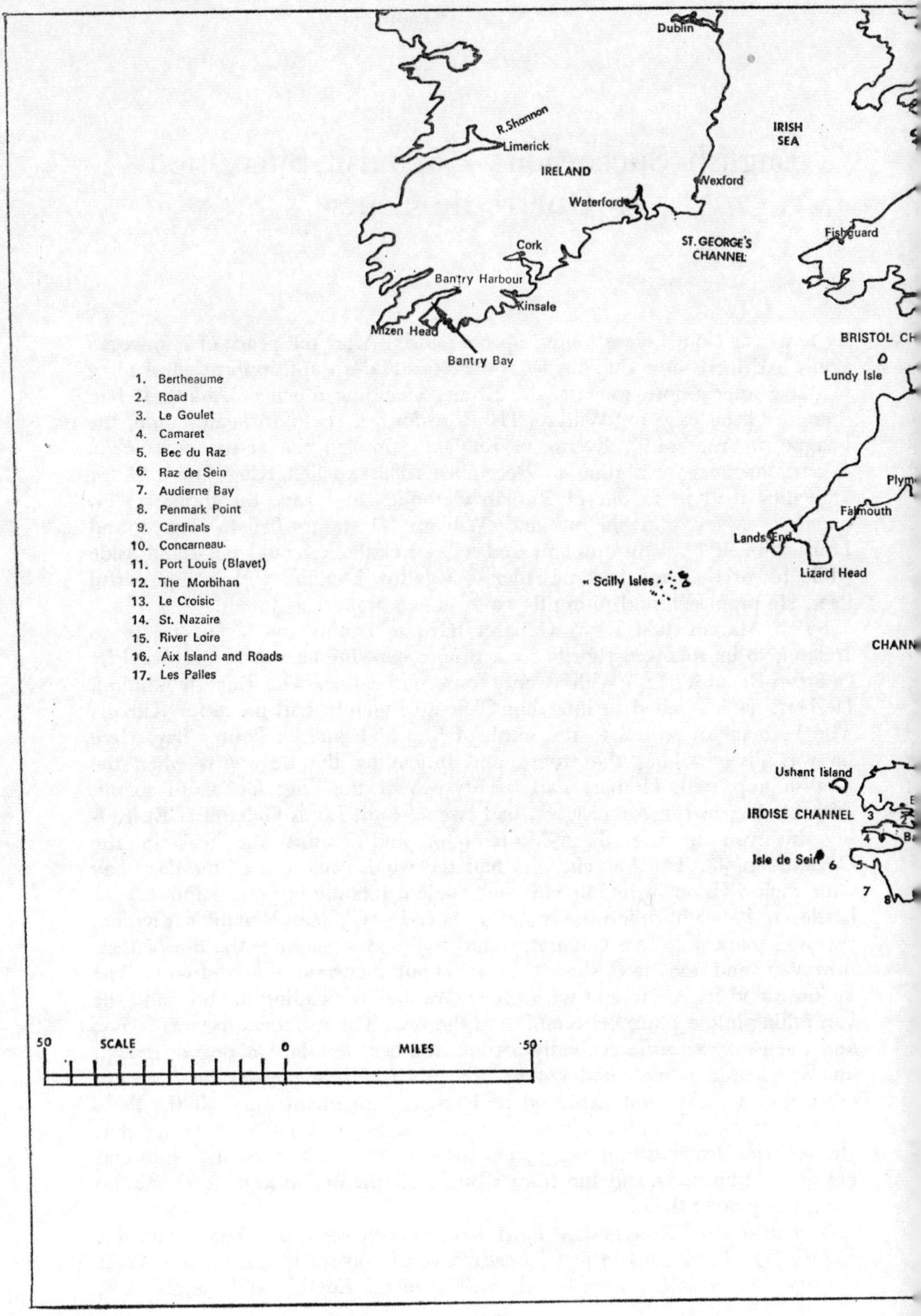
Dublin
R. Shannon
Limerick
IRELAND
IRISH SEA
Wexford
Waterford
Cork
ST. GEORGE'S CHANNEL
Fishguard
Bantry Harbour
Kinsale
Mizen Head
Bantry Bay
BRISTOL CH
Lundy Isle
Plym
Falmouth
Lands End
Lizard Head
Scilly Isles
CHANN
Ushant Island
IROISE CHANNEL
Isle de Sein
1. Bertheaume
2. Road
3. Le Goulet
4. Camaret
5. Bec du Raz
6. Raz de Sein
7. Audierne Bay
8. Penmark Point
9. Cardinals
10. Concarneau
11. Port Louis (Blavet)
12. The Morbihan
13. Le Croisic
14. St. Nazaire
15. River Loire
16. Aix Island and Roads
17. Les Palles
50 SCALE 0 MILES 50

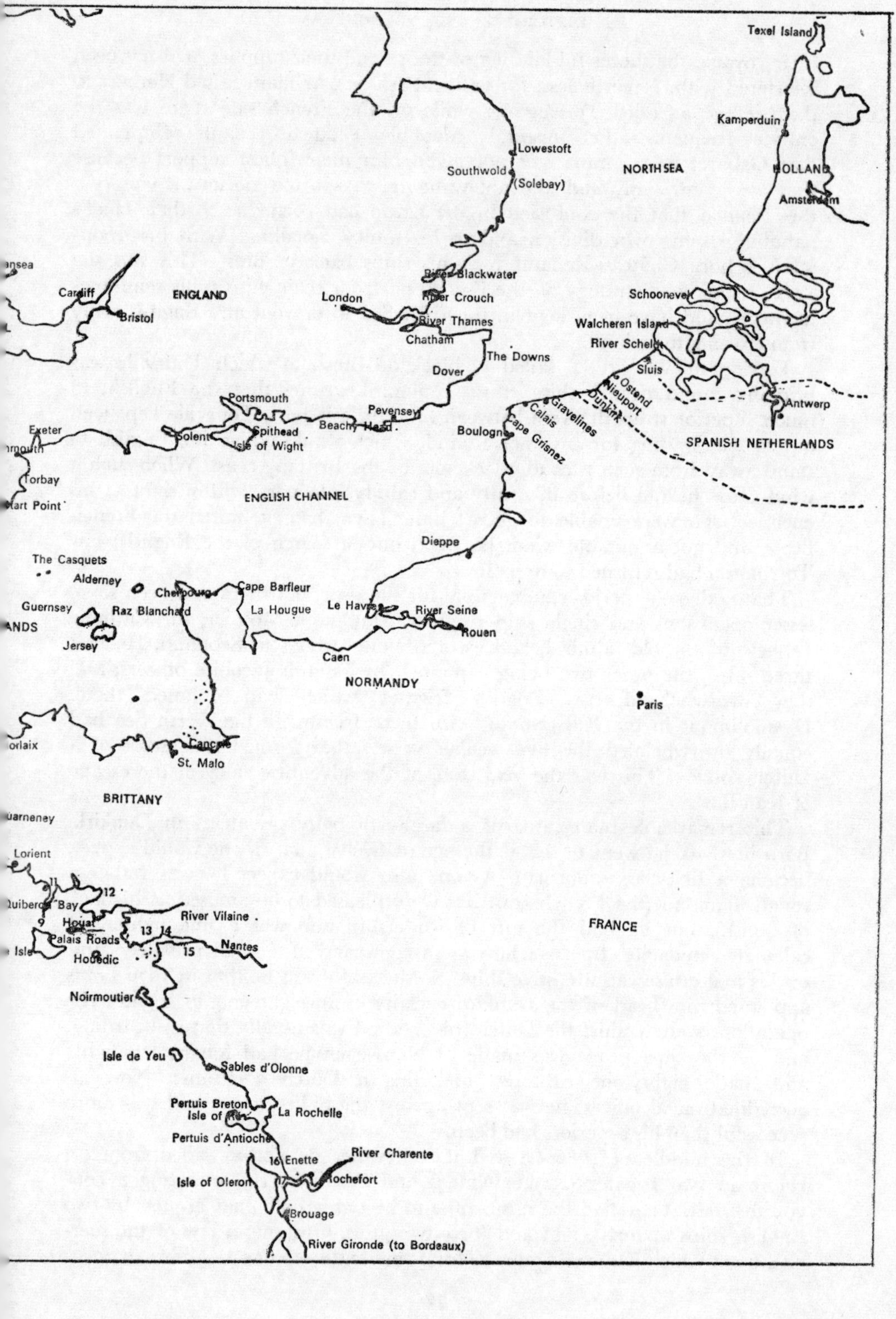

Texel Island
Kamperduin
Lowestoft
Southwold
(Solebay)
NORTH SEA
HOLLAND
Amsterdam
River Blackwater
River Crouch
London
River Thames
Chatham
The Downs
Dover
Schoonevelt
Walcheren Island
River Schelt
Sluis
Ostend
Nieuport
Dunkirk
Gravelines
Calais
Cape Grisnez
Antwerp
SPANISH NETHERLANDS
Cardiff
ENGLAND
Bristol
Portsmouth
Spithead
Solent
Isle of Wight
Pevensey
Beachy Head
Boulogne
Exeter
Torbay
ENGLISH CHANNEL
Dieppe
The Casquets
Alderney
Cherbourg
Cape Barfleur
La Hougue
Le Havre
River Seine
Rouen
Guernsey
Raz Blanchard
Jersey
Caen
NORMANDY
Paris
Cancale
St. Malo
BRITTANY
Lorient
11
12
Bay
9
Houat
Palais Roads
13
14
15
River Vilaine
Nantes
Houedic
FRANCE
Noirmoutier
Isle de Yeu
Sables d'Olonne
Pertuis Breton
La Rochelle
Pertuis d'Antioche
16
Enette
River Charente
Isle of Oleron
17
Rochefort
Brouage
River Gironde (to Bordeaux)

their voyage, the successful landing of troops and their supplies, had not been interfered with. Nevertheless, for political reasons, William raised Herbert to the peerage as Lord Torrington; while on the French side there was the only-too-frequent rather unseemly squabble. Château-Renault complained that Gabaret and Forant had not given him their fullest support because they were unfriendly and 'did not want me to win too glorious a victory': they replied that the confusion in the battle had been due to their chief's handling of his own division and to his faulty signalling. With his troops safely ashore Château-Renault took his ships back to Brest. This was the start of some bad strategy on the French part, for their ships (plus reinforcements) should have been kept in the Irish Sea to prevent any English army from crossing to Ireland.

Torrington now (1689) sailed to blockade Brest, to which Tourville was bringing round twenty ships from Toulon. Learning that the English, in much superior strength, were between him and his port, Tourville kept well out to sea, waiting for a westerly wind, which would oblige the English to stand away from such a deadly lee-shore as the Brittany coast. When such a wind came he ran before it, neatly and calmly into port, within sight of his enemies who were unable to get at him. Then, having united the French fleets, and got a suitable wind, he stood out in search of the English; but Torrington had returned to harbour.

That ended the major campaigning for the year, but there had been some lesser operations and single ship combats. DuQuesne-Mosnier, the Catholic nephew of the old admiral, had escorted some troops to Scotland. He had three ships, the other two being captained by English Jacobite officers, and they overwhelmed two English frigates which had opposed them. D'Amblimont in the PROFOND(40) with three frigates in the North Sea had equally overwhelmed the five weaker vessels that formed the escort of a Dutch convoy. This was the year, too, of the adventure story of the escape of Jean Bart.

This remarkable man came of a family of petty privateers in Dunkirk. Born in 1650, he went to sea at the age of twelve and, as one would expect, became a first-class sailorman. As one also would expect he was rather a rough diamond (the French courtiers were pleased to be amused at his lack of polish), but he had the gift of leadership and was a fine fighter, of calculated audacity. By 1674 he was in command of a small privateer, and on his first cruise captured five ships. So successful was he that in 1676 Louis appointed him head of the Dunkirk corsairs in an eighteen-gun frigate. His operations were against the Dutch, his method was usually that of boarding, and by the time peace was made at Nimwegen he had fought ten fights and made eighty-one captures, including a Dutch 30-gunner. Now an accredited naval officer, he was sent against the Sallee rovers and was more successful than his superiors had been.

In the middle of 1689 he and the chevalier de Forbin sailed from Le Havre in two frigates (RAILLEUSE(24) and SERPENTE(16)) escorting a convoy to Brest. They had the misfortune to be sighted off the Casquets by two English ships of forty-eight and forty-two guns. Ordering a few of the merchantmen which had some guns to form line and keep one Englishman occupied, Bart and Forbin made for the other – *Nonsuch*(42). But their attempt to board him failed, the merchantmen ran for it, and the two French frigates were taken by the English ships after heavy casualties on both sides, though the convoy was saved. The two commanders were imprisoned in Plymouth, but for Jean Bart stone walls did not a prison make and certainly iron bars did not make a cage. Making touch with an Ostend skipper who happened to be in the port, they got him to smuggle in a file and to put some food, water, and a compass in a boat. Then, having cut through the bars of their window, they slipped out by night, together with their surgeon and two ship's boys, got to the boat, and rowed across the Channel, Jean Bart doing most of the rowing, for Forbin had been wounded. They landed at Erquy, west of Saint-Malo, and, while Jean Bart made for home at Dunkirk, Forbin, being of the gentry, made for court. As they had been in prison less than a fortnight Seignelay was very surprised to see him and asked where he had come from. 'England.' 'But what the devil was your route?' 'Through the window, Monseigneur.' This brought Forbin his promotion to captain, and, with an unselfishness rather rare among French officers of the time, he asked for, and obtained, the same promotion for Jean Bart.

In June 1690, Tourville led the French fleet from Brest in all its glory. So to describe it is no exaggeration, for there were seventy of the line, nearly all well designed and well fitted, and in some cases beautifully decorated at bow and poop. They were manned by crews which, despite losses occasioned by the Revocation of the Edict of Nantes, contained more than a solid leaven of well-trained men, officered by gentlemen who now almost all had real sea experience[1], and commanded by an admiral whose personality, record and capacity were such that all welcomed his leadership, under which the petty jealousies that often vitiated the service disappeared. In its order of battle the fleet was as follows:

| Ship | Guns[2] | Captain |
|---|---|---|
| FIER | 80 | de Relingue (CE) |
| FORT | 68 | de la Harteloire |
| MAURE | 52 | chevalier de la Galissonière |
| ÉCLATANT | 60 | de Septèmes |
| CONQUÉRANT | 86 | marquis de Villette-Mursay (LG) |
| COURTISAN | 64 | de Pointis |

[1]One of the less experienced was Tourville's second in command, Vice-admiral Victor Marie d'Estrées, son of the unlovable Vice-admiral Jean d'Estrées. Like him he had come into the service late from the army, and owed his rapid promotion to high rank to the great influence of his family; but both in personality and in natural capacity for naval affairs he was far superior to his father.

[2]Good French authorities vary by as much as four to a dozen guns in giving the armaments of the ships: the number given in the list is what seemed the most likely, and is usually the lowest figure. Several of the ships may have carried a few more cannon.

| Ship | Guns | Captain |
|---|---|---|
| Indien | 50 | de Roussel |
| Trident | 52 | de Riberette |
| Hardi | 58 | commandeur des Gouttes |
| Saint-Louis | 56 | de Laroque-Percin |
| Excellent | 60 | chevalier de Montbron |
| Pompeux | 74 | d'Aligre |
| Dauphin-Royal | 110 | comte de Château-Renault (LG) |
| Ardent | 66 | d'Infreville |
| Bon | 52 | chevalier de Digoine |
| Précieux | 56 | de Perrinet |
| Aquilon | 50 | de Beaugeais |
| Fendant | 52 | Lavigerie |
| Courageux | 62 | de Séringy |
| Couronne | 76 | marquis de Langeron (CE) |
| Ferme | 54 | de Vaudricourt |
| Téméraire | 52 | du Rivault-Huet |
| | | |
| Brusque | 50 | de Ricoux |
| Arragant | 54 | chevalier des Adrets |
| Arc-en-Ciel | 50 | chevalier de Sainte-Maure |
| Henri | 64 | d'Amblimont |
| Souverain | 80 | de Nesmond (CE) |
| Brillant | 66 | de Beaujeu |
| Neptune | 50 | de Forbin-Gardanne |
| Sans-Pareil | 58 | chevalier de la Rongère |
| Fidèle | 52 | chevalier de Forbin |
| Diamant | 56 | de Serquigny |
| Sérieux | 56 | chevalier de Bellefontaine |
| Tonnant | 80 | marquis de Laporte (CE) |
| Soleil-Royal | 104 | comte de Tourville (V-adm) |
| Saint-Philippe | 84 | chevalier de Coëtlogon (CE) |
| Marquis | 80 | chevalier de Châteaumorant |
| Furieux | 60 | Desnos |
| Fortuné | 60 | Pallas |
| Apollon | 56 | Bidault |
| Saint-Michel | 60 | de Villars |
| Entreprenant | 56 | de Sébeville |
| Magnifique | 86 | marquis d'Amfreville (LG) |
| Content | 60 | comte de Saint-Pierre |
| Vermandois | 58 | du Chalard |
| Cheval-Marin | 50 | chevalier d'Amfreville |
| Fougueux | 58 | Colbert Saint-Marc |
| | | |
| Comte | 44 | marquis de la Roche-Courbon-Blénac |
| Vigilant | 52 | chevalier de Chalais |
| Parfait | 62 | Machault |

| Ship | Guns | Captain |
|---|---|---|
| TRIOMPHANT | 80 | chevalier de Flacourt (CE) |
| BOURBON | 62 | chevalier d'Hervault |
| VAILLANT | 58 | chevalier de Feuquières |
| DUC | 54 | de Pallières |
| CAPABLE | 54 | la Boisière |
| BRAVE | 58 | de Champagny |
| FRANÇOIS | 52 | chevalier d'Hailly |
| AGRÉABLE | 58 | La Motte-Genouillé |
| FLORRISSANT | 80 | chevalier de Cogolin |
| GRAND | 84 | comte Victor d'Estrées (V-adm) |
| BELLIQUEUX | 74 | des Francs |
| PRINCE | 56 | baron des Adrets |
| PRUDENT | 52 | Desherbiers de L'Étanduère |
| MODÉRÉ | 50 | des Augiers |
| FLEURON | 54 | Chabert |
| AIMABLE | 70 | du Magnon |
| INTRÉPIDE | 80 | Louis Gabaret (LG) |
| GLORIEUX | 60 | Belle-Isle-Érard |
| ILLUSTRE | 66 | chevalier de Rosmadec |
| TERRIBLE | 74 | Panetié (CE) |

In addition to these three divisions there were five weaker vessels – ALCYON (under Jean Bart), LÉGER, FAUCON, SOLIDE and ÉOLE – which were not put into the line of battle, but sailed to lee of it as a light squadron to assist where needed.

Tourville's orders were to try to catch the Anglo-Dutch fleet in St Helen's roads (Spithead) and destroy it, presumably in the main by use of his fifteen fire-ships. But it would have been far better if he had been got to sea earlier, even with much fewer ships, into the Irish Sea. For while he was getting his fleet up to full strength and sailing into the Channel, Rooke and Leake had broken the boom to relieve the besieged Protestants in Londonderry, and William and Schomberg had taken their army across to Ireland without interference. French strategy had not used its formidable sea-power to the best advantage. As it was, Torrington, who had been casual in taking over his command of the allied fleet, and had no look-outs in the approaches to the Channel, was surprised that the French were at sea so soon. Nevertheless, he had weighed in time from St Helens, and with reinforcements that reached him by ones and twos along the coast had fifty-seven ships with which to meet Tourville's powerful squadrons. He had the wind, and manoeuvred skilfully for several days to keep it, for his aim was to avoid action. He was the first admiral to hold the theory of the 'fleet in being' (which certain others have held since, not always advantageously) – the idea that, while one's fleet is available to make an attack, an enemy cannot attempt an invasion, or any major operation, for fear of being assailed at a disadvantage by it. But this policy alarmed the English admiralty; for Killigrew was expected soon to enter the Channel with his squadron from the Mediterranean, as

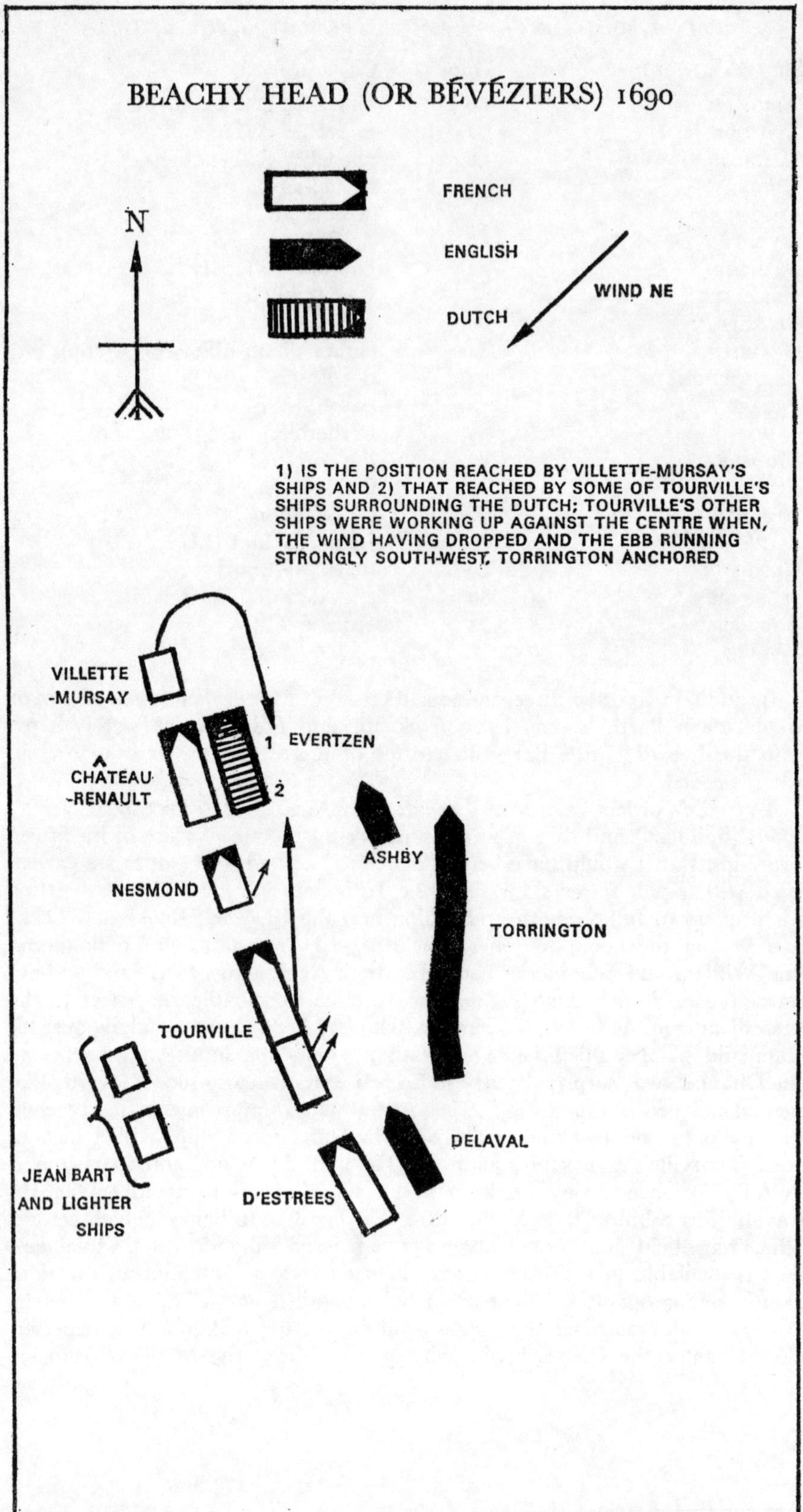

BEACHY HEAD (OR BÉVÉZIERS) 1690
FRENCH
ENGLISH
DUTCH
N
WIND NE
1) IS THE POSITION REACHED BY VILLETTE-MURSAY'S SHIPS AND 2) THAT REACHED BY SOME OF TOURVILLE'S SHIPS SURROUNDING THE DUTCH; TOURVILLE'S OTHER SHIPS WERE WORKING UP AGAINST THE CENTRE WHEN, THE WIND HAVING DROPPED AND THE EBB RUNNING STRONGLY SOUTH-WEST, TORRINGTON ANCHORED
VILLETTE -MURSAY
CHÂTEAU -RENAULT
1
EVERTZEN
2
ASHBY
NESMOND
TORRINGTON
TOURVILLE
DELAVAL
JEAN BART AND LIGHTER SHIPS
D'ESTREES

was Cloudesley Shovell with some ships from Ireland, and both of these would be an easy mouthful for the large French fleet, to say nothing of a wealthy convoy that was coming behind them. Moreover, the state of politics was such that the Government did not want a mere delaying strategy. So the queen sent orders to Torrington to fight, and the result was the battle known to the English as Beachy Head, and to the French as Bévéziers – which is a curious corruption of Pevensey.

On the morning of July 10th, Torrington ran down on his enemy. His van was formed by the Dutch squadron, twenty-two strong, under Evertzen and Callenburg. He led the centre in the *Royal Sovereign*(100) with Sir John Ashby as his second in command, and Sir Ralph Delaval commanded the rear. The wind was north-westerly and both fleets were on the starboard tack, heading north of west. The Dutch came on with great spirit and engaged at about 8 am. But they had not advanced far enough along the French line before doing so, whereby their leading ship engaged only the eighth and ninth in Château-Renault's van. That left de Villette-Mursay's division at the head of the French line, and Château-Renault signalled these ships to stretch ahead till they had room to double back on the Dutch lee-side. This they proceeded to do. It was not till two hours later that Torrington got his centre and rear into action, and even then, obsessed with the idea of keeping the fleet in being, he held the centre 'refused' – that is to say, he did not come to close quarters, but opened fire at longer range. Nevertheless, the fighting in the centre was pretty active, and it was very hot in the rear, where Delaval made determined efforts to break up the French line. In this he was unsuccessful, though he drove the FLEURON and the MODÉRÉ and the TERRIBLE out of the line, the last named because, like Coëtlogon at Bantry, its captain had got his cartridges too exposed and a lucky shot exploded them, blowing his poop away.

Meanwhile, Torrington's 'refusal' of the centre had left a gap between him and the Dutch. The leading ships of Tourville's centre were engaged with the rear ships of the Dutch line, but Tourville now began to push his other ships up into the wind through this gap so that, with Villette-Mursay now coming down also on their lee side, the Dutch were completely encircled and suffered heavily. At this stage the battle became something of a nightmare, for the wind, which had been lessening all day, dropped completely, and there was a flat calm in which some ships got out their boats to tow them, and all just drifted with the current. In the evening, as the tide began to ebb, Torrington hoisted the signal to anchor. As he did this without the usual preparation of taking in sail, he stole a march on the French, who drifted past him out of range before they brought up.

Besides the three ships in the French rear that have been mentioned, the TONNANT in the centre and the ARDENT and the PRÉCIEUX in the van had had to leave the line, and some other ships had suffered a good deal. But the French damage and casualties were nothing to those of the allies, especially the Dutch. Only one allied ship was taken in the actual battle, the *Friesland*, which had no anchor left and, drifting away among the French, had struck. (The French officers who boarded her said that hardly a foot of her was undamaged and that she was heaped with dead and dying.) The survivors were taken off, and she was sunk. But if the allies lost but a single ship in the battle, it was a different matter with the pursuit that followed. For when Torrington weighed with the turn of the tide and so drifted back up channel, and then, when some wind came, made for the Thames, there were fifteen other vessels that were too shattered to keep up – fourteen Dutch and one English – and these, standing inshore for the beaches, were all mopped up and burned by de Villette-Mursay whose ships were detached for this purpose.

The Dutch loudly cried out that they had been deserted, if not betrayed, and they had some cause. Torrington was clapped into the Tower, pending a court martial, which acquitted him, perhaps the more readily as the defeat at Beachy Head had been balanced by William's victory at the Boyne and the consequent fall of Dublin. Tourville was left in command of the Channel and had won a great victory.

Yet it was a victory without a future; for all the use that Tourville made of it was to take the little port of Teignmouth and destroy the shipping there, including a 44-gunner. For this he was much reproached in his time (Seignelay called him 'Poltron de tête, mais pas de coeur'), and has been so by historians since. It is hard to see why. Had his entry into the Channel been part of a properly planned operation, with troops ready to be taken across channel to invade England, it would have been a different story. There was no such preparation. The cause of the Jacobites in exile always reeked with undue optimism, and they had probably given the idea that the French had only to appear victorious off the south coast of England for its shores to be crowded with gentry brandishing their swords for King James. True there was much Jacobite feeling in England, but those who held it had fathers or grandfathers who had suffered fighting for Charles I, and they needed positive military support before they would rise.

For Tourville to have pursued the allies into the tricky waters of the Thames estuary and there to have fought a battle with some of his ships already damaged, and no port available for him nearer than Brest, would have been foolhardy, to say the least. He might, indeed, have attacked Plymouth to destroy not only Killigrew's ships but also the expected convoy which had both gone there for safety. Seignelay sent urging him to do this. But by that time, not only had he his damaged ships to consider, but the conditions of a long spell on board had given him a heavy sick list, as was usual at sea in those days. To enter Plymouth's fortified Sound, which contained a number of warships, seemed to him unwise in view of the state of his fleet, and probably rightly so. Therefore he took his sick and his damaged ships back to Brest, with, of course, some commercial prizes, and the victory of 'Bévéziers' remained one of prestige alone.

This war, known either as that of the League of Augsbourg or of the English Succession, contained fewer colonial operations than subsequent wars, but there were some such, apart from the usual privateering. In 1690 DuQuesne-Guiton was sent to India with six ships – the GAILLARD and the OISEAU, each of forty-eight guns, being the strongest – and some troops. On the way he surrounded the *Philip Herbert*, of sixty guns, whose captain blew up his ship after a stiff fight, and, having arrived in India and landed some soldiers, he fought an indecisive action with some Anglo-Dutch vessels off Madras. He then made his way to the West Indies, where he arrived in 1691 somewhat weather-beaten, but in time to make Commodore Wright call off an attack on Guadeloupe. The French had started well in that area by taking St Eustatius and driving the English from their part of St Christopher, thanks to a small squadron led by de Blénac. But both these islands and Marie Galante had been taken the next year by Wright. The only other oversea operation of note was that William Phips from New England, whose ships were almost all privateers, had taken Port Royal in Acadie (which was a privateer base), but had been heavily repulsed from Quebec by that old lion, Frontenac.

At the end of that year, 1690, died the marquis de Seignelay. Though he was only a shadow of his great father in capacity, in earnestness, and in a proper sense of the use of sea-power, his death was a severe blow to the Service. For he had been brought up in its administration, knew its requirements, and had sufficient of the king's ear to impress its value upon Louis, as against the military jealousy of Louvois, the Minister of War. To replace him Louis made one of his several bad appointments, Phélypeaux de Pontchartrain, who was a lawyer turned financial administrator. Cash was his main concern, and he frankly admitted that he knew nothing whatever of naval affairs. The position was a little like that when Mazarin succeeded Richelieu: since the navy was there it should be used, but when there came a strain on the purse, the navy became an expensive luxury. Yet though, as we shall see, France was later to turn her naval operations into commerce raiding by the smaller ships, it would be a mistake to gather, as one might from some historians, that the big ships were then all laid up. Too many, alas for French sea-power, were so; but there were always some kept in commission, which took part in one or two operations, though they were not enough to fight a major action.

In 1691, Vice-admiral Victor-Marie d'Estrées got to sea early from Toulon to support Catinat and his troops in taking Villafranca and Oneglia. This done, his small squadron, which was really just an escort for bomb galiots and the galleys that were to tow them into place, sailed to bombard Barcelona. Having much damaged that city, he proceeded to Alicante and demanded a large ransom to spare the town from being shelled. When it was refused he carried out a prolonged bombardment that set the place well on fire, but withdrew to harbour at the approach of a much superior Spanish fleet.

More important were the operations of Tourville. With Jacobites on both sides of the Channel, even partial secrecy was impossible. Tourville had sixty-nine ships, and it was known that the Anglo-Dutch fleet would put to sea ninety strong. The French admiral was therefore given an almost impossible combination of orders, obviously devised by a landsman. His main object was to get the large and valuable Smyrna convoy. At the same time he was to prevent the English fleet from attacking French coastal towns, or effecting landings; but in doing both these things he was to avoid any action with superior forces. He sailed north-west from Brest, with his fleet widely spread to make an interception; but the convoy had kept well out in the

Atlantic, making first for Ireland instead of for the Channel, and it reached Kinsale before Tourville could come up with it. Knowing that Russell with the allied fleet would now be after him, he headed back south-east and cut off a convoy from Jamaica that was escorted by an English 50-gunner and a frigate. He took the escorts and most of the convoy and, as the allies would now be again advised of his whereabouts, he stood right out into the Atlantic to lead them well astray before returning to Brest. He had been ordered to keep at sea till the end of August, but shortage of food and his heavy sick list brought him back in the middle of the month.

Though he had missed the big convoy he had carried out his other two orders. For he had prevented the allies from doing anything against the French coasts by keeping them running after him, and he had avoided action against their greater numbers. It had not been anything of an heroic voyage, but it had been a clever one, and Tourville, though not normally 'touchy', like most French officers, was justly angered when some at the Court and in naval administration said that he seemed to have been careful of his own skin.

More to the popular taste was the exploit of Jean Bart, who had slipped out of Dunkirk with some frigates and had taken, first two English ships of fifty and forty guns with the small convoy that they were escorting, and later a Dutch 40-gunner that was guarding a fishing fleet, which he destroyed.

The year 1692 began well enough when the chevalier des Augiers in the MAURE(64) with three lesser ships cut up a Dutch convoy, escorted by a 50- and a 46-gunner, and sank both escorts. But though the navy was later to be turned to commerce raiding, for this year it was to play the lead in what was very much a major operation.

The attempt to hold Ireland for James II had failed. After Gabaret had landed James there, and Château-Renault had taken the troops over to Bantry, and some later reinforcements had been sent in a squadron led by d'Amfreville, the fleet had given no assistance. It had been used in the Channel winning the fine, but profitless, battle off Beachy Head instead of being in the Irish Sea cutting off troops and stores from England. So at the end of 1691 Château-Renault had had to sail to take off the defeated remains of the army after its capitulation at Limerick. Now an even bigger military operation was to be tried.

It was known that Jacobite feeling had risen to a height in England; many English soldiers, too, were abroad, so England itself was to be invaded. Troops were to be collected, the infantry round St Vaast La Hougue in the north-east corner of the Cotentin, the cavalry and guns and supplies in transports at Le Havre. The warships at Brest were to be joined by those from Rochefort, and from Toulon, which d'Estrées was to bring round, and Tourville was to sail with the whole great fleet from Brest, pick up the infantry, convoy the cavalry and artillery transports from Le Havre, and take the whole lot over to England, fighting off the British fleet on the way. As the scheme was originally devised it was not in fact expected that there would be much of a fleet to fight. For the French were to sail in April, which would probably be before the English were at sea, and certainly before they had been joined by the Dutch, and it was confidently asserted by James' envoys, flitting to and from England, that a large number of the English captains would play the traitor, and would at least keep their ships out of action if they did not actually join the French. This delusion persisted up to the crisis of the campaign, but, while Jacobite sympathies and dislike of 'Dutch Billy' were certainly running high in England, they were not running as high as that, and the few captains who were really traitorously inclined were detected and relieved of their commands.

Tourville was hardly consulted about the plan, which was a bad one, drawn up by landsmen who not only had no sense of the difficulties that adverse winds and currents might present in assembling a great fleet from different parts by a given date, but did not even know the state of their own internal services, what supplies and men were available, and so forth. Even at the main port of Brest, Arnoul, the Intendant, was never disposed to hurry himself, and though Tourville sailed a month later than intended there still remained a number of ships to be manned, supplied, and to complete their fitting out, which were intended to follow him under Château-Renault. They never did. The reason that Tourville was ordered to delay a month was that it was found that the transports at Le Havre could not be ready in time. De Villette-Mursay, who was bringing the ships from Rochefort, could not get them manned till Tourville had sailed, though he was able to join him, and d'Estrées, bringing the essential reinforcement of the Toulon squadron, was so delayed, first in fitting out his ships and later by contrary winds, that he arrived in Brest only as the decisive action of the campaign had ended.

The whole plan shows the folly of what was the curse of the French service for many decades, that of separating the administration from those who actually went to sea, and of putting it in the hands of those who had little or no sea experience, with a minister who was a landsman at their head. The English administration was often faulty, and the spirit of its navy at this time, and in the next war, was by no means as good as it later became, but at least the Admiralty always contained some real senior seamen, whereby the worst blunders were usually avoided.

The idea of fighting a convoy across the Channel, and protecting the landing of troops, against a fleet that had not first been well beaten, was highly dangerous. Moreover, the tone in which the orders were couched was all but insulting. Tourville was to get to sea by a certain date without excuses, whatever the state of his flagship: he was to avoid the 'accidents' of his previous campaigns; when he met the enemy he was to fight, whatever their numbers; if, after landing the invasion force, he was to meet a superior enemy fleet he might avoid action, but he was to make sure that they were warships and not merchantmen. These orders were, of course, mainly Pontchartrain, but they were signed, and with a sentence or so in his own hand, by Louis himself[1]. Tourville's correspondence shows that he had had

[1]The king ordered him to fight a superior fleet very hard (*opiniâtre*) so that even if he were beaten the enemy would be too damaged to hinder the invading forces. Apparently neither the king nor his ministers realized that if Tourville were driven away, a mere handful of the less damaged allied ships would suffice to make havoc of the transports.

some nagging letters from Pontchartrain in the previous year, and he was quite likely to have heard that, in court tittle-tattle, Pontchartrain had questioned his courage. Though, as we have said, Tourville was less ready to take offence than most of the contentious French nobility, he was a man of his time, and the obvious reflection on his personal bravery was a smear on his honour that touched him deeply.

Having given these final orders, Louis and Pontchartrain turned the charge of the whole campaign over to the triumvirate who were to conduct the invasion: James II, who had had some good sea experience, but had now become an ageing monument of indecision, Marshal Bellefonds, commanding the troops, who had no naval knowledge whatever, and Bonrepaus, the Senior Intendant of Marine, whose sea service was limited to a few years as a lieutenant of galleys. They would send Tourville any necessary subsequent order by dispatch boats.

It was originally intended to get surprise by sailing unusually early. The delay at Le Havre prevented that, but Tourville was ordered to sail by April 25th. On the 23rd he was told to wait for de Villette-Mursay, and he eventually left on May 12th. By that time all chance of surprise had been lost. And twenty ships had to be left in Brest, for Château-Renault to complete their crews, which were a mere 2 000 men short!

The fleet of forty-four ships with which Tourville entered the Channel was as follows in its order of battle:

| Ship | Guns[1] | Captain |
|---|---|---|
| Bourbon | 68 | de Perrinet |
| Monarque | 90 | marquis de Nesmond (CE) |
| Aimable | 70 | chevalier de Réals |
| Saint-Louis | 64 | de la Roque-Persin |
| Diamant | 60 | chevalier de Feuquières |
| Gaillard | 68 | chevalier d'Amfreville |
| Terrible | 80 | de Sèbeville |
| Merveilleux | 90 | marquis d'Amfreville (LG) |
| Tonnant | 80 | de Septèmes |
| Saint-Michel | 60 | chevalier de Villars |
| Sans-Pareil[2] | 62 | de Ferville |
| Sérieux | 64 | marquis de Blènac |
| Foudroyant | 84 | de Relingue (CE) |
| Brillant | 62 | commandeur de Combes |

[1]As with the list of the fleet in the Beachy Head battle, authorities differ as to the number of guns, and the figures given may be inaccurate by a few guns in a few cases.
[2]According to some authorities the Vermandois was in line and not the Sans-Pareil.

| Ship | Guns | Captain |
|---|---|---|
| MODÉRÉ | 52 | d'Evry |
| ILLUSTRE | 70 | de Combes |
| SOUVERAIN | 80 | marquis de Langeron (CE) |
| CONTENT | 68 | chevalier de Sainte-Maure |
| ADMIRABLE | 90 | chevalier de Beaujeu |
| SAINT-PHILIPPE | 84 | chevalier d'Infreville |
| SOLEIL-ROYAL | 104 | comte de Tourville (V-adm) |
| CONQUÉRANT | 84 | du Magnon |
| GLORIEUX | 64 | chevalier de Château-Morant |
| PERLE | 52 | chevalier de Forbin |
| COURAGEUX | 58 | chevalier de la Luzerne |
| MAURE | 52 | des Augiers |
| COURONNE | 76 | chevalier de Montbron |
| AMBITIEUX | 96 | marquis de Villette-Mursay (LG) |
| HENRI | 64 | la Roche-Allard |
| FORT | 60 | chevalier de la Rongère |
| | | |
| EXCELLENT | 60 | du Rivault-Huet |
| PRINCE | 56 | de Bagneux |
| MAGNIFIQUE | 86 | marquis de Coëtlogon (CE) |
| LAURIER | 64 | chevalier d'Hervault |
| BRAVE | 58 | chevalier de Chalais |
| ENTENDU | 60 | de Ricoux |
| TRIOMPHANT | 76 | de Machault-Belmont |
| ORGUEILLEUX | 94 | Louis Gabaret (LG) |
| FIER | 80 | de la Harteloire |
| FLEURON | 56 | chevalier de Montgon |
| COURTISAN | 64 | Colbert de Saint-Marc |
| GRAND | 84 | Panetié (CE) |
| SAINT-ESPRIT | 74 | de la Galissonnière |
| SIRÈNE | 64 | DuQuesne-Mosnier |

Meanwhile, the 'triumvirate', somewhat belatedly, had learned three things: that the Dutch and English had made their junction, that they were at least twice as strong as the French, and that the expected defection of captains would not take place. They therefore sent vessels to find Tourville and order him back to the approaches of the Channel. None reached him in time.

Sighting the allies, ninety-eight strong, off Cape Barfleur, Tourville ran down on them with the greatest *élan*. The story of a dramatic council of war when all advised retreat, whereat Tourville produced the king's orders, seems to be inexact; but Bonrepaus, watching the fight from the shore, said to the others of the triumvirate that it was because the king and Pontchartrain had questioned his courage that he had attacked like a berserk (*un furieux*). True that this is probably what made Tourville engage against such odds, yet his attack was no berserk assault, but a well-designed operation. In all he had forty-four of the line with 3 132 guns with which to fight ninety-eight ships with 7 154 guns. He had the wind, which was south-west, and hoped by superior handling of ships and gunnery to do the enemy damage, and then, having carried out his orders, to use his possession of the wind to break off the battle. De Nesmond, handling his ships well, got them into action against the head of the Dutch line, which formed the allied van with van Almond in command. Russell, with Sir Ralph Delaval and Sir Cloudesley Shovell, led the centre, which Tourville's 'main battle' attacked; Sir John Ashby, with Rooke and Carter under him, had the rear against Gabaret, whose sternmost division, under Panetié, got into action somewhat late, with a gap between its ships and those ahead. The French gunnery, especially in the centre, was splendidly maintained, even though most ships were engaging two or even three of the enemy. Indeed, eight French ships, including the flagship and Villette-Mursay's AMBITIEUX, were holding off eighteen of the British line.

The battle had started at ten in the morning, but at about two o'clock the wind swung a full eight points to north-west, so that Tourville no longer had any more advantage from it than his foes, and could not readily disengage. The Dutch began to use their superior numbers to try to stretch ahead to weather the French van, but de Nesmond kept giving way skilfully to prevent their doing so, with the result that by about 4 pm the leading ships of the French van were fighting on a course almost at right angles to its rear ships. With the change of wind Sir John Ashby began to force the allied rear through the gap between Panetié's ships and Gabaret. Panetié stood off parallel to Ashby on a course that took him towards the curved-in head of the French van. Fortunately Ashby followed him for some time, which gave Gabaret the chance to bring his vessels up to starboard of the French centre, so that that much tried body could not be attacked on both sides.

It was not till about seven in the evening that, having realized their error, Ashby's ships mostly abandoned Panetié, and let themselves drift down on to the centre, trying to get at that part which Gabaret had not been able to cover. Seeing this Coëtlogon, with the MAGNIFIQUE and the PRINCE, loyally drifted down away from their place ahead of Gabaret, to give their admiral some protection. The French centre, plus some of Gabaret's ships, was now under attack from both sides, and the wind, never strong, had for some time dropped altogether, so that a flat calm reigned. Villette-Mursay sent an officer in a boat to remind Tourville of Torrington's trick at Beachy Head, and Tourville signalled to anchor so that many of the allies were carried past out of range. At this time, too, about 8 pm, a heavy mist came down and firing ceased. By now much of the sense of formation had been lost, and the ships were in a fairly general mêlée. When the mist cleared a full moon lit up the scene, and the battle was renewed. Russell tried to use his fire-ships, but the French dealt with them readily enough, though at the cost of cutting some cables and thereby losing some anchors which they could ill spare. Ashby, finding his ships separated from the rest of the allies, let them drift through the French, which cost some of them heavily till, with the mist descending again at 10 pm, the firing ceased, and the exhausted and battered French crews could snatch some rest.

They had fought a magnificent fight against odds of more than two to one for twelve hours. They had sunk one British ship and one Dutchman and had not lost a ship of their own, though some, especially the great SOLEIL-ROYAL, were very battered. On a much larger scale the action was not dissimilar to Grenville's fight in the *Revenge*, of which the English are justly proud; and the French could be equally proud of the tremendous fight that they had put up.

It was now a question of getting the fleet away and, after the strain of the battle, it is not surprising that a certain small amount of disintegration took place, especially as the retreat was made a nightmare because the wind blew only lightly, or not at all, so that ships had to anchor with each flood tide lest they should lose what little distance they had gained during the ebb; and some ships had no adequate anchor with which to bring up. France was now to pay bitterly for the economy which had rejected Vauban's proposals to make a proper fortified harbour at Cherbourg whereby nearly all would have been saved, and for accepting DuQuesne's advice to spend the money on preserving the inadequate Le Havre. Even so, practically all other loss might have been spared if Tourville had been willing at once to abandon and sink his magnificent but dreadfully damaged flagship. His unwillingness was an error, though one can understand it.

It was not till the evening of the next day that Tourville, finding that, despite getting a start of the enemy by weighing quickly in the morning, his ship had made little ground, shifted his flag to the AMBITIEUX. Then, realizing later that it was wrong to try to keep the fleet together, he signalled what was really an order for flight. He had been trying to make for the Raz Blanchard, the channel between Alderney and the French coast, a tricky bit of water for sailing ships because of the stormy cross-currents and the bad bottom if one wishes to anchor. Twenty-two vessels had got round into it and, braving its dangers, made successfully for St Malo. Some that were more out to sea than the rest took other courses. Three, including Gabaret and Langeron, went somewhat north of west to skirt the English coast and then double into Brest. Six, under Nesmond, went easterly; two of these had to drift into La Hougue roadstead, two others could get no further than Le Havre, Nesmond and the other took the route of the Armada, right round Britain, but with better fortune – for, though chased over the last part of their voyage, they reached Brest safely. Of the remainder of the fleet, Tourville directed the SOLEIL-ROYAL, ADMIRABLE and TRIOMPHANT, all in bad shape, into Cherbourg, where they beached, and took the rest to La Hougue, where at each end of the roadstead there was a battery. Here he could at least save the crews, and some of the guns and stores, even perhaps the ships – though his chances of doing so were not helped when the triumvirate kept him waiting while it made up its mind whether to approve, so that he missed the high tide that would have taken his vessels well up the mud of that very shallow anchorage.

## LA HOUGUE 1692

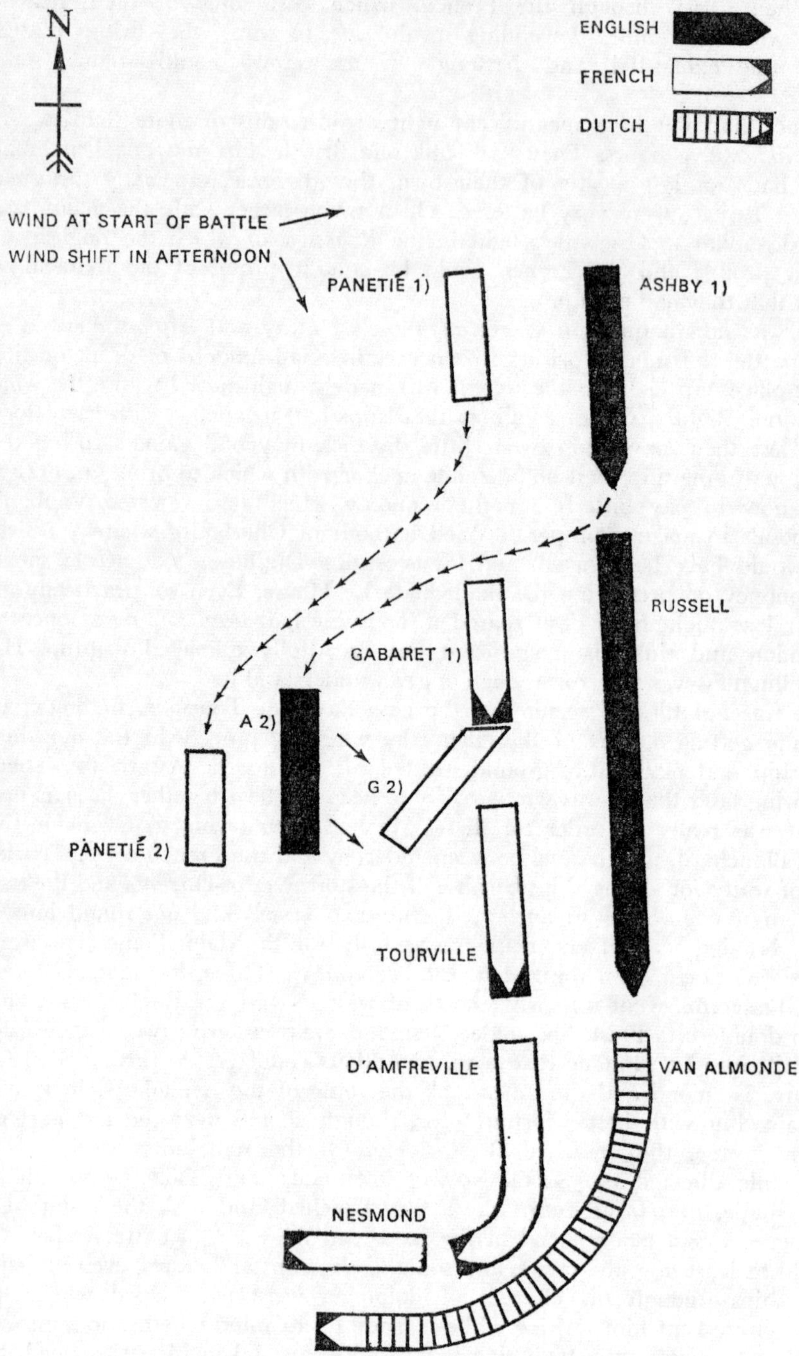


THE FIGURES 1) AND 2) SHOW FIRST AND SECOND POSITIONS. G 2) IS GABARET'S SECOND POSITION TRYING TO COVER THE CENTRE FROM ATTACK TO STARBOARD, AND A 2), IS ASHBY'S FOLLOWING OF PANETIÉ BEFORE, DROPPING DOWN TO HELP TO SURROUND THE FRENCH CENTRE.

Russell had dispatched Ashby and the Dutch after the ships that had made for St Malo, while he and Delaval turned on the remainder, and it was then, four days after the battle, that the unavoidable disaster occurred. The three in Cherbourg were attacked by fire-ships and destroyed, though they put up a stout resistance, working their guns even after they were well alight. The others – AMBITIEUX, MERVEILLEUX, FOUDROYANT, MAGNIFIQUE, SAINT-PHILIPPE, FIER, FORT, TONNANT, TERRIBLE, GAILLARD, BOURBON and SAINT LOUIS – in La Hougue were assailed by the numerous long-boats and the fire-ships of the English squadron, and though Tourville, Villette-Mursay and Coëtlogon all went out in the French long-boats to encourage their men to fight off the English, the French crews were too exhausted, disheartened, and outnumbered to do so, and all twelve ships were set on fire and destroyed. With this ends the battle of La Hougue. In the actual fight against odds the French had done gloriously and lost no vessel; but in the aftermath they had lost fifteen of the line. One of the good features of Louis XIV's unpleasing character was that he could be magnanimous. He recognized that the blame did not lie with the Service, expressed his pride that his ships had fought so gloriously against such odds, retained Tourville in his command, and next year sent him his bâton as a Marshal of France.

This honourable defeat did not stop the attack on the commerce of the allies, which cost them dear, and was sometimes carried out by some of the smaller ships of the line which had been in Tourville's fleet. Des Augiers and Forbin, with the MAURE, MODÉRÉ and PERLE, cut up a Dutch convoy in August, and captured the fifty-four and forty-eighter that were its escort. La Caffinière with the NEPTUNE, VERMANDOIS and FRANÇOIS fought a battle with four Spanish galleons off Finisterre, taking one of sixty-two guns and putting the others to flight. De Blénac with the SÉRIEUX and FLEURON gave the Tripolitans a sharp lesson off Malta; and in November Jean Bart, with three frigates, captured some twenty ships of the Dutch Baltic convoy, together with one of its escorts. The fact that La Hougue was the last fleet action of a war that lasted a further five years is apt to give the impression that the French navy was too weakened for further major battles. That is not the case. Certainly no nation loses fifteen of the line with equanimity; but the French naval forces were large and could stand the loss. The reason that there were no more battles lies in a change of policy.

The main object of French maritime forces was henceforth to be the destruction of the enemy's commerce. History has shown that, while this may be a good secondary object for sea-power, since it causes strain on the enemy's finances and resources and distraction from his objectives, yet it does not of itself bring victory. It did not do so in this war, nor in the one that followed, nor under Napoleon's Continental system, nor by the submarine attacks of the two world wars of the twentieth century. But at the time there seemed much to recommend it. The Dutch lived mainly by their seaborne trade: England's money-bags were largely filled by it. With Ireland lost, and the idea of invasion abandoned after La Hougue, the fleet had no main objective other than the slight one of supporting any French army operating along a coast. To Pontchartrain's mind it had become a costly luxury: the French had always shown a genius for commerce-raiding; by this means the navy could pay for itself. Commerce-raiding was to be the policy; and it began on a grand scale with an attack on the outgoing Smyrna convoy in 1693.

The operation was as well planned as that of the invasion had been muddled. Learning, by the usual Jacobite channels, that an allied attack on the Brest roadstead was intended, Louis hurried Tourville to sea with seventy-one of the line just before the end of May. When the allies arrived off the harbour they found the bird flown, no one knowing whither. The French fleet was in fact lying off Lagos in southern Portugal and, by flying English and Dutch colours, by having the few official visitors greeted by English Jacobites who were on board, and so forth, it was maintaining a skilful secrecy of deception. In late June, when lookouts reported the approach of the convoy, the trap was sprung.

It was not, in fact, sprung as successfully as it might have been. The convoy was enormous, and to escort it Rooke and van der Goes had twenty-one of the line and some smaller vessels. Tourville had formed a squadron of his faster sailers, twenty-two strong, to go ahead under Gabaret (LG). Between these and the convoy Rooke formed line, and in the course of a two-hour fight two Dutch sixty-fours, the *Zeeland* and the *Wapen van Medemblik*, were taken. By now the whole French fleet was in sight, and there was nothing for the escort to do but to run for it. In any case the convoy was now scattered in a complete *sauve qui peut*, some making south, others easterly for any Spanish port where they could find refuge. But instead of pursuing at his best speed Gabaret, while keeping after the Dutcn part of the escort, not only maintained his ships in line but kept them under reduced sail until his chief should come up. As a result, the victims were not overtaken before nightfall, so that the rest of the escort and many of the convoy got away. Even so the blow to the allies was severe, for what with the ships taken at sea, and those taken or destroyed in the harbours, Cadiz, Gibraltar, Malaga, whither they had fled, numbers given variously between sixty-one and over a hundred vessels were lost to English and Dutch merchants. This loss was heavy, and would have been more so had Gabaret handled his ships with dash.

D'Estrées, with the Toulon squadron, had helped the army to reduce Rosas, and English attacks on St Malo, and on Martinique and Newfoundland, had failed. Yet the attack on the Smyrna convoy was the last major fleet operation of the war, with one exception. Commerce-raiding was a matter for privateers, for single ships or groups of two or three rather than for squadrons. To this purpose the king was willing to lend his smaller ships, those of forty or less guns, and to allow his wealthy courtiers, or his own naval officers, to fit them out and operate them, in return for the payment of an agreed percentage of the prize money. So it becomes hard to separate naval activities properly speaking from the activities of royal ships privately 'leased'. Though a few of the large ships were kept in commission, most were laid up, and naval activities for the rest of the war consist mainly of operations against convoys by a group of a few ships at most. In return, the allies largely devoted themselves to attacks on French coastal places, doing damage but having no permament effect.

Indeed, the allied attacks on the coast can be dismissed briefly, for they were concerned with ports little used by the navy. After La Hougue Russell made for the ships refuged in St Malo, but considered their position and protection too strong to be assailed. A properly prepared attack on this nest of privateers was made later, and relied largely on an 'infernal machine', a ship packed with explosive. This went off in not quite the right place, and the extensive damage that it did was only superficial. So the attempt failed. A fully prepared amphibious operation against Brest in 1694 was betrayed in time for Vauban to hasten to the port and augment its defences, so that the expedition suffered a defeat not far short of disaster.

In 1694, Tourville, with a fair sized fleet, supported the army that was besieging Barcelona, but was ordered to withdraw when Russell's superior squadrons entered the Mediterranean. The leading exploit of the year was that of Jean Bart. He had been sent from Dunkirk with six vessels of medium strength to escort a home-coming convoy. He found it off the Texel, in the possession of the Dutch, who were taking it into harbour under the guard of the eight warships which had captured it. On these Jean Bart pounced like a tiger. He had the wind, and aimed at boarding. Three of his ships did so successfully, taking their three Dutch opponents. Though his other ships failed, and the MAURE was reduced to something not far from a sinking state, the remaining five Dutchmen fled, and Jean Bart brought the recovered convoy (130 sail loaded with much-wanted corn) and his three prizes triumphantly into port. For this the Dunkirk seaman was given patent of nobility. A small English convoy from the West Indies was cut up, and its fifty-gun escort taken.

Indeed, there was nothing wrong with the fighting spirit of the French navy, as was shown by the way the BOUFFONNE frigate fought off superior Dutch forces, and the stout resistance to superior English ones put up by the fever-stricken and storm-battered TÉMÉRAIRE(60) before she made an honourable capitulation. At this date, the spirit of the English service was unquestionably less good. The same fighting spirit was shown in the next year, off Pantellaria, by the CONTENT(54) and the ARDENT(42), which were overtaken by six ships under Killigrew, who was clearing the way before the Smyrna convoy. The two Frenchmen drove off three of their pursuers before being taken.

In 1696 the English made a determined effort to reduce privateering by a blockade of ports, and with success in several cases. But there were some ports that could not be blockaded owing to navigational difficulties. Such a one was St Malo, and the Malouin privateers, among whom a young man named Duguay-Trouin was prominent, reaped a rich harvest. The one naval exploit of note, however, came once again from Jean Bart. Though Benbow was blockading Dunkirk closely, Jean Bart took advantage of a misty June night to slip out in the MAURE(54) with four other naval ships and two privateers, and made for the Dogger Bank – that approach point of convoys from the north. Between it and the Texel he came up with a large Dutch convoy with an escort of five ships. These formed a close line and waited for him. Sending his two privateers to deal with the convoy, the Frenchman ran down on the war-ships and engaged. His force was the stronger for, in all, his vessels carried nearly fifty more guns than the Dutchmen. One of them was small, and was early forced to strike, but the others fought stoutly, and the action had lasted three hours before Jean Bart carried the commander's ship, the *Raadhuis van Haarlem*(44), by boarding. All the Dutch escort and twenty-five of the convoy were taken, and the French losses were surprisingly light.

It was the nearest thing to a pitched battle that Jean Bart fought in his various commands, but he was not able to carry away his prizes, for a powerful Dutch squadron was sighted in the distance. With quick efficiency he transferred all his unwounded prisoners, packing them into the smallest of the captured warships. He fired the other four together with twenty-five captured merchantmen. By now he was really hunted and was between the Dutch fleet to the north and Benbow coming up from the south; but he handled his ships cleverly and though his pursuers sighted him they were never able to come to grips. One of his courses took him as far north as Norway, and it was October before he dodged back into Dunkirk. The loss of so many of their convoy dealt a heavy blow to Amsterdam merchants.

In the colonies the war had mainly been one of privateers, such as, for instance, the exploits of Pierre Le Moyne d'Iberville and his brothers, who mopped up the English Company's settlements in Hudson Bay. Occasionally a royal ship was employed, for example, in a destructive raid that du Casse, a naval officer governing the French West Indies, made on Jamaica. His force was mainly of privateers, but he had the TÉMÉRAIRE, and it was as she was returning to France with the trophies of that raid that she was taken in the approaches to the Channel, as we have recorded (page 89). In 1696 des Augiers was sent with six of the line to tackle the English squadron in the Antilles. He failed to do so, though he used his superior force to take two Spanish galleons. There was also some fighting off Newfoundland and San Domingo.

By now both sides were ready for peace. Louis had won battles on land, but as his general, Luxembourg, was a far better tactician on the field than he was a strategist to make use of a success, the victories had led nowhere. The strain upon France had been enormous, especially upon her finances, which were in a state of combined chaos and incipient bankruptcy. The Dutch, too, were exhausted: for thirty years the little United Provinces had stood firmly, and successfully, against the assault of France; the effort had left them played out: their great period was over. In England the war was regarded as waged mainly for the policy of her Dutch king, save only that her parliamentary revolution must be protected. Let Louis renounce James II: that was all that was wanted. So negotiations were opened, but were delayed by the obstinacy of the most inefficient and distracted of the allies, Spain. So Louis decided to strike two blows to bring Spain to heel.

One of these was regular enough. Vendôme, general of the galleys, took these and some troops to capture Barcelona in 1697. He was actively supported, navally, by a small squadron of nine of the line under Vice-Admiral d'Estrées. Having endured siege and bombardment for two months, with no sign of relief coming by land or by sea, and his walls being breached, its commandant capitulated with the honours of war. The other blow was more irregular. It was an attack on Cartagena on the east side of the Gulf of Darien, the great entrepôt of Spanish riches in South and Central America. The irregularity lay in the nature of the force that was to make the attack, for it was to be a combination of ships of the line and of privateers, under the baron de Pointis (CE). But even the ships of the line were manned and fitted out on a privateering basis by a number of shareholders, though aiming at a royal objective and remaining under royal orders. De Pointis had seven of the line, with his flag in the SCEPTRE(84), and fourteen privateers, of which some went with him from France, and others, under du Casse, with motley and piratical crews, joined him in the Caribbean.

As might be expected, such a force had a somewhat chequered career. Cartagena was taken readily enough, though Pointis and some other leaders were wounded, and an enormous contribution was levied. Then an unseemly quarrel broke out between the regulars (so-called) and the local privateers over the distribution of the booty, and the latter, who contained many Negroes, proceeded to a piratical plundering of the town. The delay that this caused produced a violent outbreak of sickness (the 'black vomit') in Pointis' ships, and he left for San Domingo where he intended to water. He was cut off from that island, however, by an English squadron under Rear-Admiral Nevill. With his ships in poor state, and at least half his crews sick, he was in no condition to fight, but he managed to dodge the enemy during the night, losing only one or two buccaneers and his hospital ship, which, being full of infection, was a disastrous capture for the English. Serving the barest minimum ration of water he made for Newfoundland, where there had been desultory fighting between the French and English 'shores' throughout the war. Sir John Norris had a small English squadron in St John's, but would not leave the strong defensive position that he had taken up, believing Pointis to be the advance part of a powerful squadron known to be coming out under Nesmond[1]. So Pointis was able to struggle away for France. Approaching the Channel, he ran into five English ships under Captain Harlow, who engaged him. Pointis was able to conceal his very weak state and, some lucky shots or good gunnery having damaged the masts of two English ships, Harlow broke off the engagement. De Pointis brought his debilitated crews and their enormous booty into port.

Spain now accepted the negotiations; but before peace was signed Jean Bart had one last exploit – of an unusual kind. Poland was about to elect a king, and some Poles favoured a connection with the French royal house. So Jean Bart took the Prince de Conti to sea on a dark night, dodged through Benbow's blockade, and conveyed him to Dantzig. The prince did not get his throne.

By the treaty of Ryswick (Sept 1697) Louis received an obvious set-back. He had to surrender almost all that he had acquired since the Treaty of Nimwegen, had to allow the Dutch to garrison the 'barrier' fortresses – the strong places of the Spanish Netherlands – and had to recognize William as king of England, renouncing his support of the Stuart claims. With this

[1]Nesmond had ten of the line, with which it had been intended that he should settle the French possession of Newfoundland for good, but he arrived so storm-tossed that he thought it best to return to France.

peace ended the active service of two outstanding French naval officers.

After his withdrawal to Toulon before Russell's greater strength in 1694, Tourville had not gone to sea again. Appointed then to the command of La Rochelle and its district, he retired from the Service after the war ended, and he died in Paris in May 1701, at the age of fifty-nine. Continuously at sea from his youth, he was a first-class sailor, not merely in ability to handle a ship, but in knowing all about its construction, its gunnery, and its manning and provisioning. He was, indeed, the complete naval officer, devoted to his Service. Both Colbert and Seignelay had a high opinion of him and departed from the usual practice of the day in seeking the advice of this seaman. Nor was he above making use of his influence with them: yet much of the organization of the fine fleet that fought at Beachy Head, and many of the regulations governing the handling of ships and crews, were his work. His courage and his fighting skill were proven over and over again, and his capacity and his pleasing personality were such that even the most jealous and contentious of the naval nobility were glad to serve under him without carping or criticism. In battle he was an able tactician, in strategy he was skilful. He was capable of acting on his own, even though his action committed his commanding admiral, and he stood up to the powers that be on land rather better than other senior officers of his day.

The one query against his admission to the ranks of the great admirals is whether he was fully able to shoulder the responsibility of being in chief command of a great fleet, the greatest in numbers that France ever put to sea. This raises the question of whether it was, or was not, his fault that more was not made of the Beachy Head victory, and that the destruction of the Smyrna convoy was not complete. It may be said that it is hard to see that much more could have been done in the Channel, and that it was Gabaret's blunder that allowed many of the convoy to escape. Be that as it may, with the possible exception of the dashing Suffren (who never commanded a large fleet), he takes first place among French admirals of the sailing era. Most of his days were spent aboard ship, apparently happily so despite bad health. He married, rather late in life, and the marriage was not a success; but he left a daughter and a son, who became an officer in the army.

Jean Bart died of pleurisy in 1702 at the age of fifty-two. One who, in those days of privilege, could rise from the crude ranks of a Dunkirk coaster to be ennobled and hold a royal commission as *chef d'escadre* must have been a man of no ordinary personality. He was a consummate seaman, a dashing, almost reckless, fighter, whose favourite method was that of boarding, yet he was able accurately to weigh up risks and possibilities. His cleverness in using wind and tide amounted to genius. Whether he could have made a success of commanding a fleet, or even the division of a fleet, in a major battle is, perhaps, open to question; but as the commander of a group of a few ships of medium or light weight he was unsurpassed. Even in comparison with his great privateer contemporaries, Forbin and DuGuay-Trouin, and the subsequent Surcouf, he remains the commerce-raider *par excellence*.

With this peace there passed, too, a great period of the French navy. It is the tragedy of that service that, whenever it has reached a peak of efficiency and merit, it has been betrayed by its administrators or its Government. Richelieu created a fleet: his successor, troubled by the Fronde, allowed it to fall away to nothing. The great cardinal's follower, Colbert, built up a navy so good and so strong that it could meet the three other maritime powers of Europe combined, on equal terms, and might even have proved superior to them. The ships were splendid, the corps of officers was skilful and spirited, the gunnery good and the organization sound. It could have obtained the mastery of the seas, with all that that means. But Colbert's second successor did the reverse of support the service that it was his duty to nourish. After 1693 it was allowed to run down to a large extent; and though, as we shall see, there was something of a revival with the start of the next war, yet, before that ended, it was allowed to run down almost completely. The main objects of sea-power were abandoned in favour of profit-making attacks on traders, and the mastery of the oceans passed into other hands.

Before the ink was dry on the Ryswick treaty, the clouds of war began ominously to rise above the horizon. The poor Charles the Sufferer of Spain clearly could not live long, there was no direct heir, and the Spanish dominions were vast. Besides Spain and the Indies, there were the Netherlands (the modern Belgium), and the Kingdom of Naples, Sardinia, and the Duchy of Milan – and these gave Spain the controlling influence in Italy. There were several claimants, all claiming through the distaff side, and these claims were complicated by renunciations and unfulfilled conditions. The leading claimants were Louis and the Emperor, the one on behalf of a younger grandson, the other for a younger son, since it was clear that Europe would not stand aside and allow all the vast empire to pass into the hands of either the French or the German monarch. William, now beginning to ail, was busy keeping his Grand Alliance together, but was ready to negotiate. For he knew that his alliance might not move without the support of the English fleet and subsidies, and that the English Parliament was suspicious of his foreign policy as being designed for the benefit of his native country, and might refuse to endorse it. Louis, too, was surprisingly ready to negotiate, for it seemed that at last he had learned his lesson. So he and William agreed on a Partition Treaty that gave Spain, the Indies, and the Netherlands to the Hapsburg, the French claimant being satisfied in Italy. At this point, in 1700, the poor king Charles died, and by will he left his throne and dominions to the French claimant.

The temptation to Louis was great. The Emperor, who wanted Italy, had not accepted the Partition Treaty, the Spaniards, apart from the separatist Catalans, clearly wanted the French claimant, the English Parliament was likely not to support William. In a famous scene Louis appeared before his Court with his grandson and announced: 'Messieurs, voici le Roi d'Espagne'. The prince was hurried to Spain and received with acclamation, French troops overran the Netherlands and turned the Dutch out of the barrier forts, the Emperor growled but made no move, the English Parliament refused to support its king's policy: Louis appeared to have 'got away with it'. For a year all was comparatively quiet. Then Louis was so impolitic as to allow his generosity to run away with him. In late 1701 old James II died, and Louis recognized his son as king of England. This was a flagrant breach of the treaty of Ryswick, and the survival of England's parliamentary revolution was threatened. By his actions in Italy, and by a commercial treaty whereby France was given a share in the monopoly of trade in the Indies, Louis had shown that, to him, the Spanish Empire was to be a French satellite. War was declared and, in 1702, the dying William III could see his Grand Alliance in motion.

England had employed the few years of uneasy peace in strengthening her already strong navy. France had not. In 1699 Pontchartrain had been succeeded as Minister of Marine by his son, Jérome; and the younger Pontchartrain had followed his father's policy of a tight-fisted disparaging of the service that he administered. He was, to boot, a most unpleasant person whom it was hard to serve with enthusiasm, or even with loyalty. In addition to her frigates, France had more than ninety ships of war; but some were elderly, most were ill-maintained, and only a few were in commission. Nor, owing to an excessive economy, were the dockyards adequately supplied to commission many more. The Spanish navy had almost ceased to exist.

For France, the naval side of the war opened with a disaster. In 1701 the Spanish 'flota', the ships containing the year's bullion from the Indies with much other precious cargo, had to be brought home, and Coëtlogon was sent with a few ships to provide an escort. At the end of the year Château-Renault, who had succeeded Tourville in the rank of vice-admiral which he had so long and so jealously solicited, was sent after him with thirty hastily commissioned ships. Coëtlogon, finding that, with war threatening, the Spaniards would not risk sailing, had given them up as a bad job and returned home, while Château-Renault, on reaching the Antilles, discovered that some of his ships were so ill-found as to be leaky, that he had a heavy sick-list, and, as he would have to wait for the Spaniards, that he would run short of stores. So, reducing some of his ships to the minimum of supplies and crews necessary for the return voyage, he sent them back to France, and it was with fifteen of the line only, his flag in the FORT(76), that he eventually escorted the 'flota' – which was specially valuable as it now contained two years' accumulation of treasure.

On reaching the Azores he learned that war had begun and that Rooke was already off Cadiz in force. He therefore proposed to make for a point south of Ushant and to run down the coast so that the 'flota' could be diverted for safety into the nearest French port, should the enemy heave in sight. The Spaniards refused, no doubt fearing that their treasure would be too great a temptation to the cupidity of the French. It was then agreed to make for Vigo. Here there is a deep bay, with an entry only about three-quarters of a mile wide and commanded on either side by a battery. Reaching this harbour, Château-Renault ran a boom of spare spars and chains across the entry, saw that the batteries were manned by Spanish militia strengthened by some seamen, stationed a few ships inside the boom, and sent the rest to the bottom of the bay, where he proceeded to unload the specie. In this the local inhabitants appeared singularly uninterested and they were not over-willing to help, so that, though the bulk of the bullion was got away on pack animals, much of the valuable cargoes remained when Rooke appeared on October 22nd with the allied fleet, twenty-seven strong. They were returning from an attack on Cadiz that had been a failure when they heard from an indiscreet boast of the French consul at Lisbon that the galleons had reached Vigo.

In order to take Cadiz, Rooke had been given a number of troops. These were promptly landed and carried the southern Vigo battery, whereupon the English second in command, Hopsonn, and the Dutch van der Goes led an attack on the boom, their ships being in two lines ahead. Though the fire from the other battery and the French ships inside the boom so damaged the *Torbay* that Hopsonn had to change his flag to the *Monmouth*, the allies broke through. Finding himself after some fighting to be hopelessly outnumbered and outweighted, Château-Renault ordered the captains of warships and galleons alike to set their vessels on fire. He might have done better for his country to have fought it out and sunk or damaged as many allied ships as possible; for his order seems to have created panic conditions and was incompletely executed, some captains not firing their ships at all, and in other cases the allies being able to board the vessels before the flames had got real hold.

None the less, there was a startling scene of gunfire and of flaming and exploding ships. But when it all ended the FORT(76), ORIFLAMME(64), PRUDENT(64), SOLIDE(56), DAUPHIN(44), and two frigates were destroyed[1]: the SUPERBE and ESPÉRANCE(70), SIRÈNE(60) and VOLONTAIRE(42) were hard ashore, and the PROMPT(76), FERME(74), BOURBON and ASSURÉ(66), MODÉRÉ(56) and TRITON(42) were taken by the allies. (It is perhaps the only naval battle in which tobacco has played a part, for a fire-ship sent against the *Torbay* had a cargo of snuff, whose griping smoke caused a temporary panic among part of the crew.) The casualties on both sides were heavy; but the allies carried off a number of undestroyed galleons which, even after the landing of the royal bullion, still had so much valuable cargo on board that it was reckoned fully to have balanced all the losses to privateers in the previous (English Succession) war.

Such dishonour as there was in this disaster, though not its magnitude, had been balanced by an action fought on the other side of the Atlantic. DuCasse, who had entered the navy by way of privateering and was given a local rank of CE in the Antilles, had the task, with six ships, of escorting a convoy of troops to Cartagena and of taking a new governor to Mexico in 1702. As the Governor was taking his wife, with her complement of duennas and maids of honour, all of whose modesty required a separate cabin on board, DuCasse had an unusual problem. However, he reached Porto Rico without mishap, sent on thence two ships with the governor and his entourage, and with the remaining four sailed to escort the transports to Cartagena. He was running along the north coast of South America towards the end of August, with a north-easterly wind, when he was sighted and pursued by the English West Atlantic squadron of seven of the line, com-

[1]Fricambault, captain of ORIFLAMME, fired her magazine and blew himself up with her.

manded by Vice-Admiral Benbow, with his flag in the *Breda*(70). DuCasse had the HEUREUX(68), PHÉNIX(60), AGRÉABLE and APOLLON(50) and a thirty-gun *flûte*. Benbow's ships carried a total of nearly 400 guns.

DuCasse ordered his convoy to make all sail, and formed line, keeping between it and the enemy. For several days the action and pursuit continued, for whenever Benbow got his ships into position, some of his captains either fired a broadside or two and then stood off, or even refused to come within range to engage. When the APOLLON, damaged in masts and rigging, lagged astern and was in danger of being surrounded, DuCasse boldly came about with the rest of his line to disengage her, as he did again when the *flûte* was specially threatened. The *Ruby*(40), which alone gave Benbow proper support, was so shot up as to be driven out of action, and the flagship itself was much damaged.

At last Benbow, who had a wound that eventually proved mortal, was obliged to stand away for Jamaica. He was a rough-tongued man who had come into the navy through the hawse-pipe, and his defaulting captains were of the kind that were in the service to fill their pockets and were unwilling to fight seriously unless a profitable convoy could be taken – as DuCasse was clearly determined should not happen. There was a court martial which sentenced two of them to death (as it would have done a third who died during trial), and dismissed another from the service with imprisonment. But the fact that this is a dishonourable page in the history of the English navy does not alter the other fact that DuCasse, who knew nothing of these captains' cowardice, conducted a fine fighting retreat and thereby wrote a most honourable page in the history of the French service. He had done much the same as Cornwallis was to do when faced by Villaret-Joyeuse some ninety years later[1].

The French had also made an appearance in the waters, unusual for them, of the Adriatic, where Forbin with some frigates had been sent to cut off supplies sent by sea to the imperial armies in Italy. He did this with such ruthless disregard for the rights of neutrals that he had to be recalled; but his successor, DuQuesne-Mosnier, lived up to his family's reputation with an unusual exploit. Learning that the imperialists had formed an army supply depôt at Aquileia, some miles up a river from the Gulf of Trieste, he took his boats thither and destroyed it in a good surprise attack. Of course, with the outbreak of war the privateers and commerce raiders had been let loose, and Saint-Pol Hécourt, who had taken Jean Bart's place in command of the little squadron of four lightweight vessels in Dunkirk, had several successes. He cut up a convoy escorted by the *Salisbury*(50), which he took and added to his little squadron; he cut up another off the Dogger Bank, though in subduing its escort one of his ships, the ADROIT(30), was sunk by the explosion of the vessel it had grappled: he cut up a Dutch convoy, and, in 1704, an English one, but lost his own life in doing so. The galleys in Nieuport had their last little success, for sallying out in force against the Dutch *Licorne*(56), which was completely becalmed, they overwhelmed it and set it ablaze. A more regular engagement took place off

[1]DuCasse wrote to Benbow: 'I had no hope but to have supped in your cabin. As for your cowardly captains, hang them, for, by Heaven, they deserve it'.

A FIRST-RATE SHIP OF THE LINE

A 30-GUN FRIGATE UNDER FULL SAIL (*by Ozanne*)

PLATE VI

## THE 'CORSAIRS'

JEAN BART

FORBIN

DUGUAY-TROUIN
*(by Ozanne)*

Lisbon in 1703, when Coëtlogon (LG), taking five of the line from Brest to Toulon, came upon a large convoy escorted by five Dutchmen under the veteran Roemer Vlacq. These formed line between their convoy and the French and put up a truly heroic resistance, especially on the part of their admiral (for Coëtlogon's ships were much superior in strength), so that, though all were eventually taken (two in a sinking state), their convoy was saved.

But there had been little or no fleet action. The admiral of France at this time was the comte de Toulouse, a royal bastard aged twenty-six. He was utterly without experience, but instead of treating his office as a mere appanage he raised his flag at Toulon, and was given vice-admiral Victor d'Estrées, now Maréchal de Coeuvres, as his 'adviser'. In 1703 they had taken the Toulon squadron to sea to support military operations in Italy, but had been obliged to return to port when Sir Cloudesley Shovell appeared in the Mediterranean in superior force.

Clearly if France was not to leave the sea free to her enemies, where any operation more than a darting commerce raid was concerned, she must put out a fleet in strength, and her flag-officers were able to persuade her rather unwilling king and his minister to that effect. So in 1704 the comte de Toulouse arrived in Brest, was joined by the ships from Rochefort, and took the whole force round to Toulon, sighting, but not engaging, Rooke on the way. The Englishman had landed the Austrian archduke and an army at Lisbon (for, by the Methuen Treaty, Portugal had now allied herself to England), and had sailed on to attack Barcelona. This Toulouse should have put to sea to thwart; but the Toulon ships were not ready, nor had Pontchartrain taken any particular pains to see that they should be so. Therefore it was not till late in July that Toulouse was able to leave harbour, commanding the last major fleet that France was to put to sea for many a year. In order of battle it was:

| Ship | Guns | Captain |
|---|---|---|
| ÉCLATANT | 66 | de Bellefontaine |
| ÉOLE | 62 | marquis de Mons |
| ORIFLAMME | 62 | Château-Renault |
| SAINT-PHILIPPE | 92 | d'Infreville (CE) |
| HEUREUX | 72 | Colbert de Saint-Marc |
| RUBIS | 56 | de Renneville |
| ARROGANT | 56 | Desherbiers de l'Étanduère |
| MARQUIS | 56 | Patoulet |
| CONSTANT | 68 | comte de Sainte-Maure |
| FIER | 90 | marquis de Villette-Mursay (V-adm) |
| INTRÉPIDE | 84 | DuCasse |
| EXCELLENT | 60 | La Roche-Allard |
| SAGE | 58 | Montbault |
| ÉCUEIL | 68 | d'Arigny |
| MAGNIFIQUE | 90 | de Belle-Isle-Érard (CE) |
| MONARQUE | 84 | Chabert |

| Ship | Guns | Captain |
|---|---|---|
| FURIEUX | 58 | marquis de Blénac |
| VERMANDOIS | 60 | comte de Béthune |
| PARFAIT | 74 | marquis de Châteaumorand |
| TONNANT | 90 | comte de Coëtlogon (LG) |
| ORGUEILLEUX | 72 | du Palais |
| MERCURE | 50 | chevalier de Lannion |
| SÉRIEUX | 60 | Champmeslin |
| FLEURON | 54 | chevalier de Grancey |
| VAINQUEUR | 86 | bailli de Lorraine (CE) |
| FOUDROYANT | 104 | comte de Toulouse (adm)<br>comte d'Estrées (v-adm) |
| TERRIBLE | 102 | de Relingue (LG) |
| ENTREPRENANT | 58 | comte d'Hautefort |
| FORTUNÉ | 54 | de Bagneux |
| HENRI | 66 | de Serquigny |
| MAGNANIME | 74 | de Pointis (CE) |
| LYS | 88 | de Villars |
| SCEPTRE | 84 | chevalier d'Ailly |
| FENDANT | 58 | de la Luzerne |
| ZÉLANDE | 60 | de Ferville |
| SAINT-LOUIS | 60 | chevalier de Beaujeu |
| ADMIRABLE | 92 | comte de Sèbeville (CE) |
| COURONNE | 76 | de Champigny |
| CHEVAL-MARIN | 44 | de Pontac |
| DIAMANT | 58 | D'Oroigne |
| GAILLARD | 54 | chevalier d'Osmond |
| INVINCIBLE | 68 | marquis de Rouvry |
| SOLEIL-ROYAL | 102 | marquis de Langeron (LG) |
| TRIDENT | 56 | chevalier de Modène |
| CONTENT | 60 | chevalier de Phélypeaux |
| MAURE | 54 | Saint-Claire |
| TOULON | 62 | DuQuesne-Mosnier |
| TRIOMPHANT | 92 | de la Harteloire (CE) |
| SAINT-ESPRIT | 74 | DuQuesne-Guiton |
| ARDENT | 64 | d'Aligre |

Rooke's attempt to promote a rising in Barcelona had failed, but just as, two years before, he had made good his failure at Cadiz by taking advantage of the lucky chance that Château-Renault and the treasure fleet had gone into Vigo, so now, having joined Shovell in the Straits, he heard that the Spaniards had almost criminally neglected to supply Gibraltar with an adequate garrison. Rooke had some troops, and these were landed under the Prince of Hesse-Darmstadt, while Byng's squadron bombarded from the sea. The Spaniards made as stout a resistance as could be expected from a mere 150 men, plus the crew and guns of a small French privateer;

but they could not hold long against such odds, and one of the strongest strategic positions in the world, the key to the Mediterranean, had passed into English hands unless the count of Toulouse could recover it. This he was sailing to do, while Rooke, having landed his marines to increase the garrison, and supplied them with all the stores and ammunition that he could spare, stood off eastward to intercept him. When the fleets got in touch the French went into Velez-Malaga to water, and, checked by the inshore current and lack of wind under the land, could not get to sea again for two days. Rooke, with the outer, or ocean, current and a little wind, took advantage of this to get back his marines, though not his spare ammunition, from Gibraltar.

Battle was joined off Malaga on August 24th 1704. The French were in line with an easterly wind abeam, steering south-westerly. Rooke had the advantage of the wind and bore down on them. Shovell commanded the allied van in the *Barfleur*(90) with Leake as his second in the *Prince George* (90); Rooke led the centre in the *Royal Catherine*(90) with Dilkes as his second in the *Kent*(74); the rear consisted of the Dutch squadron under Callenburgh and van Wassenaer, with a few English ships under Byng in the *Ranelagh*(80). The fleets were roughly equal, for, though the allies were a little more numerous, they were short of powder and shot, while the French carried the larger crews, had at least an equivalent total weight of metal, and, under their lee, had their galleys which could give a tow, or other help, to a ship in trouble.

Shovell did not keep station well, and came into action about ten in the morning rather far ahead of Rooke, leaving something of a gap between van and centre. Villette-Mursay signalled his ships to stretch ahead, to round the English van, but Shovell's vessels kept up with the French and thereby increased the space between themselves and their centre. Into this Toulouse (or more probably d'Estrées, who was really in chief command) endeavoured to work; but Rooke made all sail and closed the gap, forcing the French back into line, though some of his ships suffered damage and heavy casualties in the process. The practice of fighting in line was still comparatively new and, apart from the two manoeuvres attempted, its tactics were not yet (nor for a long time) fully understood. So the battle became a bloodthirsty slogging match, with now and then a ship on one side or the other paying out of line to make good some grave injury, but usually getting back into action.

At four in the afternoon a shell landed in the poop of Villette-Mursay's FIER, exploded some reserve ammunition, blew away most of the poop, caused heavy casualties and a fire, and so forced the 'admiral' of the French van to pull well out of line. The other ships of the van took this to be a deliberate manoeuvre and also pulled out, re-forming line on him. Firing between the two vans therefore ceased. Leake wished to go into the gap thus left to get to leeward of the French centre, but Shovell preferred to bring the van ships about to give direct support to Rooke who seemed under heavy pressure. Thereat the engagement in the centre gradually ceased, though that in the rear continued till nightfall. Each fleet then drew off to lick its wounds and make essential repairs. The Dutch *Graaf van Albemarle* sank during the darkness.

The French claimed a victory, and the claim had this much justification – that they had caused the heavier casualties and had done the greater damage. But the battle had in fact been most bloodily indecisive. Had they known it, a victory could have been won the next day, for the wind changed to give them some advantage, and a number of Rooke's ships were almost, or even entirely, out of powder and shot. The French had lost rather unusually heavily in flag-officers[1], but the remainder were summoned to a council of war. The dying Relingue sent a last letter urging attack, and Pointis also urged it. No one else did so. D'Estrées summed up the pros and cons, and Villette-Mursay advised against it. At the age of seventy-two he was no longer the stout fighting officer that he had been under Tourville twelve years before, and was probably badly shaken by the damage done to his own ship, which had suffered more than any other. Yet, perhaps because of his seniority and reputation, it was his advice that carried the day: it was decided certainly to remain in line and fight if the English gave battle, but not to attack them. Councils of war seem always to take the more cautious course.

Rooke put up a bold front to conceal his weakness but, of course, did not engage; and the two fleets drifted apart, Rooke to return his marines to Gibraltar before sailing to base, and the French to go to Toulon. Though Te Deums were sung for a French victory, the aftermath of this indecisive slaughter was, in a way, a defeat. For the object had been to recapture Gibraltar before its defences could be made good; but the Rock was still in the hands of the English, who were busily strengthening their position.

As the realization of this sank home, and as defeats on land imposed an increasing strain on resources and finances, the fleet once again seemed to be an expensive luxury that achieved nothing. So once again it was decided to lay up most of the big ships, and to rely on the profitable commerce raiding: a policy of chasing an attractive phantom instead of facing a stern reality. For unopposed, and with the Dutch too exhausted to play anything but a minor part, the English were to take advantage of it to give themselves a mastery of the oceans that lasted over 200 years. It is odd that France failed to notice that, when she put a large fleet to sea in 1704 and so compelled the English to concentrate their forces, her privateers were able to operate more freely than they had done before, or were to do later, and the Spanish treasure convoy came home without the least interference. It should not be thought, however, as one might gather from French naval historians, that the idea of having a fleet at sea was contemptuously dismissed. On the contrary, it was intended to fit out a fleet of full strength, some seventy ships, in 1705; but it was found that the arsenals were too run down to equip it. Royal credit was no longer good, and it was to some extent perforce that privateering became the main maritime weapon because the *armateurs*, the bodies of merchants who financed privateers, had the money to fit out the ships, which were, of course, smaller and less costly to equip than the large vessels of the line. Indeed, as the war proceeded it was only in privateers that naval officers and crews could make a living, for the pay,

[1]Relingue, Belle-Isle-Erard, and the Bailli de Lorraine killed; also several ships' captains were *hors de combat*.

even of those at sea in the few ships that were kept in commission, became sadly overdue.

Meanwhile it was decided to reduce Gibraltar by siege, and Pointis was sent with a small squadron to blockade it. While he went to Cadiz for water a convoy was slipped in, and he was thereupon ordered to blockade closely through the winter. He represented that then, if a superior force should appear, he might be caught against a lee shore, whereas if he was allowed to cruise in the approaches he would be able to intercept any convoy that was not escorted by a force so superior that he would have to retreat. His representations were disregarded and what he feared came to pass. In March 1705 a gale blew eight of his thirteen ships well out to sea, and the remaining five, as a reward for their good seamanship in keeping to their station, were trapped close inshore with no room to manoeuvre by a superior force under Leake and Dilkes. The ARDENT(64), the ARROGANT and the MARQUIS(58s) struck; Pointis with the MAGNANIME(80) and the LYS(88) beached, and the ships were set afire by their crews. Shovell went on with a combined force to take Barcelona. The next year, the French and Spaniards made a joint land and sea effort to recapture it; but thirty ships were the most that could be equipped, and they had to retire to Toulon and let the siege be raised when forty-four allied ships came on the scene.

Indeed, in 1707, Toulon itself was in the greatest danger, being attacked on land by the Duke of Savoy and from sea by Shovell. The navy played a leading part in the defence, for sailors manned the guns ashore, and some ships turned themselves into floating batteries to keep off Shovell's bomb vessels. It was a tense situation, though eventually the defence triumphed and the assailants withdrew. But the bombardment had destroyed the SAGE and the FORTUNÉ by fire, and had done so much damage to the arsenal that Toulon was out of use as a naval base for some time. Then in 1708, over-riding such little aid as the French could give the defence, Leake and his squadron took Minorca, with its fine harbour of Mahon, almost opposite Toulon. The English were clearly masters of the Mediterranean.

It was well before this that commerce raiding had become the national maritime policy. Its great advocate was Vauban, who had an honorary flag-rank in the navy as a reward for his work in coastal fortification. To him it was at once a means of replenishing the exhausted royal coffers and of bringing low two enemies who depended much on their seaborne trade. He advocated that the raiders should sail, not singly, but in small squadrons, operating as Jean Bart had done in the previous war. Large fleets were 'vanity'. This did in fact lead to the capture of several British ships of the line on convoy duty when they had to fire on two or three smaller adversaries at once while being also ready to repel boarders, since boarding was the main French method of attack – a method in which they employed almost reckless dash and bravery and were often successful. But though part of the considerable profits that they made went into the royal coffers when the raiders had been royal ships that were leased, most of them went into the pockets of merchants, or of officials and officers acting as private persons. Moreover, while Vauban was undoubtedly a great military engineer, he had no great vision of sea-power, which is to control the oceans; and as far as there could be control in the days of sail, it was the English who exercised it. The privateer made, so to speak, destructive sallies upon the besiegers' camp, but where the sea was concerned it was France that was besieged.

The main centres of privateering were Dunkirk, specially planned for this purpose, and St Malo, a tricky port to enter for those who did not know it, and one impossible to blockade because of the dangerous currents and neighbouring coast – though the Malouin privateers were always likely to be attacked by English ships using such shelter as the Channel Islands provided, and they did indeed suffer losses in this way.

The leading names among the privateers are Forbin, the old colleague of Jean Bart, who acted now as a naval officer and now on his own account, and DuGuay-Trouin. This latter, who rivals Jean Bart in Success and in French estimation, came of a family of Malouin merchant shippers and, with his combination of dash and of personal charm of manner, was a sort of Claude Duval among privateers, quite unlike the rough Dunkirker. Born in 1673, he had gone to sea when sixteen, and his quick eye for a chance in attacking merchantmen had brought him a privateer command when only eighteen. Like Jean Bart he had made some dashing and sensationally successful attacks, and, like him, he had on one occasion been overwhelmed and taken prisoner and had escaped from Plymouth, though rather less sensationally than the Dunkirker. He was therefore well known when he resumed his career with the outbreak of war in 1702, now leading a group of three ships or so, with the same success. In 1706 he was sent to assist in repelling a wrongly anticipated attack on Cadiz, and fought off a squadron of six Portuguese of the line *en route*. It was only after this that he was given a royal commission as captain (together with a patent of nobility) for he had sailed under royal orders as a privateer.

It is, in fact, difficult to say how much the activities of the larger privateers should, or should not, be treated as naval affairs. When, later, Du-Guay led his famous expedition against Rio de Janeiro, it was a privateering concern; yet the king himself was among the *armateurs*. The king's written conditions were that he would lend such and such ships in good condition and supplied with guns, spare parts, and munition for a nine-month campaign. He would also supply the officers, seamen, and soldiers, who would be subject to naval discipline, and he would provide their pay at harbour rates; but DuGuay (which means the other *armateurs* who were financing the expedition) was to pay the expense of the actual fitting out of the ships, the additional seagoing pay, and the officers' table money, also for all provisions. His Majesty would accept captains for three of his ships who would be chosen by DuGuay and would give them temporary naval rank. The king would not expect compensation for any ships lost in battle or by the hazards of seafaring, nor for any consumption of ammunition or spare masts and sails, and the ships should be used entirely as DuGuay wished; but there are complicated and strict conditions as to the sharing of any prizes, booty, or other profits of the campaign, of which His Majesty was to receive one fifth, while if there were no profits the *armateurs* were to lose their money. Truly the Crown and its navy went into business in the privateering racket!

There were some notable events in the mass of minor commerce raiding. In 1706, Forbin, sailing from Dunkirk with two small ships of the line and two frigates, attacked a Dutch convoy. Its escort put up a fight and saved the convoy, but one of the escort blew up, a second was sunk and a third taken. Next year, with his flag in the MARS (60), and six other small ships of the line and some corsairs, Forbin attacked a convoy leaving the Downs under escort of three English seventy-gunners. After a stiff fight two of the English ships were taken, and meanwhile his corsairs had cut off more than a third of the convoy. He completed his highly successful cruises from Dunkirk by cutting up the Dutch 'Muscovite' convoy and, knowing that the approaches to Dunkirk would be packed with ships waiting to entrap him, bringing his booty home by sailing round the north of Scotland and west of Ireland. Later he joined DuGuay-Trouin in an attack off the Lizard on a large convoy carrying troops to Portugal to support the Austrian archduke in his attempt on the Spanish throne. DuGuay had his flag in the LYS (74) and Forbin in the MARS (60): they had fourteen ships, all but one being of the line, but all acting as privateers, and DuGuay, only recently given a captain's commission, was in command, though Forbin was a CE. The escort was five of the line varying in strength from eighty to fifty guns. Three of these were taken, and the *Devonshire* (80) blew up after a strenuous resistance. Sixty of a convoy of eighty sail fell to the victors. Du Quesne-Mosnier, with five ships from Brest (fitted out with Spanish money, for the intendant at Brest had had to fall a year behindhand with wages for his dockyard hands), cut up another troop convoy.

These convoy losses had a grave effect on the allied campaign in Spain and contributed to the archduke's cause coming to grief in the battle of Almanza. While it is understandable that Dutch convoys were weakly escorted, since the United Provinces were so strained that they could not carry out their treaty obligations to England of keeping a certain number of ships at sea, it is odd that important English convoys were not more strongly guarded. For England had ample ships, which were largely kept concentrated in good-sized squadrons, though the intelligence of those days was such that she must have known of the impoverished state of French dockyards and of the laying up of the French fleets. Of course the French had losses too; for a number of privateers were taken, as were some commerce-raiding royal ships, such as the HASARDEUX, the SUPERBE, and the TOULOUSE.

Beaten on land by the genius of Marlborough and Eugene, and forced back inside their own frontiers, the French were willing for any desperate venture. So in March 1708 an attempt was made to put the 'Old' Pretender, James III, on the throne of Scotland. With seven ships, none of great strength, his flag in the MARS, and three transports carrying troops, Forbin left Dunkirk for the Forth. The Scots, especially those of Jacobite leaning, were dissatisfied with the recent Act of Union, and a good sized conspiracy had been worked up. But it had become known, the Dukes of Atholl and Gordon had been arrested, and Byng was after Forbin in force. The French commander, having reached the Firth, learned this, and was faced with the awkward choice of sacrificing his ships to put the prince and the troops ashore, when the country might or might not rise in his favour, or of getting away if, and as best, he could. One or two of his ships had parted company in bad weather, the SALISBURY, fighting to protect the rest, had been taken, so he ran for it, and got back to Dunkirk. Ministers of War and Marine each threw the blame on the other, and Forbin, who had been so unwise as to describe the whole plan as chimerical, was made the scapegoat, and sent to his country home in disgrace. It was a poor reward for forty-four years' strenuous service. This expedition did, however, serve one useful purpose. In view of the threat to the homeland the British fleet that was waiting for the 'flota' was called home, and du Casse, who had become a sort of specialist in coping with the Spanish treasure ships, was able to run his convoy of them safely into Passages.

France was, indeed, in so grave a state that in 1709, with a grim winter and a bad harvest, she of all countries was threatened with a corn shortage. The precious grain could be obtained from the Turkish Empire, but with Toulon helpless it had to be left to a corsair, Jacques Cassard of Nantes, and his privateers to bring it home. And he successfully dodged the allies in getting two convoys through. So it is not surprising that the one lightening of the gloom, DuGuay-Trouin's capture of Rio de Janeiro, has attracted great acclamation.

We have already seen the terms under which this piece of royal buccaneering was arranged. The Malouin sailed from La Rochelle in June 1711 with seven of the line, his flag in the LYS(74), six frigates (two being of forty guns), two privateers and a bomb vessel; and he had on board over 2 000 troops. The Portuguese were ready for him, for an ill-planned raid had been made, and failed, in the previous year. Favoured in his approach by an east wind and some mist, he came on in line ahead, the MAGNANIME(74) leading, with little damage from the batteries at the entrance, and drove the Portuguese warships easily before him to the base of the huge bay. Here lay the actual harbour with a narrow entrance, covered by Cobra Island and with powerful forts at each side. The town itself had no defences, but there were several batteries. Under fire, DuGuay occupied Cobra Island and mounted on it his mortars and his heaviest guns to engage the forts. Then, a summons to surrender having been rejected, he landed his troops. They were attacked by the Portuguese forces, but repelled them. At this the governor fired most of his ships and abandoned the town and the forts, retreating inland to await reinforcements. But on a threat that the town would be destroyed he agreed to pay a heavy ransom. This being collected, DuGuay withdrew; and though in a violent storm the MAGNANIME and the FIDÈLE(60) were lost without trace, while the AIGLE frigate struggled into Cayenne only to sink there, the expedition repaid 92 per cent to those who financed it. DuGuay-Trouin himself was later promoted to CE though he had never commanded anything but privateers.

A smaller but no less spirited action was fought in the same year off Vado, south of Leghorn, by four ships commerce-raiding under a *capitaine de frégate* named l'Aigle. He sprang upon three ships of the British Mediterranean squadron which were keeping watch upon the coast and caused them heavy casualties and damage before the arrival of the rest of the squadron obliged him to break off the action and retire to Leghorn. One of his ships, the PEMBROKE (a capture from the English), was a bad sailer and was overtaken. But she made a most determined fighting retreat, fairly beating off the *Exeter* with heavy loss, and holding off other attackers before struggling into Spezzia. L'Aigle was killed a few weeks later in action with a Dutch frigate.

In the colonies as on land, things went ill. True the privateers in distant seas made some rich prizes, and both Cassard and du Casse conducted fairly profitable raids on the West Indies. Yet the solid gains went to the English. The French East India Company was ruined, almost all French ships in the West Indies had been taken, the key points in Acadie (Nova Scotia) had fallen into English hands, Newfoundland had been held only with difficulty, and Canada had been saved only by a storm off Sept Isles in the St Lawrence that brought disaster to Admiral Hovenden Walker's squadron before it could seize an almost defenceless Quebec. Every French overseas possession lay at England's mercy, and, despite a last minute victory at Denain, the French armies had been beaten in Europe – except in Spain.

But the European situation had changed. The Emperor Leopold had died and had been succeeded by his elder son, Joseph. He too died in 1710, and the Archduke Charles, claimant to Spain, had become Emperor. England had less wish to see the Hapsburg Empire of Charles V re-created than she had to see a Bourbon on the throne of Spain: that country was clearly going to be most difficult to conquer, and England had got all she wanted. A Tory anti-war government had come into power, Marlborough was recalled with his great victories uncompleted, pourparlers were opened, Holland was too weak to do anything but follow her ally while protesting, and France was so strained that Louis was willing to make peace on almost any terms. So in 1713, by the Treaty of Utrecht, France agreed that the thrones of France and Spain should be for ever separate, that the old Pretender should be disowned, that the Emperor should have the Spanish Netherlands and share Spain's Italian possessions with Savoy, that England should retain Gibraltar and Minorca (which meant the command of the Mediterranean), and France surrender to her Nova Scotia, St Christopher, all Hudson Bay rights and forts and, subject to the retention of some fishing rights, Newfoundland. And it was England, not France, who had the Assiento, the breach of the monopoly of Spanish trade in the Indies. Finally, and most offensive clause of all, the forts and harbour of Dunkirk were to be destroyed by Frenchmen under English supervision. It was said that the Dunkirk raiders had taken some 1 600 ships worth, with cargoes, over thirty million: England was not prepared to allow that 'nest of vultures' to survive. The terms were a humiliation, but France could do nought else but accept them. Her navy was in ruins; her oversea possessions lay at the mercy of the new dominant sea power: the ambitions of Richelieu and the realizable dreams of Colbert had all vanished into the Ewigkeit.

CHAPTER VIII

# Starvation Years and the Austrian War

Louis XIV, le Roi Soleil, died in 1715. He had lived and reigned so long that both his son and his elder grandson had predeceased him, and the throne passed to his great-grandson, a child of five, as Louis XV. Though he was to be as much an autocrat as his predecessor, he lacked his zest for work, and grew up to be lascivious and lazy. For a while there had to be a regency, the next prince of the blood, the Duke of Orleans, being the regent, with an advisory council.

France was utterly exhausted by the strain and expense of the long wars, and nowhere did that exhaustion show more than in the navy. Under the two Pontchartrains it had fallen far below the numbers of Tourville's day. The great ships lay at their moorings, unoccupied and slowly rotting: the crews were all paid off: the officers were ashore, some in poverty, for their pay was in arrears; the arsenals were all but empty. Yet it would be wrong to accept the impression given by French naval historians that the navy was left in neglected decay. Even when things were at their worst DuQuesne-Mosnier (CE) sailed with one of the line and a frigate to show the flag off the Barbary ports; and if his force was pathetically small, it nevertheless reminded the pirates that France still had naval ships and conveyed a hint that they would be wise to keep their depredations within limits. Moreover, the count of Toulouse, the royal bastard who had been given the admiralty of France as an 'appanage', but who had actively commanded from 1702 to 1706, retained his interest in the service of which he might have been only a nominal head, and, early in the regency, both he and one of the vice-admirals, Victor d' Estrées, presented memoranda to the Council which formed a strong attack on Pontchartrain's administration. Indeed, in nineteen years Colbert had spent 216 million livres in building up the service, and in roughly the same time the two Pontchartrains had spent more than twice as much in running it down. One wonders where the money went.

At any rate, this attack led to Pontchartrain's dismissal, though the family influence was such that this was accomplished only by giving the office to his son, the comte de Maurepas. As this gentleman was then a youth of fifteen, the navy was to be administered by a council. It consisted of the vice-admirals, d'Estrées and Coëtlogon, the general of the galleys, Tessé, and two other naval officers, plus Renau d'Eliçagaray (Petit-Renau, the naval constructor) and three naval intendants, with Toulouse having the right to attend when he wished. And though, for some time, their representations to the Council of Regency as to what was wrong and what funds were required brought the invariable reply that 'The Council will provide this as soon as it can', they gradually got things straight, though on a scale much reduced from that of former days. Supply did not make up for arrears in stocking the arsenals, much less for those in building new ships, but it was enough to keep the much reduced fleet in trim.

The general foreign policy was one of peace, which, for France, was vitally necessary. Had the regency taken the opportunity to put her chaotic and virtually bankrupt finances in order, the period of peace would have been a benefit as well as a necessity; but this would have upset many vested interests, and anyway the regent was a philandering spendthrift, incapable of promoting such a policy. The peace meant keeping on friendly terms with England; but the statement that England bribed Dubois, the regent's Foreign Minister, and the marquise de Prie, mistress of the short-lived chief minister, the duc de Bourbon, for the express purpose of keeping the French fleet small, is right off the mark. Any such money that passed was a *douceur* to irenics; for England was equally anxious to avoid war, because the parliamentary (Whig) revolution depended on maintaining the Hanoverian succession, which was unpopular, and war might provide the Jacobites with foreign support. In any case it would have taken France many years to build up to England's naval strength. England, having only a sea frontier, could devote the bulk of the money for her forces to her navy. France had both land and sea frontiers, and put the safety of the land one first. An enemy army could invade, an enemy fleet could do no more than raid a port or so. Louis XV was so little interested in his navy that in all his long reign he spent just two whole days in one of its bases, and that the least important, Le Havre; the French people outside the ports were equally uninterested. The general attitude to it was that of Berryer in the Seven Years War, when his attention was called to its needs: 'Monsieur, when the house is on fire one does not bother oneself about the stables'. The result was to be the loss of an empire.

The regency ended with the death of Orleans in 1723, and, after three years with the Duc de Bourbon as chief minister, the boy king handed the conduct of affairs to the aged cardinal Fleury. It was to his credit that he was pacific-minded, but less so that he was a mean man, who tried to run the state on a shoestring. In fact her natural resources were so great that France had recovered fairly quickly from her exhaustion, and commerce had revived. The trade with the West Indies soon became good, to the benefit of the Atlantic ports; and later the apparently moribund East India Company became positively prosperous. Royal ships had to sail to the Antilles taking out officials and relief garrisons and bringing home others. This was almost all the actual sea-training that naval officers could get, for the East India Company ran its own armed vessels. The Navy Council came to an end with the regency. It had done good work, especially with the colonies, founding New Orleans, fortifying Louisbourg, and developing the port of L'Orient. A few months later the young comte de Maurepas took over his office.

He was only twenty-two, was chiefly known for a facile flippancy in making epigrams, and, in the words of one angry intendant, 'did not know the colour of the sea nor how a ship was built'. Yet he proved a minister who did much with rather little. The 'Barbaries' were again becoming a nuisance, and in 1727 an expedition was planned against them. Because of the muddling of its leaders it came to naught, but the next year Grandpré (CE), with two ships of the line, some frigates, galleys, and bomb vessels, went against Tunis and Tripoli and obtained reasonably good satisfaction and release of slaves, though he could do no more at other ports than make a bombardment that did some damage. In 1730, the navy budget being nine million livres, de Maurepas presented a report on its expenditure. The galleys, which had still to be maintained though their day was well past, would take 1 600 000. The colonies, which were lumped under 'Marine', would need 1 700 000. The marines and the *plume* (the administrative services) would need 2 100 000; and, after a sum had been put aside for hospitals, stores, and lighthouses, there would be left two and a half million for building new ships and refitting existing ones. It was not much; but with it he hoped to maintain a fleet of fifty ships of war, excluding really small vessels such as brigs and corvettes[1]. As England maintained over 160 ships, with others building or in dock, this meant that, where she was concerned, the French fleet would not sail proudly forth to meet the foe but would have to dodge out of harbour and proceed on whatever was its object knowing that it would possibly be pursued by, and have to fight against, superior forces.

It was largely from this that there grew the doctrine of 'the mission' that was to dominate, and often bedevil, French naval activity for almost the rest of the century. A fleet sailed with a mission, a positive task to carry out, and should do its best to avoid any action likely to prevent it from carrying out that mission. Taken to a little short of its extreme this meant that a fleet would fight an indecisive action from which it could slip away to carry out its object, rather than strive for a victory as a result of which, though triumphant, it might have to return to harbour. From this, plus the lesson of the writings of tacticians, came a preference for engaging from leeward. (It was pointed out that ships which obtained the wind and then ran down on the enemy were likely to be much damaged by the enemy's broadsides before they could bring their own broadsides to bear – as happened with Rooke at Malaga.) A corollary of being able to disengage in order to carry out one's mission was that one fired to damage the enemy's masts and rigging so that he was too crippled to follow. The English preferred to get to windward so that they had the choice of attack, and to fire at the enemy's hulls and batteries.

As a result of Maurepas' report the king required that ships of 'classes' and frigates, to the number of fifty-four in all, should be maintained. This meant building about three ships a year, and left less than half a million livres with which to maintain stores in arsenals. Maurepas managed to get a little extra money to keep the ports up to date, but in 1744, with war obviously on the doorstep, when he asked for an extra twenty million

[1]The French ships were still graded into five classes, but the fifth class – of 30–39 guns – was now completely confused with the frigates, and it was not reckoned that the fourth class – 40–49 guns – could take a place in line of battle, except in an emergency.

livres he was given only six. A navy so financially starved could put in very little sea time, for exercises at sea meant expense in stores, equipment, and pay of crews, for which the funds were just not there. Young officers were therefore given training in such things as hydrography which made the service a highly scientific one; but, with actual ship-handling limited to a few exercises in harbour, this could not give the experience and the confidence that would have come from voyaging.

Nor was the service a united one. The *plume* looked on the *épée* (the fighting officers) as mere brainless fighters who carried out its intellectual plans, while the *épée* resented being run by landsmen who had little knowledge of seafaring. Privilege, too, made for disunity, for the French noble was more proud than ever of his birth and scorned the *officiers bleus*, those who had come from merchant service stock. Even among the nobles degrees of birth counted, and it cannot have been over-acceptable to LGs of long service that the marquis d'Antin was pushed over their heads by being given the reversion of the first vacant vice-admiralty at the age of only twenty-two, because he was the stepson of the count of Toulouse, admiral of France and royal bastard – even though d'Antin proved a notably able officer. None the less there was a good deal of family tradition in the service, and among the officers we find Gabarets, Nesmonds, great-nephews of DuQuesne, the grandson of Tourville and of Jean Bart, and so on for the first half of the century. The ships that Maurepas built were mostly seventy-fours of first-class design.

Apart from the voyages to and from the West Indies there were a few small, but more active, expeditions. In 1731 DuGuay-Trouin (CE) was off the Barbary ports in the ESPÉRANCE (72) with three others of the line, 'showing the flag', as were de Court la Bruyère (LG) in 1734 and the marquis d'Antin (CE) in 1737. In 1733, in the affair of the Polish Succession, when king Stanislaus Lesczynski, father of Louis XV's queen, was driven by his Saxon rival to a last stronghold in Dantzig, La Luzerne (LG) with eight of the line and five frigates went to the Baltic. He achieved nothing; for he had insufficient forces, and refused to put his squadron to great risks at the demand of Count Plélo, the French soldier ambassador to Denmark, who landed with a few troops and was killed. In each of the next two years DuGuay-Trouin was ordered to fit out the Brest squadron for the Baltic and began to do so, full of his old energy despite the gout, only for the orders to be countermanded. Death brought his inspiring career to an end in 1736, and Toulouse followed him to the grave in the next year. The great office of admiral then went to his son, the duc de Penthièvre, who was aged twelve!

The years of uneasy and sometimes broken peace were now drawing to an end. By 1739 Britain was at war with Spain over the working of the Asiento, which the British had abused; and in 1740 the Hapsburg emperor, Charles VI, died, to let loose the eight years' war of the Austrian Succession. As he had no son and it was doubtful if her sex entitled his daughter to inherit all his dominions, he had devoted his diplomacy to persuading all the powers to agree to a Pragmatic Sanction, whereby she should succeed in peace. No sooner had he died than Frederick the Great of Prussia seized

Silesia, entirely unscrupulously, and the Elector of Bavaria advanced his claims. As he was a traditional French ally and the Sanction had already been broken by Prussia, France joined in to obtain a share of what was going, as did Spain, to drive the Hapsburgs from Italy. Since England kept her word in the Sanction (because of her Hanoverian connection, which dreaded Frederick) war between England and France was inevitable – though it did not break out for a further four years.

To protect the colonies and ships of her ally, Spain, France sent out a fleet twenty-two strong to the Antilles, under the marquis d'Antin (v-adm). Its presence did serve, to a certain extent, to halt English operations; but its usefulness was brought to nought by yellow fever, which devastated its crews. So d'Antin[1] brought all but six of his ships back early in the next year, losing the veteran BOURBON, which was too decrepit to be at sea at all, with almost all hands when in sight of Ushant. Some of the ships which he left behind were attacked by the English under the (alleged) belief that they were Spaniards, falsely sailing under French colours, and in 1743 the same thing happened, with the same excuse, off Gibraltar to some ships returning from India. Of course, when war on land started in 1740, the revived French commerce felt itself very exposed, and Fleury had allotted an extra three million to the navy, which Maurepas used wisely on repairs and on new constructions. And when Fleury died in 1743 (aged ninety!), Louis decided to do without a chief minister, leaving each department to its own devices. This left Maurepas with all too short a time in which to prepare for a naval war that must come. It began the next year, 1744, with a battle fought before war was declared.

This battle was in more than one way curious. In 1743 a Spanish fleet of twelve ships, taking troops to Genoa for use in the Duchy of Milan from which Spain was trying to expel the Hapsburgs, was pursued by a stronger English fleet and took refuge in Toulon, where it was promptly blockaded. France and Spain were allies: Britain was at war with Spain but not, as yet, with France. In answer to a Spanish appeal Louis XV ordered the Toulon squadron, which had been put on a war footing, to escort the Spaniards to Genoa, but not to fight the British except in defence of the Spaniards. An odd position. The Toulon fleet was sixteen of the line, commanded by de Court la Bruyère (LG), a distinctly veteran officer, for he was aged seventy-eight. His flag was in the TERRIBLE(74) and his second in command was a Gabaret with his flag in the ESPÉRANCE(74): the twelve Spaniards were commanded by Don José Navarro in the *Real Felipe* (110). De Court, having seen how ill equipped and ill trained were the Spaniards, had proposed that they should be mixed among the French for their better protection; but Navarro had refused. The allies therefore sailed with Gabaret commanding French ships in the van, de Court commanding French and two Spaniards in the centre, and the *Real Felipe* leading a Spanish rear. De Court was, of course, ready to fight, for his orders were plainly an instruction to trail his coat.

The English were lying in wait for them off Hyères. They were under the command of Vice-admiral Mathews in the *Namur*(90), a person so

[1]This promising officer died in the next year, 1742, aged only thirty-two.

stormy in temper that the Italians had nicknamed him *Il Furibondo*. His second, who disliked him intensely, was Vice-admiral Lestock in the *Neptune* (90) and Rear-admiral Rowley in the *Barfleur*(90) commanded the other division. Their order of sailing was that Rowley's division formed the van, Mathews had the centre, and Lestock the rear. They were twenty-nine strong, and there was no real disparity between the fleets, except for the inefficiency of the Spaniards – and that was to be balanced by the conduct of Lestock. De Court had issued strict orders for station-keeping, but the Spaniards had not been sufficiently skilful to maintain station, and there was a gap between the last French ship and their leader, *Poder*, and again between their fifth ship and the remainder of their line. Mathews had the wind and came down on the Spaniards, in such a way that Rowley's van stretched very little beyond the head of the French centre, while Mathews himself was opposite the Spanish tail of the centre, and the leading ships of the rear (Spanish) division. This attack, mainly on the Spaniards, was probably an attempt to preserve what had now virtually become a fiction, that the English were fighting only against Spain. The dispositions meant, however, that most of the French van had no ships opposite to them: moreover, Lestock, who was both astern and to windward, sulkily remained so, and made no attempt to get his division into action.

On Mathews opening fire on the Spaniards, the French guns and those of Rowley's vessels also came into action, though the Anglo-French gun duel never became intense, and indeed the two lines were at medium-long range apart from each other. Anxious for closer action, and to set an example for it, Mathews closed the Spaniards with his flagship and expected the English line to form on him; but only his next ahead and astern, and the *Berwick*(70) under Captain Hawke from Rowley's division, did so. The Spanish line gave, the captain of the *Real Felipe* was shot down, and its admiral promptly left the deck with a minor wound. At this the command was assumed by the French liaison officer in the Spanish flagship, Captain de Lage de Ceuilly, who managed to restore the Spanish line and, with the superior weight of guns of four ships, to do considerable damage to Mathews and to one of his supporters, the *Marlborough*. To balance this, Hawke had out-gunned the *Poder*, which had to strike.

After some time de Court signalled to his unengaged van to come about, in order to put the English between two fires. The van took a long while to do this, partly because the signal was either not seen or not understood at first; partly because two or three of Rowley's ships had intelligently stretched forward to prevent this manoeuvre; and partly because the wind was only light. Eventually they got round and, followed by de Court, bore down to the relief of the *Real Felipe*, which had by now almost half its crew *hors de combat*. At this threat Mathews also signalled his fleet to go about, and with one or two desultory shots at the remaining Spaniards, ended the action. The *Poder* was too shattered for her prize crew to keep her up with the fleet and she was retaken, but had to be sunk next day. The fleets were then still in sight and when the Spanish *Hercules* fell astern, some English ships made for her, but stood away again when de Court bore up.

The English then made for Minorca, some of their ships, notably the flagship and the *Marlborough*, being very damaged: the Spaniards, also damaged, returned to Cartagena, and de Court to Toulon. On the whole the allies had rather the better of it, the road to send troops to Genoa now being open; but the action was in fact indecisive. It was not only oddly fought, but it had an odd aftermath, for the air was filled with recriminations on both sides. The Spaniards complained that they had been ill supported, and to please them de Court was removed from his command, a poor reward to the old man for long service and for having dealt at least adequately with a battle in strange circumstances. The English had a series of courts martial, in which Lestock was rewarded for his disgraceful conduct by being acquitted on a technical point, and Mathews was rewarded for his stout fighting by being dismissed the service, also on a technical point. An amazing judgment. War between France and England was then formally declared.

With war obviously coming, France had prepared to strike a blow direct at England by taking advantage of the still quite strong Jacobite feeling while the bulk of the English troops was operating in Europe. Prince Charles Edward Stuart was brought from Italy, a formidable army under Marshal Saxe was collected at Dunkirk with some troop transports, and de Roquefeuil (CE)[1] was to bring other transports round with the Brest fleet, twenty-six of the line strong. The idea was good: the planning was less good. For, Dunkirk having been reduced under the terms of the treaty of Utrecht to a small harbour, the troops would have to embark from the open beaches to be made famous, nearly two hundred years later, by another, and highly emergency, embarkation of an army. The landsmen planners had not allowed for the difficulties of this, nor for the time required, nor for the danger to the transports with no proper harbour and on a possible lee shore if foul weather should arise.

It did so arise. Roquefeuil's fleet, not over well fitted out because of shortages in the arsenals, was largely scattered by violent gales; some transports were driven ashore. The British had had time to collect a superior fleet in the Downs; and the attempt had to be abandoned. It is a grave reflection on the lack of continuity in France's haphazard policies of that time, and of her government's failure to think ahead, that no effort was made to renew it, and no plan was made to support the prince when he sailed on his own the next year. That France should have been unwilling to fit out another major expedition, so that he had to sail in two privately hired ships with no more aid than merely a few officers, some arms and some money, is just understandable. But surely there should have been some plan to support him quickly with troops if he gained the initial success that justified

[1]Being aged eighty he was even more of a veteran than de Court. Most French flag officers at this time were at least in their sixties, for, with a reduced fleet and a long period of peace, promotions had become very slow. There were very junior officers aged forty and a man reaching the rank of captain, or of CE, might wait twenty, or even thirty, years for his next step.

it. As, in fact, for much of a year he had all the firths and sea-lochs of west Scotland in his control for French ships to dodge into, such a plan might have altered the course of history. In its absence Prince Charles's early and alarming successes came to nothing, and his cause, and France's opportunity, ended early in 1746 with the rout at Culloden.

Other activities in 1744 were the blockading of Admiral Hardy's squadron in Lisbon by the veteran Rochambeau (CE), who withdrew precipitately at the news that English reinforcements were coming; the scattering by a gale of sixteen ships under Gabaret that had sailed to see the Havana convoy home; and the capture of the *Northumberland* (70) by two ships under captain the comte de Conflans.

With one exception fleet movements ceased with 1744. The reasons were chiefly financial. Money was desperately short, the navy being allotted only what could be spared from the army, and sending a properly equipped fleet to sea was expensive. Further, once the idea of invading Britain had been dropped, there seemed to be no object for which a fully assembled fleet was needed, while desertions were frequent and the arsenals were largely short of supplies because of the lack of funds. There was, however, a good deal of sailing to and fro by single ships, or a few ships escorting a convoy, for unescorted merchantmen were being mopped up by English cruisers. These sailings led to single ship battles, or to fights between small groups of ships attacking or escorting a convoy; and in these actions the French nearly always fought with gallantry (showing that there was nothing wrong with the spirit of the service) even though sometimes with lack of success owing to want of training and experience. Losses in single ship actions worked out about even between the two nations. Conflans got a West Indies convoy through against odds in 1746, and later that year, in the NEPTUNE (70), attacked an English convoy, taking one of the escort, the *Severn*. But there was no doubt of the English domination, by sheer weight of numbers. They were able to land troops for an attack on L'Orient. A farcical panic ended this assault on a defenceless port, but the landing of the troops and their subsequent re-embarking were able to be conducted at leisure, without even a threat of interference by French ships; and they were able to burn the beached ARDENT (64). The one operation resembling a fleet movement was made in connection with Louisbourg.

When Nova Scotia had been ceded to England by the treaty of Utrecht a number of its settlers had moved to the adjacent Ile Royale (or Cape Breton Island) and had created a town and fortified harbour at Louisbourg, which commanded the entry to the St Lawrence and so was the main defence for Quebec and Canada. This was taken by a volunteer army from New England with the help of a naval squadron, and an expedition was sent in 1746 to retake it. The duc d'Anville (CE and general of galleys) was in command of ten of the line with some lesser vessels and a large number of troop transports. The usual ill fate that often awaited the French, even when the English were absent, was lying in ambush. It took the forms, first, of a gale which scattered the expedition and drove some ships and some transports back to France, and, second, of a violent epidemic that killed the commander and more than decimated crews and troops. In this situation, with food also running short, the expedition could only stagger back home. Over 2 000 men had died on the outward passage, and a further 800 did so on the return voyage.

Indeed, French medical arrangements for the care of men aboard ship had been, and were to continue to be, almost non-existent. As we shall see, this huge convoy was not the only naval operation to be handicapped or even ruined by an appalling sick-list. It is not until the next century that the medical service becomes satisfactory.

The following year, 1747, saw two remarkable convoy actions. De la Jonquière de la Pommarède, who had just been promoted to CE after forty-nine years' service, had been appointed Governor-General of Canada and was taking out thirty transports, with three weak vessels of the line and two frigates as their escort, his flag being in the SÉRIEUX(64). Delayed by weather, he was joined by the INVINCIBLE(74) and JASON(50), commanded by Grout, chevalier de St Georges, an East India officer who held a temporary royal commission, and was taking a number of East Indiamen and a few other ships out to the Orient, where his two fighting ships would reinforce the armed East Indiamen operating there. Off Cape Ortegal they were sighted by an English fleet of fourteen of the line under Vice-admiral Anson, with his flag in the *Prince George*.

The odds were hopeless: there was nothing to do but to run for it, or to fight it out as a sacrifice to save the convoy. La Jonquière chose the latter course, formed line with all his ships, including five Indiamen, and though two of these disobeyed signals and fled with the convoy, he put up a tremendous resistance. The English came down in a general chase, but they did so rather slowly (whereby the convoy got a good start), and then attacked the French line from both port and starboard. The Indiamen soon struck, so did the JASON, but the forty-gun GLOIRE and the fifty-six-gun DIAMANT fought for three hours before surrendering. The former had her captain killed and most of her crew then out of action, and the latter sank soon after she was taken. It was not till after that that La Jonquière struck in the SÉRIEUX, with every officer wounded. The INVINCIBLE lived right up to her name, for it was not till yet later, when she had fired her last shot, that the flag was lowered on her battered remains. The English officers made no secret of their admiration for the defence offered, and rightly, for though the ships and some 800 men were lost, the convoy was saved.

This battle in May was followed by a similar one in October. A large convoy was to leave for the Antilles, and the biggest escort that could be found was provided for it under Desherbiers, marquis de l'Étanduère, a veteran officer with much fighting experience, which had included being at Rio with DuGuay-Trouin. When a few days' sail out into the Atlantic he was seen 300 miles off Cape Finisterre and pursued by Rear-admiral Hawke in the *Devonshire*(64) with fourteen ships. Sending the convoy on ahead de l'Étanduère formed line under topsails across the path of the English in the following order:

| Ship | Guns | Captain |
|---|---|---|
| INTRÉPIDE | 74 | comte de Vaudreuil |
| TRIDENT | 64 | marquis d'Amblimont |
| TERRIBLE | 74 | comte du Guay |
| TONNANT | 80 | marquis Desherbiers de l'Étanduère (CE) |
| MONARQUE | 74 | de la Bédoyère |
| SEVERN | 50 | du Rouret de Saint-Estève |
| FOUGUEUX | 64 | du Vignau |
| NEPTUNE | 70 | chevalier de Fromentière |

Hawke signalled for a general chase and the English ships came up in any order, attacking the rear of the French line and passing down it as other ships got into action, engaging from both port and starboard. The first shots of a stern battle were fired at 11.30 am. It was not till 4.0 pm that a French ship struck – the NEPTUNE, which had had the fire of all English ships as they came up, with her captain and first lieutenant killed, more than half her crew *hors de combat*, and all her masts down. At 5 pm the FOUGUEUX, TRIDENT, MONARQUE, and SEVERN, the first two with only one mast standing, also submitted; and at 7.0 pm the TERRIBLE, surrounded by six of the enemy, lowered her colours. It was now all but dark and Vaudreuil (who had fought his ship stoutly and, being at the head of the line, had been the last to be assailed and so had kept his ship almost intact), though he had every excuse to run for it, did nothing of the sort. Instead, making sail, he brought the INTRÉPIDE round, and bore down to support his admiral, who, with his foremast down, was still holding out in the TONNANT. He managed to get a hawser to him in the gloom and under fire, and towed him out of the battle in the dark. Steering an unusual course, they lost Hawke during the night, and both got safely into Brest. It was a great achievement with which to end an action that wrote a gallant page in French history and deserves to be better remembered. For, like that of La Jonquière, it was a defeat that was more glorious than many victories; and, again, it saved the convoy, for Hawke was too delayed and too damaged to pursue. (Suffren, who was captured as a lieutenant in the MONARQUE, always stated that it was one of the finest actions he had seen.)

Captain Dubois de Lamotte had done less well when bringing back a West Indian convoy with a weak escort for, signalling a *sauve qui peut* when chased, he lost forty-seven ships of his charge. He made up for this when taking out another convoy in the MAGNANIME (64) with a frigate, for he managed to hold off a superior force while getting the ships into Cap Français. (The MAGNANIME, under jury rig after a storm, was taken the next year just before the war ended.)

But apart from European waters and the Atlantic there was another area of activity, so remote, in those days, from communication with Europe that it is best treated as a separate entity – the Indian Ocean. Here, as far as France is concerned, the small size of her fleet meant that royal ships rarely appeared. The fighting had to be done by the East India Company's vessels. These were built for it, up to a point, and were pierced with many gun ports; but their prime object was to carry cargo, and this took first place in their design. As a result, in most of them the lower battery was barely usable, and so was not armed. This meant that, despite a number of ports, a ship would carry only thirty or so guns, and these would be only eighteen-pounders or less. The ships were also slow, heavy sailers.

After a struggle for existence in the seventeenth century, during which a station was set up in Madagascar, the Île de France and the Île de Bourbon (Mauritius and Réunion) were occupied, Chandernagore was leased, and Pondicherry built, the company began to flourish round about 1720. In 1725, needing a port on the Malabar coast, it sent a squadron under Pardaillan to take the fortified Maihi, which was renamed Mahé after Mahé de Labourdonnais, who had devised the plan by which it was captured.

This remarkable man was born in St Malo in 1699, and went to sea as a ship's boy at the age of ten. During several long voyages he got priests and officers to teach him mathematics and navigation, and some military engineering, and at the age of twenty obtained a second lieutenantcy in the Company. A daring open boat voyage to get help for his stranded ship and her cargo brought promotion and made him a marked man; and in 1735 he was appointed governor of the Île de France. This was then a decrepit, lazy, unambitious sort of watering station for ships; but the energetic Labourdonnais altered all that. He built a harbour, a dockyard, and a building slip at Antongil Bay, and armed and trained some Indiamen and their crews as a fighting squadron. With war coming he was ready for it. At this, however, Fleury, hoping to keep the peace, not only denied him the two royal ships that he had been promised, but recalled the squadron that he had created. Only after it had sailed did orders come for him to retain it – too late. There never was a worse administrative muddle, and his plan to sweep English commerce from the sea was ruined. He had to start again from the beginning. Personally he designed the means for converting the unarmed merchantmen that arrived, and he supervised their conversion. He put the island on starvation rations to build up a reserve of stores. With the arrival of the ACHILLE(74) and his royal commission as 'frigate-captain' (equivalent to the English commander) he went to sea. His making a fighting fleet out of nothing had been a prodigious achievement.

In addition to the ACHILLE he had eight Indiamen of which six carried only thirty guns, and the other two had less than forty. The English commodore, Peyton, had four sixty-gun ships of the line, and two lesser vessels. With his much heavier armament, to say nothing of his more handy ships, Peyton had distinctly the superior force. But he was a feeble sort of commander, and in the action which the two squadrons fought when they met off Negapatam Peyton used his superior mobility, having the wind, to fight at a distance, which he could do because of his greater gun power, and did not allow Labourdonnais to get anything like close enough to board. Yet at the end of a long indecisive cannonade Peyton stood off, leaving open the road to Madras and Pondicherry. As Labourdonnais was short of stores and ammunition he put into the latter place, and was now, rather to his displeasure, under Dupleix, its governor, who was using clever diplomacy with the natives to found a French empire in India. Labourdonnais was hesitant about carrying out Dupleix's orders to attack Madras; but after Peyton had reappeared, and then, on sighting Labourdonnais, had again bolted, and having received positive instructions from the Pondicherry Council, the French sailor led his fleet to the English headquarters of Madras and took the place.

Now ensued a sordid quarrel, which is not really a naval affair, between the two French leaders. Labourdonnais had held Madras to a ransom which its governor, Morse, accepted, and to that end he gave Labourdonnais what amounted to a bribe. Dupleix refused to accept the agreement that had been made, and retained Madras, for he wished to expel the English from the subcontinent. Trying to get his way, Labourdonnais stayed too long on the coast and was caught by the monsoon, which wrecked two of his ships and damaged the others. He struggled back to the Île de France with them only to find his recall awaiting him. But in four months, having started from scratch, he had driven the superior English out of the Bay of Bengal and taken their chief depot in India. It was a great performance. Yet on his return to France he found that his reward for it was to be thrown into the Bastille for disobeying orders and perverting funds. He had certainly been impetuous, and by his defiant quarrel with Dupleix had thrown away much of the position in affairs that his patriotic energy had gained for him; but he deserved better treatment than imprisonment. Unless they were of exalted birth France did not treat her successful servants well, if once they made a slip. After rather more than two years he was declared to be innocent, and he emerged from prison broken in health, angry and bitter. He died not much later in 1753.

The Peace of Aix-la-Chapelle, which ended the war in 1748, was a restoration of the *status quo ante bellum* as between England and France. British armies had been defeated, but Britain had established domination at sea. French privateers had been busy, but British shipping losses had been no more than those of France, which, as Britain's merchant marine was much greater, meant that that of France had lost disproportionately – had, indeed, been hard hit. So Britain returned Louisbourg, and France returned Madras. Dupleix was furious; but Madras could not have been retained, for Boscawen was in the Indian Ocean with a powerful squadron and would have taken the place if the treaty had been delayed. The colonial rivalry between the two powers, and Britain's desire to seize the French possessions, were now obvious. The treaty was clearly only a breathing space. France's financially starved navy had suffered heavy losses, but had shown fine fighting spirit. War was to come again in a few years. Was France going to employ the interval to build up good fleets? The sad answer is that she was not.

CHAPTER IX

# The Seven Years War

Maurepas' first step when the war ended was to get rid of the galleys. This was long overdue: it had been clear for at least forty years that the galleys had had their day, and only a mixture of vested interests with the dislike of terminating a service that had a historic record had preserved them. Their abolition freed money for building ships, and their convict crews could be used on labour in the dockyards. His next step was to work out a plan for the future of the navy. Its war losses, plus the fact that some ships were too worn out to be worth repair, had reduced it to thirty of the line, of which only nine were in commission. Unfortunately his plan for bringing the service back to strength almost coincided with a witty epigram that he made on Madame de Pompadour, the reigning mistress; and, as reigning was very much the word, he was incontinently dismissed in 1749. He was the third, and last, of the Pontchartrain 'dynasty', and was much the best of them. Not a great Naval Secretary, but France might have done worse, and was to do so.

His successor was de Rouillé, comte de Jouy, an intellectual, and experienced in several branches of the Administration, but not at all in that of the navy. Where ships were concerned he could do no more than follow Maurepas' plan as far as meagre supplies of money allowed. He founded the Académie de la Marine for naval studies in 1752, and he established a gun foundry on the Loire in 1754 so that France no longer had to buy her cannon in England or Sweden. But though he began to get the navy back to its pre-1744 strength, he could not find the funds to arm and commission the ships, much less to give the service the deep-water sailing and the battle practice that it needed; and there was no money available for the highly necessary fortification of key positions in the colonies. Nor could he do anything to end two other evils from which the navy suffered. One was the unwillingness of the noble officers to accept the *bleus* from the middle-class merchant marine – a strange attitude when they had before them the example of such men as Jean Bart, DuGuay-Trouin and Cassard. The other, and greater, was the rivalry between the *épée* and the *plume*. The latter, operating the practices that are now loosely called Parkinson's Law, had so increased their numbers that, when Rouillé went out of office, there were 563 of these administrators to 940 active service officers, and they had so infringed on all the duties of the latter (except that of going to sea!) that the active service officers were reduced to performing just one or two purely routine duties such as taking an occasional guard. The position was not unlike that of English education in the 1950s and 1960s, when the heads of schools and their staffs were handicapped by the interference of a host of administrators, many with little or no practical experience, who were themselves trying to carry out the instructions of councils and committees whose ideas only too often came from irresponsible theorists. It is not surprising that during these years between the Austrian Succession war and that which followed it the spirit of the French navy deteriorated gravely.

Yet *élan* and determination were badly needed, for a war against England, which must obviously be mainly a maritime one, was surely coming. Colonial friction and rivalry were intensifying. In the East, the French were the aggressors, for Dupleix in India, acting nominally as the ally of native princes, was warring against the British East India Company and forcing that body to become a military power against its will. Though the two countries were at peace, a state of warfare had existed between their India Companies since 1751. In the West the British were the aggressors, for the New Englanders were afraid of the French in Canada and of their aims to control the valleys of the Ohio and Mississippi rivers; and from 1754 the British were taking advantage of every 'incident' on a most loosely drawn frontier to promote reprisals that soon were little if anything short of open war. In the Atlantic, too, the English were to start attacking French ships before war was declared.

It was in the middle of 1754, as the outlook started to be minatory, that M Machault d'Arnouville took over the Naval Secretaryship. He was a vigorous man who saw the danger that threatened and aimed at once to increase the fleets to a total strength of eighty of the line, when he would reckon always to have sixty in commission with twenty refitting. He proposed, in the event of war, to give temporary commissions, whether the regular officers liked it or not, to suitable merchantmen. He had made rather more than a good start with this programme when war was formally declared in 1756; but on the eve of that war, instead of concentrating all her energies on the colonial struggle, in which, thanks to her financial neglect of her navy, the odds were already against her, France must needs take one of the most unwise steps in diplomatic history. This was to yield to the subtle statesmanship of Kaunitz and carry out a diplomatic revolution whereby she allied herself to the hereditary enemy, the House of Austria, in an attack upon Frederick of Prussia. France was therefore to be fighting two separate wars, from one of which she stood to gain little, when the inevitable colonial one would have been fully all that she could have managed with her hands free elsewhere.

The step was the more unwise in that, without any formal declaration, England had already started the war in the previous year, 1755. If the fighting in India was due to French aggression, and if England could plead in mitigation of her military action in North America that France was making a claim to vast territories for which she had not enough soldiers and colonists for even a token occupation, nothing could excuse her action at sea. Because of the threatening situation on the Canadian frontier, France had sent out reinforcements and supplies. These were carried in eleven ships of the line which, to make room for troops and stores, were armed only *en flûte* with twenty-two or twenty-four guns apiece. Their escort was three of the line under Dubois de la Motte (CE) with his flag in the ENTREPRENANT(74). The English sent Boscawen with a powerful squadron to intercept them. Thanks to the fogs off the Newfoundland banks he missed them; but those fogs caused three French ships to become separated from the rest – the ALCIDE(64) and the LYS and the DAUPHIN ROYAL, which were armed only *en flûte*. Boscawen came up with these and, on their refusal to salute the British flag, attacked them. The DAUPHIN ROYAL, being a fast sailer, got away: the other two were taken after a stout resistance against heavy odds. (By an odd freak of coincidence the ALCIDE's captain became Boscawen's prisoner for the third time, the other two occasions having been in the Austrian Succession war, in 1744 and 1747.)

At this news Louis recalled his ambassador, whereupon the English seized all French merchant ships that they could find in British harbours or on the seas. Moreover, when Dubois de la Motte, having landed his troops and stores, brought his ships back to France, he had had to leave behind him, for repairs, the elderly ESPÉRANCE; and she, *en flûe*, was taken by three English ships of the line, after a resistance so determined that the captors sank her as too damaged to be worth taking into harbour. By these seizures of war and merchant ships, whose crews were imprisoned, France had lost thousands of seamen. In return, Comte du Guay, protecting the Channel ports against possible raids with a small squadron, took an English corvette. If these doings in 1755 did not create a state of war one would like to know what did so: yet war was not formally declared.

The reason that France refrained from such a declaration was probably the depleted state of her arsenals, and the lack of money to buy the supplies needed to fill them. The British had about a hundred of the line ready for sea and ample stocks for them: France had forty-five of the line, by no means all of them fitted out. It had needed all Machault's drive to get the ships of Dubois de la Motte and du Guay to sea. A major expedition, however, was prepared, and well prepared, at Toulon, for the spring of 1756. Its object was the capture of Minorca, that island so strategically important to the British; and an army under the duc de Richelieu was embarked in a large number of transports, escorted by the Marquis de la Gallissonière (LG) with twelve of the line. Though the object of the expedition was kept commendably secret, the fact of its fitting out could not be concealed, and the English Government should have been able to deduce its objective. That government, however, was under the direction of the duke of Newcastle, who was a party politician impure and far from simple, but little of a statesman and less of a strategist. Even when the aim of the expedition was clear, he was sufficiently distracted by the assembly of some troops on the north coast of France to fear an invasion, though there was no sign of any transports or escort for them, and therefore sent out Admiral Byng to Minorca with a squadron which, even when he had collected the ships already in the Mediterranean, would be only thirteen of the line instead of being sufficient in number to overwhelm the French. Yet it was vital that Byng should gain a complete victory, since Minorca was in no state to hold out for long, the garrison being below strength, and the governor and other officers being quite haphazardly on leave.

La Galissonière was a hump-backed little man, of poor health, but of good ability and experience. He came of a family of naval officers, had himself been forty-six years in the service, had been to sea in most of such fleets as the French had put forth in that time, and had been for two years a forceful Governor of Canada, responsible indeed for establishing the frontier posts whose existence had largely contributed to causing the war. In the absence of good charts of Minorca, he landed his troops at Ciudadela, at the west end of the island, and made no attempt to cut off the few English ships in the roadstead at Mahon, which was at its east end. General Blakeney withdrew his troops into the fort of St Philip there, and sent asking the meaning of this landing in time of peace. Richelieu replied that he was behaving just as the English king had done to French ships. The troops once landed, La Galissonière cruised off the island to protect their transports and re-embarkation, as his instructions strictly enjoined. His ships, in order of battle, were:

| Ship | Guns | Captain |
|---|---|---|
| ORPHÉE | 64 | bailli de Raimond d'Eaux |
| HIPPOPOTAME | 50 | de Rochemore la Devèze |
| REDOUTABLE | 74 | commandeur de Glandevès (CE) |
| SAGE | 64 | Durevest |
| GUERRIER | 74 | Villars de Labrosse |
| FIER | 50 | d'Herville |
| FOUDROYANT | 80 | marquis de la Galissonière (LG) |
| TÉMÉRAIRE | 74 | de Beaumont Lemaître |
| CONTENT | 64 | de Sabran Grammont |
| LION | 64 | marquis de Saint-Aignan |
| COURONNE | 74 | de la Clue Sabran (CE) |
| TRITON | 64 | Mercier |

with some frigates.

Byng had his flag in the *Ramillies*(90), his second in command being Rear-admiral Temple West in the *Buckingham*(70). The fleets were approximately equal, the French having the greater gun-power, especially as the weak *Deptford*(50) was not brought into the English line till near the close of the battle. The enemies sighted each other on 19th May, but did not come to action till the next day. There was a good deal of manoeuvring, skilful on both sides, as a result of which Byng, having first got the wind, next got into the position he wished, and came down on the French at about 2 pm. His plan of attack was clever enough, for he came down 'lasking', that is to say in line but on a diagonal or at an angle to the French, so that his ships were less exposed to their fire without being able to reply than they would have been coming down head on. He aimed for his leading ship to strike the French line only a little ahead of its centre, so that its van would be for some time left out of action while he concentrated

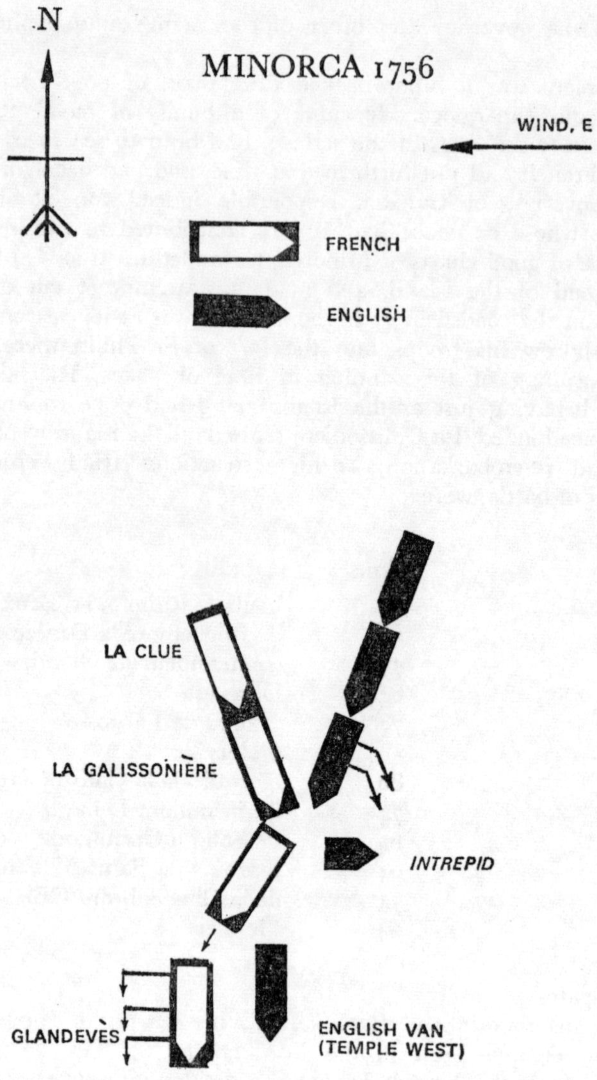


THE POSITION WHEN THE DAMAGED *INTREPID* PAID OFF, FORCING THE ENGLISH IMMEDIATELY ASTERN TO BACK SAILS. LA GALISSONIÈRE IS SHOWN ATTEMPTING, UNSUCCESSFULLY, TO WORK INTO THE GAP THUS CREATED, AND GLANDEVÈS AS (LATER) STANDING OFF TO RE-FORM TO LEEWARD on its centre and rear. La Galissonière, seeing this danger, backed topsails, so that the French awaited the attack under little more than steerage way. This countered Byng's plan, and its ruin was completed by his leading captains who, misunderstanding his intentions, made sail to stretch ahead, so that the *Defiance*, which led, should come up with the ORPHÉE.

In the absence of a fuller code of signals than then existed, Byng could not correct this and, having come up with their opposite numbers, each sportsman taking his bird, as it were, the English van ships ran headlong down on their opponents under heavy fire and suffered much damage in the process, the rest of the ships, committed to their lasking line (which Byng would not break), coming into action one by one. Once brought up broadside on to the enemy their van ships did well enough, and first the SAGE and then the GUERRIER stood out of line to leeward. The firing had begun at 2.30, and an hour later Glandevès took the rest of his division to leeward and re-formed his line. Temple West's van seemed to have gained a partial victory, but were too damaged to be able to close the re-formed French and complete it; and meanwhile things had gone very wrong for the English astern of them. For good French gunnery against the *Intrepid*, sixth in the English line, had brought her foretopmast down, fouling the foresail, so that she came up into the wind unmanageable, and not only got heavily pounded but held up the rest of the British line, which was further thwarted by damage to the seventh ship, the *Princess Louisa,* which also brought up to windward. This stoppage left a large gap astern of the fifth English ship into which La Galissonière proceeded to work with his next ahead and astern (FIER and TÉMÉRAIRE), drubbing the *Intrepid* as they passed, in order to get the English van between two fires. But seeing the powerful *Revenge, Trident,* and *Ramillies* making sail to cover the *Intrepid* and close the gap, La Galissonière desisted and paid off again. The position now was that the French van was virtually out of range, a bit ahead of the English van and to leeward of its own centre and rear, which were drawing ahead to fire on the English van, while the English had re-formed line, the *Deptford* taking the place of the *Intrepid*, and were drawing up on the French. At this La Galissonière also stood away to leeward and re-formed line on his van. Thus the fleets were out of effective range.

It was now a little after 5 pm, and the fleets remained in sight till dusk, when the English brought to. La Galissonière had handled his ships well, better than his opponent, in an action that was tactically somewhat clumsy, and the French gunnery had been good. The English had suffered the greater damage, but the battle had been indecisive. It rested with Byng whether to renew it, for La Galissonière's instructions were entirely defensive – to husband his fleet in order to cover the troops and their transport – and Byng turned the battle into a defeat by his decision to withdraw to Gibraltar. This left Minorca to its fate, and cost him his life. For, apart from the loss of Minorca, things had gone badly for England in America and India, the mob screamed for a scapegoat, and though the court martial which, strictly correctly, sentenced Byng to death also recommended royal clemency, Newcastle was not the man to defy public opinion by advising his king to exercise it.

Meanwhile, the British in fort St Philip having capitulated, and a French garrison having been left in Minorca, La Galissonière re-embarked the rest of the troops and returned to Toulon. Here he asked to be relieved of his command. He was a sick man; and he died of dropsy later in the year. Indeed, there would have been no Toulon fleet for him to lead, for it was paid off, except for five ships, in order to save expense! Incidentally, and almost absurdly, war had been formally declared three days before the battle of Minorca.

The year had been a successful one for the French, but in 1757 two administrative changes were to alter the whole conduct of war affairs. In England a new government was formed by a coalition between Pitt and Newcastle. The duke was to deal with all the Parliamentary jobbery, while Pitt, whose previous period in office had been brief and without real power, was to deal with the war. He gave it a force and direction that were not to be equalled by any British leader in time of war for nearly two hundred years. In France, the most notable of the past year's successes had been the capture of Minorca, thanks to the way d'Argenson at the War Ministry and Machault for the navy had worked together to provide the most efficient combined operation that France produced in the days of sail. Their reward was to be dismissed, in February, at the bidding of Madame de Pompadour; for her mistress-ship was displeased at their want of obsequiousness to her.

As Minister of Marine, Machault was succeeded by the marquis de Moras who continued his predecessor's policy of introducing *officiers bleus* to make good gaps in the officer corps, and, being also controller of finance, was able to find funds for the service. He lasted only one year, resigning in 1758 as no longer willing to endure the utter disorganization into which governmental affairs had fallen. In his place, and for the first time, a genuine naval officer was appointed. This should have been a good step to take, but unfortunately the man picked was the least suitable of the various flag officers available, the marquis de Massiac (LG), a dandified person, more concerned with his perfumes and his peruke than with his ships and their arsenals. After five months the absurdity of his incompetence became obvious, even to the Court, and he was dismissed in favour of Berryer. This man was a narrow-minded police administrator who knew nothing of the naval service, considered it, indeed, as a wasteful excrescence, a man capable of the worst Pontchartrainry. It is small wonder that, under such administrators, when faced by an opponent of the calibre of Pitt, the French navy fell upon disaster.

Not that this happened at once. Indeed, in 1757 some things, at least, were as successfully handled as could be wished. Machault had realized the urgent need to strengthen the colonial position and, before his fall, had arranged to reinforce the Indian Ocean, the Antilles, and Canada. Captain Kersaint de Coëtnempren in the INTRÉPIDE (74) with two others of the line and a *flûte* had sailed for the West Indies at the end of 1756. In early 1757, having raided British trading posts in West Africa on the way, he reached San Domingo to pick up a convoy. As his road was barred by three English liners, he came out to give them battle. In this his ship was badly damaged aloft, and he had to return to harbour. But the English had also suffered, so that they could not maintain a blockade and had to stand away for repairs. The road being thus left clear, Kersaint brought his convoy home safely.

Early in the year the chevalier de Beauffremont (CE) in the Tonnant(80) with four others of the line took troops out to San Domingo, and went on to Louisbourg. Here he was joined by Dubois de la Motte, recently promoted to LG at the age of seventy-four(!), who had his flag in the Formidable(80) and brought with him eight other ships of the line. Though he sent two of these on to Quebec, whence, having landed troops and stores, they returned to France, Dubois was joined by four more of the line, under Durevest, from Toulon, which had exchanged some shots with a superior force under Saunders, as they ran though the Straits of Gibraltar. The fleet at Louisbourg was now sixteen strong, plus some frigates. The comte d'Aché (CE) had also sailed, in May, for the Indian Ocean in the Zodiaque(74), preceded by a number of East Indiamen carrying troops and stores. (He should have sailed earlier, with three of the line and some frigates, but had been delayed by a ludicrous running-aboard collision in getting out of harbour, and two of the liners had then been diverted to other service. The accident makes a sad commentary on the deterioration of discipline among French officers, for it was due to one of the frigate captains having spent the night ashore and not being back in time to weigh his ship.) The English had planned an attack on Louisbourg, and Vice-admiral Holbourne had sailed from Halifax, with fourteen of the line and some troops, to make it. He was surprised to find the enemy in such strength, and stood off to wait for reinforcements. Dubois de la Motte, perhaps made hesitant by his great age, did not take the initiative and come out to attack him. When Holbourne's reinforcements arrived, so did a furious storm, which wrecked one of his ships of the line on the shore, and so damaged most of the others that there was nothing to be done but to take them back to England.

In all this, things had gone well for France, but there were flies in the ointment. One was that, by reinforcing her colonial positions, she had left herself unable to protect her own coasts. In September, Hawke, with a large fleet and a number of troops, entered Aix Roads. He easily occupied the island of Aix and razed its weak fortifications; and he could readily have taken Rochefort and destroyed its dockyard, for it was all but defenceless. What stopped him from doing so was the timorousness of the general commanding the troops, who just funked attempting the operation for which he had been sent out. What *should* have stopped him should have been the presence, or appearance, of the Brest fleet. But the Brest fleet consisted of little more than the hulls of ships; for they were short of stores, short of crews, short of guns, and could not put to sea even for an attack which had been projected on the nearby Channel Islands. Conflans, who was in command, complained with justifiable bitterness that there were at sea a large number of privateers carrying a large number of seamen and a large number of guns, while he could not complete the manning and armament even of his own flagship.

Worse than this was the return of Dubois de la Motte. Realizing that the storm-battered English fleet from Halifax was no longer a threat to Louisbourg, and being short of provisions with also a heavy sick list, he wisely decided to bring his ships back to France. The return voyage was a purgatory. The weather was the North Atlantic at its worst, the sickness proved to be typhus and spread like a bush fire. The condition of the fleet became so bad that when it ran across the English seventy-four *Vanguard*, Dubois felt obliged to call off the DIADÉME(74) which was near enough to attack her, and some ships were so short of fit men that they could not come properly to an anchor when the stricken squadron at length crept into Brest. Over 6 000 invalid seamen were put ashore, and they spread the disease throughout the town so that it was almost threatened with depopulation. Of course the dockyard was all but out of action for weeks. A lost battle would have been a lesser disaster.

The year 1758 was to show how grave was the real position. It had been decided to maintain a reasonably adequate fleet at the vital point of Louisbourg. The Toulon squadron was therefore to replace Dubois de la Motte. It was under the command of de La Clue Sabran (CE), an officer of no remarkable capacity, and not particularly respected, for it was felt that he owed his promotions and appointments to his having been a protégé of the comte de Toulouse, during that prince's admiralty of France. His orders were to go first to the West Indies, and then on to Louisbourg. The dockyard at Toulon was so ill-provided, and desertions were so frequent (for the seamen's pay was long overdue), that it had taken La Clue nearly a year to fit out his ships. Even so, when he sailed near the end of 1757 with his flag in the OCÉAN(80), he was by no means at full strength. He had six of the line: others were to follow. Learning that Osborne and Saunders were in force in the Straits, he put into Cartagena to await these reinforcements. Two liners and a frigate reached him in January; and near the end of February DuQuesne, great-nephew of the famous admiral, arrived with the FOUDROYANT(80), the ORPHÉE(64), and the ORIFLAMME(50). Wishing to get out of harbour on the morrow, La Clue signalled them to anchor in a nearby bay: DuQuesne, instead of obeying, sent in a preference for cruising off the port, and La Clue, instead of having his orders carried out, casually left him to do so.

This questioning of orders by the captain and failure to enforce them by the admiral cost France dear. For during the night a strong wind blew DuQuesne's three ships away from the coast into the track of the English fleet, which had come round from Gibraltar, and kept La Clue in harbour unable to help them. They ran for it on different courses. ORIFLAMME tried, unsuccessfully, to mislead the British, but managed to get away and, eventually, to reach La Clue in Cartagena. The pursuit had started at day break. ORPHÉE was overtaken by the evening, and struck after no great resistance. At four in the afternoon the *Monmouth*(64) had come up with the FOUDROYANT, and that much more powerful ship should have been able quickly to drive off her weak opponent. That she failed to do so was because so many of her seamen deserted the guns and ran for safety into the bowels of the ship that her powerful lower battery was only partially in action. Her mizzen mast was brought down by *Monmouth*'s fire so that other English ships came up with her; at 1 am her mainmast also fell, and she struck. With two of his ships taken, La Clue's mission was clearly 'off', and in April he returned to Toulon. He had been away nearly half a year, and had got no nearer than the east coast of Spain to Louisbourg.

To that dangerously threatened harbour had been sent captain the marquis Desgouttes with the Prudent and Entreprenant (74s), the Célébre, Bienfaisant, and Capricieux, all three armed only *en flûte*, and five frigates. Three of these Desgouttes sank across the entrance to the port to bar it. Attempts were made to reinforce him. The Raisonable (64) was taken by two English ships on the way, but a little division of two 64s, two others armed *en flûte*, and an Indiaman with some frigates under Captain Duchaffault de Besné arrived in May. The harbour entry being already barred, they landed their troops and went on to Quebec. Boscawen, with a formidable force, came on their heels to Louisbourg in June, and made a full-scale attack. The Entreprenant, Capricieux, and Célébre were set on fire by the British shells, boats came in on a dark night and cut out Prudent and Bienfaisant, the bulk of whose crews were manning defences ashore, and, in mid July, this key to Quebec and Canada had to surrender. The navy's desperate efforts to keep the place supplied and reinforced had been in vain.

With Louisbourg taken, there was no point in Duchaffault's ships remaining at Quebec, where, indeed, they were in danger. So in September they made for France. Seventy miles north-west of Ushant they came into touch with Boscawen. who was returning with six of the line, and there was an exchange of gunfire which never became a close engagement owing to heavy weather. This may have been as well for Duchaffault, who had a big sick list. So his ships all got back to France except the Belliqueux (64) which had an unusual fate. The night after the cannonade with Boscawen the weather became very bad, the ships separated, and the Belliqueux was blown up the north Cornish coast. She anchored off Lundy, but her cables gave and she was carried helpless up the Bristol Channel. Her captain, remembering that in the last war an English frigate under flag of truce had obtained water from Brest, approached Saumarez in the *Antelope* with a request for aid. The English were not disposed to be so chivalrous, and Belliqueux was politely but firmly made a prize.

The African trading posts at Sénégal and Gorée fell during the year, and the French coast itself was attacked. A British fleet, with an army, came to take St Malo in June. It destroyed much shipping, but did nothing else owing to the incompetent handling of the troops. A further attempt in September brought a second British failure, for the same reason plus the vigorous defence of the duc d'Aiguillon, Governor of Brittany. But in between these two landings Howe had got ashore at the newly created port of Cherbourg and had destroyed everything maritime without interference; and though military incompetence had twice saved St Malo, on all these three occasions the British had been able to make their landings as they chose while their ships had lain off the coast in perfect security.

The situation had become desperate. England could not be reckoned to make a military mess of every landing, it was only a question of time before Canada and other overseas possessions fell into her hands, she had swept French trade from the seas, and her money bags were providing the means to keep Frederick of Prussia in the field on land. It was now clear that her naval superiority made it impossible to hold her, much less defeat her, in the colonies. Cost what it might, she must be dealt with in her own country. To that, therefore, 1759 was to be devoted, and the duc de Choiseul, a man with the determination to do this, had just been recalled from Vienna to take over Foreign Affairs and to be, in effect, the Chief Minister.

Even in modern days of first-class meteorological reports and of power-driven ships, which can to a large extent defy wind and tide, to take a large army oversea in the face of an active, let alone a superior, naval enemy is an enterprise involving both meticulous planning and, even so, considerable risk. To do this in days of sail, subject to unpredictable vagaries of wind, required not only combined planning to meet a number of eventualities, but a naval force which was highly efficient, and united in its team work. The French navy, sadly inferior in numbers, was not such a force. The parsimony which had led to the arsenals being all but empty, and to the ships being laid up and the crews paid off – only too often without any payment – except when a squadron was urgently needed at sea, had prevented any training, and knocked all willingness to serve out of the seamen. The seagoing officers, too, frustrated by having their wishes and proposals rejected by a contemptuously powerful corps of administrators who were little more than landsmen, and riven not only by Service rivalries and class prejudices but by petty jealousies among themselves, were anything but an efficient team[1]. Nor was the commander of the Brest fleet a man likely to make them so. The maréchal comte de Conflans-Brienne, vice-admiral of the Levant, had seemed a good appointment; for his record as a captain in the Austrian Succession war had been that of a firm fighter and a successful tactician. Now, after fifty-three years in the Service, he was past his best. Even as a young officer he had been officially described as 'courageous and able in his profession, but too disposed to stand on his dignity in requiring the respect due to one of his birth as a descendant of the former titular kings of Jerusalem'. He was ever ready to write to Berryer complaining of the lack of men, of equipment, and of supplies, but much less ready personally to do anything about it. If ever suitable for the highest command, he was so no longer; though his-

[1]One officer had left to serve abroad, saying that one injured France less by being a bad citizen than a good courtier. Another refused a command in Toulon, saying that his life was the king's but his honour was his own; and he would not serve to command seamen to whom promise after promise had been broken, so as to leave them destitute.

**THE EARLIER EIGHTEENTH-CENTURY ADMIRALS**

MAHÉ DE LABOURDONNAIS
*(by Vangelisty)*

DESHERBIERS DE L'ETANDUÈRE

LA GALISSONIÈRE

PLATE VIII

## LATER EIGHTEENTH-CENTURY ADMIRALS

COMTE D'ESTAING

COMTE DE GUICHEN

COMTE DE GRASSE

VILLARET-JOYEUSE

torians have perhaps been over-ready to load him with obloquy for having made a disastrous failure of an impossible task.

While the vital expedition was being planned the British had already got to work. In spring, a squadron carrying a small army appeared off Martinique. The landing forces were repelled by a stout resistance, and the British attack was shifted to Guadeloupe. A little after its departure de Bompar (CE) arrived with eight of the line and some frigates. He spent some weeks haggling with the Governor of Martinique, who was unwilling to lend forces for the rescue of the neighbouring island, and when Bompar at length arrived there he was, as usual with French squadrons at that time, too late. Guadeloupe had fallen, after a rather poor resistance (wherefore its governor was committed to prison), and with it had fallen the lesser islands, Marie-Galante and La Désirade. Bompar therefore took his ships on to San Domingo whence he later returned to France, reaching Brest on an occasion when the blockading squadron had been blown clear, and in time for his crews to be used to fill gaps in the manning of Conflans' fleet. Canada, of course, was without naval defence. The British fleet that took Wolfe's army up the St Lawrence to its victory at Quebec had no other enemy to contend with but the navigation of what was to them an unknown river. And in July Rodney had appeared off Le Havre and subjected it, without interference, to a two-day bombardment that did immense damage.

Meanwhile the planning for the invasion of Britain had gone slowly ahead. The people responsible for it were Marshal Belle-Isle, Berryer, and Choiseul: the last two merely concurring in what was really the marshal's plan; so, as might be expected, the plan was very much that of a landsman. The navy received no consultation worth mention, and the ways in which winds, currents, and tides might ruin timing and create all sorts of difficulties were just ignored. So, largely, was the predominant presence of the British fleets. Two good-sized armies were to be collected. One, based on Vannes and commanded by the duc d'Aiguillon, was to embark from the little ports and beaches of the Morbihan – the south-east corner of Brittany. The other was to assemble round Ostend and embark from that port and its neighbouring beaches. The duc d'Aiguillon's army was to be taken well to the west of Ireland and to land in the Clyde. The escort was then to round the north of Scotland, pick up the Ostend army and land it on the Essex coast by the mouths of the Crouch and Blackwater. A distraction was to be provided by some light-weight ships carrying troops which were to slip out of Dunkirk under a renowned privateer named Thurot, and effect a landing in Ireland. The Toulon fleet was to come round to join Conflans. It seems to have been left to a last moment decision whether the whole fleet of combined Brest and Toulon squadrons should escort the expedition, as Conflans urged, or whether this important task should be entrusted to a mere five of the line under Bigot de Morogues, while the main fleet came into the Channel to distract the main British forces.

From any army point of view the plan was good enough. With some forces in the colonies, and others in Germany, Britain had few troops at home. Though France had long ceased to put any trust in the airy-fairy promises of Jacobites, and d'Aiguillon was in fact instructed on landing in Scotland not in any way to proclaim the Stuart cause, yet there was still some Jacobite disaffection, and it was stronger in Scotland than elsewhere: the French would not be entirely unwelcome. As to the other army, the Essex coast was an easy one on which to land: its creeks made it hard to defend; and it was within short marching distance, over country with no natural obstacles, of London itself. But to get the troops to these two landing places was another matter. The English, always intensely sensitive to any threat of invasion, and knowing, in those days of poor 'security', that some such scheme was afoot, already had their squadrons on guard, from Boys in the Straits of Dover via Rodney in the Channel to Hawke off Brest. Boscawen watched in the Mediterranean. The Toulon fleet would have to run the gauntlet of his superior strength in the Straits of Gibraltar, and then it would have to get past the stronger Hawke into Brest; and, of course, assuming it got so far, a pursuing Boscawen would have been on its tail. The Brittany army had not been made to cover the extra distance for an assembly at Brest, whence its escort could at least have taken it to sea when a favouring wind had blown the blockading fleet off its station, but, being in the Morbihan, gave the escort some days' additional sail from Brest to reach it, during which, and during embarkment, the ships were likely to be attacked. If, by amazing good fortune, it reached the Clyde and landed the troops, the escort had then to negotiate some of the most stormy waters in the world in rounding Scotland, and yet arrive in sufficient trim to pick up the second army at Ostend, and embark it, presumably without interference, though only a short distance from the English base in the Thames estuary!

It was to be a summer campaign, but there had been such difficulty in fitting out and manning the Toulon fleet that it was not until August that it could sail. La Clue was still in command, though he should surely have been removed after the grave error of judgment that he had made at Cartagena. His flag was in L'OCÉAN(80) and he had twelve of the line, though two were only fifty-gunners, plus three frigates. His orders were to go to Cadiz, where he would get further instructions. He told his captains this, but made no other arrangements, such as an alternative rendezvous in case of accident, and issued no orders as to what was to happen if he had to fight Boscawen. He hoped to get a favourable wind and a dark night during which he could run past Gibraltar unobserved. He was lucky with the wind and the darkness, but he was seen at dusk by one of Boscawen's vedettes. Knowing this, he decided to alter course for the open Atlantic during the night, and at 2 am lit his stern lanterns for a short while, made his change of course, and then extinguished them, taking it for granted that his ships would all see this, would understand, and would follow him. The OCÉAN was a fine sailer, but some of his ships were not, and five which had straggled rearmost did not see his lights and change of course and continued their route for Cadiz, into which, after cruising outside to wait for him for a day, they entered. At daybreak, therefore, La Clue found himself with only just over half his fleet. He took in sail to wait for the truants, and even closed some ships sighted at 8 am only to find that they were those of Boscawen, who, though partially refitting his ships in Gibraltar, had got the canvas on to them with a speed incredible to the French navy of that depressed day, had slipped his cables, and had stood out hot-foot in pursuit.

La Clue ran for it before the wind, forming line in the following order: TÉMÉRAIRE (74), MODESTE (64), REDOUTABLE (74), SOUVERAIN (74), OCÉAN (80), GUERRIER (74), and CENTAURE (74). The speed of ships fleeing in line ahead is that of their slowest sailer, in this case the SOUVERAIN. So at 2.30 pm the fastest ships of Boscawen, who had signalled a general chase, came up with the CENTAURE. Her captain, de Sabran Grammont, did his best to behave like the escort of a convoy, to hold off the enemy to give his charges a chance to get away; and he fought his ship with great determination. He was engaged by each English vessel as it came up, and eventually he was surrounded, while the other pursuing English vessels reached up the French line as far as the SOUVERAIN. La Clue fought his powerful OCÉAN well, and so damaged Boscawen's ninety-gun *Namur* that that admiral had to stand off and shift his flag to the *Newark*. The OCÉAN herself was hard hit, with ninety dead and a hundred wounded, including La Clue himself. At 7.30 pm, after five hours' struggle against odds, the CENTAURE struck. She had put up a fine fight covering La Clue's retirement, but was completely dismasted, and so shot about that it was only by jettisoning her guns and anchors that her captors were able to coax her into Gibraltar. If she earned glory from this battle, she was the only French ship that did so. The firing ceased with nightfall, but the pursuit did not. In the darkness the captains of the GUERRIER and the SOUVERAIN, with complete lack of loyalty to their admiral and their colleagues, slipped away by violent alterations of course. Both eventually reached Rochefort.

So in the morning, with the enemy still close on him, La Clue found himself with only four ships. He was fairly near the south coast of Portugal, and made for it, aiming at a creek a little west of Lagos, where there was a battery. He hoped that the combination of protection from this and the neutrality of Portuguese waters would call a halt to the pursuit. Boscawen was not to be stopped by such small considerations as violation of neutrality and the guns of a battery that, in fact, were not very energetically served: he followed the French into the bay, and opened fire. With the hopelessness of their position, the fight had gone out of the French, and they thought only of getting ashore. OCÉAN, which with REDOUTABLE had been run aground, struck, on fire and with a number of her crew still aboard. REDOUTABLE, on being attacked, did the same. MODESTE and TÉMÉRAIRE which had anchored in the shallows, abandoned ship as the English approached. Boscawen took the TÉMÉRAIRE to Gibraltar and set the other three ships ablaze. The Toulon fleet could be written off.

Of course Portugal protested at this violation of her neutrality, so Lord Knowles was sent out to make a humble apology in the presence of the diplomatic corps. The victory was cheap at that price. La Clue had been got ashore, and when he had sufficiently recovered from his grave wounds he wrote a long self-exculpatory report, blaming the disaster on the captains who had separated from him, especially those who had lost him in the night when they came through the Straits, and he took no account of his own failure to give adequate orders and signals. The ships in Cadiz were told to collect the survivors from Lagos, which they did, and then to come north to join Conflans. This highly dangerous order they were unable to carry out because of the English blockade of the port, and when at last they did get away, in January of the next year, there was no longer a Conflans for them to join. So they sailed back to Toulon.

With the disaster to La Clue the invasion scheme should have been promptly abandoned. But France had gone so far with preparing, and the situation was so desperate a one calling for a desperate remedy, that it was decided to continue. Thurot was to sail with his little diversionary expedition: Conflans was to bring the main fleet out of Brest, sail to the Morbihan, free the transports, which were blockaded there by some fifty-gun ships and frigates under Commodore Duff, escort them west of Ireland to the Clyde, pass round Scotland and run down the North Sea, collect the second army at Ostend and deposit it in Essex. If, as was likely, the fleet could not dodge the British it must fight them off as well as it could. To have conducted such a campaign would have tested to the utmost a Nelson with a fleet officered by his 'band of brothers' and manned by crews trained and hardened by years of sailing and fighting at sea. Conflans was no Nelson, his officers were anything but a band of brothers and were entirely inexperienced in working their ships as a fleet, while the crews, mostly collected quite recently, barely knew their ships or each other, were without adequate gun-drill, and, as to a good proportion of them, were not even real deep-water sailormen. (De Guébriant, CE, considered that he had not above twenty to thirty true seamen in the whole 750 of his ship's company.) In short, what had been a most dubiously possible plan had now become a dangerously reckless gamble.

It was not expected, either by the planners or by Conflans, that he would get clear into the Atlantic, or even round to the Morbihan, without a battle; and, while still in harbour, he issued lengthy orders, and confident ones, to cover eventualities in fighting it. At long last, on November 14th, a date dangerously late with regard to the probable weather – a gale having blown Hawke away from his blockading station to the English coast – Conflans weighed anchor and made slowly for the Morbihan. His fleet was:

1st Division

| Ship | Guns | Captain |
|---|---|---|
| Soleil Royal | 80 | Maréchal de Conflans (V-adm) |
| Orient | 80 | chevalier de Guébriant Budes (CE) |
| Glorieux | 74 | Villars de la Brosse |
| Robuste | 74 | marquis de Vienne |
| Dauphin Royal | 70 | chevalier d'Urtubie |
| Dragon | 64 | Le Vassor de la Touche |
| Solitaire | 64 | vicomte de Langle |

2nd Division

| Ship | Guns | Captain |
|---|---|---|
| TONNANT | 80 | chevalier de Beauffremont (CE) |
| INTRÉPIDE | 74 | Chasteloger |
| THÉSÉE | 74 | Kersaint de Coëtnempren |
| NORTHUMBERLAND | 70 | Belingant de Kerbabut |
| SUPERBE | 70 | Montalais |
| ÉVEILLÉ | 64 | chevalier Laprévalais de la Roche |
| BRILLANT | 64 | Keremar de Boischateau |
| | 3rd Division | |
| FORMIDABLE | 80 | Saint-André du Verger (CE) |
| MAGNIFIQUE | 74 | Bigot de Morogues |
| HÉROS | 74 | vicomte de Sanzay |
| JUSTE | 70 | Saint-Allouarn |
| INFLEXIBLE | 64 | chevalier de Caumont |
| SPHINX | 64 | chevalier de Coutance-Laselle |
| BIZARRE | 64 | chevalier de Rohan |

with five frigates and corvettes.

On the evening of November 19th, he was seventy miles south-west of Belle-Isle, with a strong wind blowing and a big sea rising. He altered course northward for the Morbihan, and took in sail in order not to reach there before daylight. By then a full gale was raging and the seas were becoming mountainous. Commodore Duff's ships, which had been blockading the transports in the Morbihan, had sighted the French and were seen beating urgently out to sea before they were cut off. Conflans made a signal for a general chase, and was swooping down on this easy prey when his frigates and mastheadmen reported a large number of sail of the line coming up from westward. This must be Hawke! That admiral had stood back from England for his station off Ushant as soon as the weather allowed and, finding the Brest roadstead empty and knowing of the army awaiting embarkation in the Morbihan, had at once made for that area carrying all the sail that he dared in such weather. Conflans at once signalled for his fleet to form line on him, and for the rearmost to make all sail, and steered northward for Quiberon Bay. This is formed by the long Quiberon peninsula and its continuations the islands of Houat and Houëdic ending in the group of rocks known as Les Cardinaux. (It is after these that French historians have named the battle known to the English as Quiberon).

Navigationally these waters, with their reefs and shoals lying off an inhospitable shore, are most dangerous. Conflans wrote later that he expected to get all his ships into the bay before the English could come up with them and that the English would not dare to follow into such dangerous waters in such weather; if they did so, by keeping inshore he would have the advantage of the wind in choosing how to fight them, while they would have the Cardinals and the alarming Du Four shoal on their lee. The first part of this, that he did not think the English would follow into the bay, is true: the second part sounds like an afterthought, to give a better colour to his actions. Hawke had twenty-three of the line, so that in numbers of ships and strength of armament the fleets were not so disparate. But the difference in quality was great. Conflans knew it only too well: gone were the confident fighting orders; something not far short of a panic flight took their place.

Hawke had signalled a general chase, and at 2.30 pm, as the SOLEIL ROYAL, an excellent sailer, was rounding the Cardinals, the French rear, some eight miles astern, was being engaged by the leading British vessels. By going about at once Conflans could probably have extracted them from immediate danger, but he carried on into the bay, his fleet, and the English, tumbling pell-mell after him. The scene was dramatic in the extreme: a howling gale, huge seas breaking over the vicious rocks, and the great ships struggling at once to engage the enemy and save themselves from wreck. It is not possible to describe in detail the confused action that was fought. Most ships were engaged even if, in a few cases, only briefly and in passing enemies, but, except that Hawke's ship, the hundred-gun *Royal George,* was distinguishable by its size and its admiral's flag (as was the SOLEIL ROYAL), few captains knew whom they were fighting at any time. Once well into the bay, and seeing that his plans had all gone wrong, Conflans went about to try, too late, to save his rear. But his signals to his fleet were obeyed by only a few captains. Some, for lack of seamanship or sea room, could not do so, others, like de Beauffremont, disobeyed for lack of trust in their admiral, and struggled amidst the fighting to get to the open sea. The French admiral had lost control of his fleet.

Kersaint, whose record was that of a fighter, opened the lower ports of the THÉSÉE to get his heavy battery into action. His vessel was rolled to leeward by a sudden fierce squall, the raging waters seethed through the gun ports, and the horrified onlookers had the terrible sight of a great ship foundering instantly while under sail. Only a mere twenty-two upper deck men, who managed to grasp pieces of floating wreckage, were saved. A little later the SUPERBE, shattered on the water-line, mainly by Hawke's broadside, and trying also to use its lower battery, suffered the same fate. At about 4 pm the FORMIDABLE, which had borne the main weight of the English assault on the French rear, with her CE and his flag-captain, who was his brother, both killed, hauled down a helpless flag. Soon after this captain Lord Howe, in the *Magnanime,* who had been fighting THÉSÉE until a collision with another British ship turned him away, battered the already damaged HÉROS into submission, though a prize crew was not sent aboard because of the high sea running. Conflans, seeing all lost, had gone about once more, to try to lead his surviving ships south-eastward for the open sea, but collisions with first one and then another of his own ships had forced him off his course. Others of the French, unable to get to seaward, had run for it to the Morbihan, for during the battle, in the furious westerly gale, the fleets, willy-nilly, had been carried eastward out of the bay proper.

By five o'clock on that winter day of overcast storm it had become all but dark, and Hawke signalled to anchor. The signal, by gunfire, was not immediately noticed by all, and some desultory firing between retreating

French ships and the British continued; but by five-thirty all the English ships still in or near the bay had brought up, few knowing who or where was his neighbour, or being certain, even, of the fate of the day. At dawn the gale was still raging. Hawke found all his fleet in sight and more or less together, though *Essex* and *Resolution* had dragged on to the treacherous Du Four shoal. But Conflans found himself alone, and cut off from the open sea by nearby English ships, while the few French ships still in sight to the south were making sail to get clear. He therefore cut his cables and ran the Soleil Royal ashore near Le Croisic, close to the surrendered Héros, also aground there. There was nothing better that he could do.

The dispositions of the French fleet, if they can be called such, were now as follows: To the north-east, hopelessly cut off from the open sea by the British, were seven ships; ashore near Le Croisic, at the mercy of the enemy, were Soleil Royal and Héros; Formidable was a wreck in enemy hands, Thésée and Superbe were sunk, Juste, badly damaged in the battle, with her captain and his second in command killed, had struggled to sea only to be wrecked with the loss of almost all hands near the mouth of the Loire for which she was making[1]; de Beauffremont and the remaining eight, by no means all in company, were running for the shelter of Aix Roads and the protection of the Rochefort batteries. Hawke did nothing that day save to try, vainly, to get his two ships off the Du Four. But five of the Frenchmen trapped off the Morbihan, having lightened themselves somewhat, took advantage of the high tide and the mass of water that the gale had piled into the bay to bump their way into safety over the bar of the river Vilaine, where no ship of their size had ever been before. Inflexible was wrecked in the process.

The next day Hawke, having dispatched some ships under Keppel after de Beauffremont (who was found too far up the Charente to be attacked), made first for the Soleil Royal. The few remaining aboard promptly fired her, and the English arrived too late to do more than carry off a trophy or two. They themselves set fire to the Héros, which was too firmly aground to be got off. Hawke then made for the Morbihan, at which the two ships still remaining outside the Vilaine followed their fellows over its bar. He spent several days vainly trying to devise some means of getting at them, and in claiming the crew of the Héros as prisoners – a claim rightly rejected, for, though the ship had yielded, a prisoner who has not given parole is always entitled to make his escape. He then took his fleet back to England. Counting the wrecked Juste and Inflexible the battle had cost France seven of the line, of which one had passed into the hands of the British, in return for the two English ships left to break up on the Du Four. The remaining ships were scattered, most of them unable to go to sea without refit. The hope of invading England was gone for good: what spirit had been left in the service was broken; the disaster was complete.

Conflans did not realize fully what he had done. He wrote a long self-satisfied, self-exculpatory report. He had done all that could be wisely ex-

[1]The three sunk ships were all Breton manned, and the loss of almost all of their crews plunged the whole of Brittany into mourning.

pected in the circumstances, the superiority of the enemy had been the trouble (which was true), but everyone had fought jolly well, though it was sad that some officers had been killed; of course, the expedition must be considered as 'off'. A letter to the duc d'Aiguillon, somewhat similar in tenor, seemed to have as its most important point the risk that a man of his age and dignity had to undergo in clambering out of his ship over her stem by way of spars and cables to reach the shore. Only when he found himself generally execrated did the vice-admiral alter his tone. Then – his signals had not been obeyed, he wrote, the hindmost ships had failed to crowd on sail, in particular he had been deserted and not supported by his second in command. Indeed, as someone said, if anybody gained any honour out of the battle, de Beauffremont certainly did not. That officer, ignoring the true charge that he had abandoned his chief, retorted that it was not his fault, he had not been in command, the responsibility for the disaster lay with the vice-admiral, who had made the mistakes.

The task had been impossible, but the aftermath does not make a pretty picture. Nor does the fate of Thurot (see p 129). Understanding that Conflans would certainly sail before he actually did so, Thurot had slipped out of Dunkirk in the middle of October in the MARÉCHAL DE BELLE-ISLE (44) with three frigates and two corvettes, carrying a few soldiers. His passage northward was stormy and trying, and his ships were distinctly weather-worn when, having rounded Scotland, they arrived off the Ulster coast in January 1760. The tempers of the officers were now as worn as their ships. The naval men and the commander of the troops had never liked sailing under the orders of a mere privateer, however successful and able, and they were at loggerheads with him. One of the frigates and a corvette had parted company during a storm, and the frigate was lost with all hands; and now the remaining corvette sailed for home without so much as by-your-leave. Thurot landed his troops, who took Carrickfergus. But they were too few to attack a large place like Belfast, and all they could do was to revictual the ships from the neighbourhood before being re-embarked. Sailing south of the Isle of Man the French were found by three English frigates. Though weather-worn, they were collectively equal in strength to their opponents. But two of the French frigates ignored Thurot's signals and deserted him. He was killed, and the MARÉCHAL DE BELLE-ISLE was taken. So, after a mere show of resistance when overhauled, were her two absconding consorts.

Nor had matters gone any better in the Indian Ocean. We have seen (p 105) that d'Aché (CE) sailed for India in May 1757. He typified both what was best and, emphatically, what was worst in the French service. On the one hand, he knew his business as a sailor and was as courageous as one could wish in combat. On the other hand he was obstinate, indolent, ill-tempered, unwilling to take orders, and so haughty that he deemed his appointment to be below his dignity, as he was to command merchants (the East Indiamen who were to make up his squadron), and he accepted it only because it was the road to promotion. With him sailed the new governor and commander-in-chief in India, Count Lally-Tollendal, the eager, hot-tempered son of an Irish Jacobite. Even before the start the two got on so ill that d'Aché offered his resignation: it is a great pity that Moras did not accept it; but he was new in office, and did not want further to delay an already postponed sailing. The Indiamen, with the troops, had left in advance and, had d'Aché made a reasonably quick voyage, he would have arrived when the position in India was most propitious. For France's main concern was first to expel the English from the southern part of the subcontinent, the English troops had gone north to deal with Suraj-ud-Daula's seizure of Calcutta, and their ships were in the Hooghly to support them. His passage, however, was the very reverse of rapid. He early captured an English merchantman, and, since his haughty *noblesse* was by no means above a little commerce, sailed slowly so as not to part company with her. He put into Rio, for he had scurvy aboard, and his stay there was not shortened by his selling his prize and its cargo. He then steered an absurdly southerly course to avoid the equinoctial winds off the Cape of Good Hope, took in sail frequently, and on reaching the Île de France and Bourbon spent almost three months in exercising the Indiamen who were to form his squadron. By this time the relations between him and the fuming Lally were worse than dire. He finally reached his station off the Coromandel coast at the end of April 1758, having managed to take all but a year over the voyage. Commodore Stevens, with ships to reinforce Vice-admiral Pocock, had sailed three months after him, and arrived five weeks before him.

D'Aché's fleet consisted of the ZODIAQUE(74), a fine royal ship, and eight Indiamen, each armed with approximately fifty guns. Pocock, with his flag in the *Yarmouth*(64), had only seven ships of fifty to sixty-four guns; but as they were all royal vessels, and d'Aché had so many Indiamen which were awkward in battle, especially in using any of the heavier guns, the two fleets were not unequal. The French went at once to attack Cuddalore (Fort St David) and drove two British frigates there ashore, whereat Lally, anxious to get to his headquarters at Pondicherry, went ahead in one of the Indiamen. He was still in sight when Pocock appeared, but he ordered his ship's captain to ignore d'Aché's signal of recall. (This filled the enmity between the two French chiefs to its brim.) D'Aché stood out in line of battle, and Pocock, who had the wind, came down in line to fight him. The Indiamen ahead of the ZODIAQUE did well, and the VENGEUR drove the fifty-gun *Salisbury* out of the line. Astern of the ZODIAQUE, however, the weak MORAS was driven out of action, and the timorous DUC DE BOURGOGNE kept almost out of range. As the English line began to straggle so as to open a gap, d'Aché wore his ships to make for it, and in so doing brought a concentration of fire on the *Yarmouth*; but the manoeuvre put his own line into disorder and the two fleets drifted apart. The French had had rather the better of this indecisive battle, Pocock having suffered a good deal of damage; but when the French moored off Alam Parva on their way to Pondicherry one of the Indiamen, the BIEN AIMÉ, was driven ashore by the strong winds. Her captain replaced that of the DUC DE BOURGOGNE, to whom d'Aché meted no other punishment for his cowardice; but Pocock brought the three lagging captains of his rear to court martial, two of them being removed from command.

As his ships needed much repair aloft, the French gunnery having followed its customary practice of aiming at masts rather than hulls, Pocock could not keep Cuddalore open from the sea, and Fort St David surrendered in early July. This should have shown d'Aché the valuable part played by the mere presence of his fleet on the coast.

But d'Aché had been kept cruising off Fort St David till its fall only by being threatened with arrest by Lally, and because Lally had paid his almost mutinous seamen out of his own pocket. Now that it came to the question of a complete siege of Madras by land and sea d'Aché refused to co-operate, saying that it was his business to cruise off Ceylon to capture the trade; and he sailed. A formal order from Lally brought him back, but he still refused to make for Madras on the plea of shortage of victuals. While this unseemly squabble continued Pocock reappeared, having refitted. D'Aché, somewhat short-handed, stood out to meet him and, after some days of manoeuvring, the English, having gained the wind, came down to attack in line off Karikal. As before, Pocock had fewer guns and fewer, but better, ships than the French. Another indecisive, but this time a brief and rather hot, battle followed. D'Aché's leading Indiaman, the COMTE DE PROVENCE (54), was set afire, but was covered and disengaged, just in time, by the DUC DE BOURGOGNE. The ZODIAQUE had her wheel and tiller smashed and went to leeward out of control: fire also broke out aboard her. With his two weakest Indiamen, the forty-four gun MORAS and CONDÉ, also standing out of line, and knowing that facilities for repair were negligible, d'Aché broke off the action and returned to Pondicherry. Pocock was again too damaged aloft to pursue.

Lally now urgently reverted to his main object, the siege of Madras. D'Aché would have none of it. His ships needed supplies and refitting: the materials were not available at Pondicherry; he would sail for the Île de France – 'that Cythera of the French navy'. Since it was known to all that the Île de France was, if anything, shorter of supplies than the Coromandel coast, his proposal was plainly dictated by another reason, and that reason was clearly his hatred of Lally. It is, of course, the gravest illustration of the parsimony of the French administration, and of its lack of understanding of naval needs, that a squadron had been sent to so remote a station as the Indian Ocean without the least arrangements having been made for its maintenance there. But in similar trying circumstances Labourdonnais had made do somehow, as Suffren was to do later. There was room in India for only one major European power, and that power would be the one that could hold the sea. D'Aché was not a man of any strategic vision, nor had he any patriotism to equal, much less override, his personal dislikes and conveniences. Ordered to stay until mid October, he asked for a general council which, to his annoyance, bade him remain till mid September. At this he got the opinion of his own officers that he could not refit at Pondicherry, and sailed on the night of September 2nd, leaving Lally and Pondicherry to look after themselves. His arrival at Port-Louis horrified the Governor of the Île de France, who was barely able to keep the island's population fed. He found there three royal ships under Froger de l'Éguille (CE) who had brought troops and two million of money. D'Aché retained half this, just sending a frigate to Pondicherry with the remainder, and with a whole nineteen soldiers of the reinforcements on board!

It was a great pity that he did not hand the command to the more enterprising Froger de l'Éguille. True that the conditions at Port-Louis were desperately inadequate for caulking and that therefore the process of refit was slow; true that victuals were so short that ships had to be sent to the Cape of Good Hope to raise food; but the delay need not have been so lengthy that it was not till mid July 1759 that d'Aché sailed again for India. During the long interval since his departure Lally had attempted the siege of Madras, but, with the place open to the sea, he had failed. He was now short of money, growing short of troops owing to the wastage of campaigning in a tropical country, and so he was forced back on to the defensive round Pondicherry. The fate of France in India now depended on d'Aché.

As he had been reinforced, so had Pocock, but the French still retained the superiority of about a hundred guns and eleven ships to nine; though as only four of the French ships were royal and the others were Indiamen there was no real inequality of strength. The fleets met off Porto Novo, both on the port tack with a northerly wind, and battle was joined at about eleven in the morning. It followed much the same course as d'Aché's previous two actions. The ACTIF, a royal sixty-four, was set afire by the *Elizabeth*(64), but Froger made sail and covered her with the MINOTAURE (74); Stevens, in the *Grafton*(68), had similarly covered the fifty-gun *Newcastle* and the *Tiger*, which had had the worse of deals with the MINOTAURE and the Indiaman, SAINT-LOUIS; and the *Salisbury*(50) had been hard hit by another Indiaman, the FORTUNÉ(54). But the toughest of these slogging matches had been between the two flagships, and Pocock's *Yarmouth* had given the ZODIAQUE such a hammering that, when d'Aché went below to have a severe wound dressed, and his flag-captain was killed, the next in command stood away to leeward. The rest of the French not unreasonably followed to form line on their flagship and, as the English were again too damaged aloft to close, the battle ended after five hours' slaughter. It was again indecisive, but d'Aché was free to make for Pondicherry. He arrived there on September 15th, landed some troops and 800 000 livres of the million he had retained from the funds that had been sent out, announced that he had thereby done his task, refused to hear any arguments, and left again. A coaster was sent after him with a letter, signed by everyone of importance in Pondicherry, charging him with the responsibility for the loss of the colony. At this he returned, landed more men, announced that his fleet was in no state to fight the British, and sailed for the Île de France for the last time. He had thrown away an empire.

He spent a year at the Île de France doing nothing but suffer heavy damage from a hurricane, and then, in December 1760, he sailed for home. Pondicherry, besieged by land and blockaded by sea, yielded to starvation a month later. Its defender, Lally, put in the Bastille for two years, was then executed after a trial that was 'a monument of iniquity'. D'Aché, who had deserted him and who was one of his accusers, ended his days as a vice-admiral. Truly, to serve France during the reign of Louis XV was a chancy business. During these campaigns an attempt to give indirect support had been made by sending a ship of the line and two frigates to cruise off St Helena and intercept the British India trade. Owing to quarrels between the ships' captains they achieved precisely nothing. But a soldier in India, the comte d'Estaing, disgusted with the tergiversations of the fleet, fitted out an Indiaman, the COMTE, and a smaller ship at his own expense, and throughout 1760 carried out some highly successful privateering raids on the English company's stations in the Persian Gulf and the East Indies. On the strength of this he was allowed to transfer from army to navy, with the rank of CE – a step which aroused some jealous displeasure among his new colleagues.

At home, the year 1760 had been chiefly notable in naval affairs because of schemes for getting the six vessels of Conflans' fleet out of the Vilaine, where they lay on the mud: these produced unseemly wrangles between officers of the navy and of the India Company, both of whom put forward suggestions. Eventually, early the next year, all six, in pairs, were got out in advantageous conditions of tide and wind, and sometimes of fog, past the English blockade and into Brest. Berryer, the Minister of Marine, had so little faith in what was left of his Service that he proposed to sell up such few stores as remained in the arsenals. The degradation of the French navy seemed as complete as its disaster.

But now a new life was to be breathed into it. Choiseul, who had already added the Ministry of War to that of Foreign Affairs, in 1761 took over the Secretaryship of the Navy from the deplorable Berryer, so that France had for the first time a single minister for both its armed services. Already, by astute diplomacy, he had succeeded in lightening France's unduly heavy commitments to Austria, and he was now busy forming the Family Compact between the Bourbon monarchies of France, Spain, and Naples. It was too late for it to have any successful effect in this war. Indeed, the notorious inefficiency of Spain would probably have made it of little effect if it had been formed earlier – as was to be shown by the failure of the great combined campaign in 1779. He also set to work to re-create the navy, with what results will be seen later. For the moment, such French vessels as crept to sea, while occasionally successful in holding off English ships that came upon them, were usually taken; and what was left of the French colonies lay at England's mercy.

Martinique fell to a fleet and army brought by Rodney early in 1762. St Lucia and Grenada followed in a few weeks. With these went the nests of privateers that they had harboured. For, whatever may have been the fate of the official fleets, the French had shown their usual genius for privateer attacks on commerce. The English losses had been large, usually of ships whose owners would not delay for them to sail in convoy; but though an occasional merchant pulled a long face, in the main the losses were laughed at, for England had so much shipping, and her mastery of the seas was so complete, that, despite the privateering, her trade had flourished. Nor did the privateering have any effect on the main course of the war: in 1761 25 000 Frenchmen were prisoners in England, in France the English prisoners were a mere 1 200. Canada had gone, India had gone, and the only reason that Louisiana, San Domingo, and the Mascarenes (Île de France and Bourbon) were left was that, with Spain entering the war under the Pacte de Famille, the Spanish colonies were a simpler and more profitable prey. The few remaining French oversea possessions could wait.

What saved them was a political change in England. Pitt, guessing the secret war pact that lay behind the new Bourbon alliance, wished to anticipate it by declaring war on Spain. His colleagues, though they were later to find themselves obliged to do this, refused to follow him; and he resigned. This split the Government, and the new young king, George III, concerned chiefly with recovering powers of the Crown that Parliament had somewhat unconstitutionally seized, was able to get rid of its remnant under Newcastle, in order to bring to office his own tutor, Lord Bute, who sought peace to be free to further his master's wishes. Yet before peace was signed the moribund French navy gave one last kick. The captain chevalier de Ternay, who had shown much enterprise in getting the stranded ships out of the Vilaine, sailed with two of these ships, ROBUSTE(74) and EVEILLÉ(64), and two frigates for an unexpected raid on Newfoundland. He had troops which he landed, and he and they carried out a destruction of fisheries and their shipping that cost the English over a million. But even this success was incomplete, for six English of the line came on the scene and Ternay had to flee, leaving the troops ashore to be made prisoners. Though pursued, he got safely into Corunna.

The peace, negotiated by the end of 1762, was signed the next year. Spain and France both lost heavily by it. France handed back her one conquest and success, Minorca, gave up all claims to Canada and anything east of the Mississippi except New Orleans, was allowed to retain some depots in India, to be used purely as trading posts, and yielded her conquered West Indian Islands, except for Martinique, Guadeloupe, and St Lucia, which were returned to her with the little fishing islets of St Pierre and Miquelon by the Newfoundland banks. It was a heavy price to pay for the parsimony and neglect that had so debased her navy that it was inefficient as well as inferior. Yet from England's standpoint the peace was not a good one. Having France at her mercy she should have exacted even more. As it was, she had taken so much that a desire for revenge was inevitable, the means were left to France to work for it, and Choiseul was already building the new fleets that were to take it.

CHAPTER X

# Reform: Recovery: Revenge

When the peace was signed the French navy still had some forty ships of the line and about ten frigates; but these were scattered in the various ports, in various states of disrepair, and the arsenals were empty. Yet already Choiseul had started to prepare for revenge upon England. To meet her many more ships were needed, and there was no money with which to construct them. Some could be built out of the annual naval budget, in the usual way; but that would not increase the size of the fleet with anything like sufficient rapidity. Sensing that the country at last realized that the disasters of the war were largely due to the inadequacy of the navy, and that that inadequacy was partly due to its own lack of interest in the Service, he felt that an appeal to national pride might produce the ships or the funds for them. The result must almost have exceeded his hopes. A number of the provinces had their Estates, a kind of provincial Grand Council, and the Estates of Languedoc met in 1762. To them Choiseul tentatively made his first appeal. With acclamation they voted to provide a ship of eighty guns. Other estates and bodies followed suit, wherefore the following magnificent addition was made to the fleets:

The BRETAGNE, 110 guns given by the Estates of Brittany

The VILLE DE PARIS, 90 guns, given by the City of Paris

The LANGUEDOC, 80 guns, given by the Estates of Languedoc

The SAINT-ESPRIT, 80 guns, given by the Order of the Saint-Esprit

The ZÉLÉ, 74 guns, given by the Receivers-General of Finances

The BOURGOGNE, 74 guns, given by the Estates of Burgundy

The MARSEILLAIS, 74 guns, given by the Marseilles Chamber of Commerce

The DILIGENT, 74 guns, given by the Controllers of Postal Services

The SIX-CORPS, 74 guns, given by the Paris guilds of merchants

The CITOYEN, 74 guns, given by the Court Bankers and the Treasurers of Supplies to the Forces

The UNION, 64 guns, given by a large group of country gentlemen

The UTILE, 54 guns, given by the Farmers-General of taxes

The FERME, 54 guns, given by the Farmers-General of taxes

The FLAMAND, 54 guns, given by the Estates of Flanders

The BORDELAIS, 54 guns, given by the City of Bordeaux and its Court of Lawyers

The ARTÉSIEN, 54 guns given by the Estates of Artois.

Besides the gift of these ships there was a huge general subscription that brought in thirteen million livres. The shipyards could again be busy, very busy. It was not only with the size of the navy and the replenishing of its arsenals that Choiseul had to deal. He had also to breathe into the service a new spirit and a new efficiency. For it was clear that between the last two wars, that of the Austrian Succession and that of the Seven Years, the quality of the officer corps had gravely deteriorated: there were no La Jonquières and l'Etanduères to make courageous last stands against odds. It was clear, too, that sheer inefficient handling of the fleets and of the ships had greatly contributed to the disasters.

A new classification of ships and their armament was made. They were already in five classes. The first three of these were rated as ships of the line. First rates of ninety guns and over were three-deckers armed with 36-pounders, 24-pounders, and 12-pounders in their respective three batteries. The second rates were two-deckers of between seventy and ninety guns, armed with 36-pounders and 18-pounders. The third rates, also two-deckers, were of over fifty but under seventy guns, mounted with 24- and 18-pounders. This standardization of guns, which was carried to other equipment, made the arming or refitting of a ship, at whatever port it might happen to lie, a much more simple matter than heretofore. Brest and Toulon were still as they had been built in the days of Colbert and Vauban, seventy or so years before. Work was put in hand to rebuild them almost completely, and so to bring them up to date. The East India Company, on the verge of bankruptcy, lost its monopoly in Eastern waters, and traded its great port of L'Orient back to the Crown in 1770, to make an Atlantic station for the navy between Brest and the roadsteads off Rochefort. Provision was made for other ports, including those in the Île de France and the West Indies, to cater for a naval ship or so in case of emergency.

As to officers, in 1764 Choiseul remodelled the *gardes*, the corps of cadets into which intending officers were entered and in which they were trained, and he provided for the entry of some good quality commoners as well as for the sons of the nobility. This did not, unfortunately, break down the attitude of social superiority with which the noble *rouges* treated the *bleus*. His success in getting rid of the dead timber among the senior officers, too, was only partial: privilege and custom were rooted too deep. (For example, the comte de la Rochefoucauld, whose last command was that of a ship of the line in 1760, stayed in the service going through its various grades of promotion till, by mere seniority, he became one of the vice-admirals in 1782, and drew his emoluments till his death two years later.) But Choiseul did succeed in virtually ending the rivalry between the *épée* and the *plume* with his great ordinance of 1765. By this the *plume* was, technically, abolished and a new and much smaller corps of administrators was created from its ranks, with their former powers cut right back. The naval officer commanding in a port was now the equal of the Intendant and not his subordinate; and, under him, each captain was master in his own ship. His duties and those of his officers were fully and severely prescribed. So were details of uniform and pay. This change was not at first liked: the seagoing officers wondered if too much was being thrown upon them, and the administrators naturally regretted their lost supremacy; but it worked. In the same year (1765) the corps of naval constructors was created. In this year, too, the navy went into active service again: some of the few ships that had been kept in commission were sent to deal with the Moroccan pirates, who were again giving trouble. In command was Duchaffault de Besné (CE) in the UTILE(52) with five frigates and corvettes and two bomb vessels. Sallee was heavily bombarded, and the squadron then sailed north and bombarded Larache. Here something of the old inefficiency showed itself. Boats were sent in and set fire to a ship that could be seen up the river. The business was muddled and they had to go in again. Seeing another vessel further upstream they made for it and rowed into an ambush, coming under heavy fire from both banks. Three hundred officers and men were killed or enslaved.

In France the reforms and reorganization continued. Choiseul himself gave up the Ministry of Marine in 1766; but as he handed it to his cousin, the duc de Praslin, who was his alter ego, there was no change in policy or cessation of drive. Naval medical schools were established in 1768 so that something could be done to cope with the appalling losses from illness that were, so far, the accompaniment of any long voyage. The next year the Marine Academy was remodelled. That the service was revitalized was shown by the voyage of Bougainville[1] in the BOUDEUSE frigate in 1768, which made him the first Frenchman to sail round the world. The explorations of Kerguelen followed a few years later.

All this inspiration of a new vigour, all these reforms, were made with one object, that of revenge upon England. There were other things far more necessary for France: the reorganization of her clumsy system of autocratic government and especially of its chaotic and needlessly overstrained finances, the removal of some of the more grave of her social inequalities. They were all but ignored in the desire to get her own back for the humiliating peace of 1763. Perhaps no European government, not even that of France after 1871, was so set upon the idea of sheer revenge as that of Choiseul, until the advent of the very different Adolf Hitler. As an astute man the French Minister had realized the truth of Montcalm's prophecy that if the French were driven out of Canada England would have trouble with her American colonists, who, with no enemy now to fear, would resent the charges on their trade that were made to help to foot the bill for the war that had protected them. As is well known, England did have that trouble, and was tactless and unintelligent in dealing with it. Choiseul probably saw it coming, and realized that, if the trouble should lead to rebellion, it would need some support to keep the rebellion alive and thereby to distract and draw oversea English Forces.

But he knew that even his revived and enlarged navy was not likely to get the better of England in a most expensive war of long-distance transatlantic campaigns. It must be upon the British Isles themselves that the

[1]Bougainville was another instance of soldier turned sailor, and a more suitable one than some. For his personal interest was Higher Mathematics, on which he had written a notable book, and this enabled him to become a skilful navigator. Of course other Frenchmen had actually been round the world, but only working as seamen or traders in ships belonging, for at least part of their travels, to other countries.

blow must be struck. He therefore had a number of spies, Jacobites and English-speaking Frenchmen, who made surveys of parts of England, and who proposed various schemes of invasion. The king, too, unknown to his minister, was acting on the same lines, through a spy organization controlled in strict secrecy by the comte de Broglie. For Louis XV had an odd habit of conducting secret diplomacy behind the backs of his ministers, and even of counter-working them, despite the apathy that had become his leading characteristic.

Choiseul was encouraged by the increasingly faction-ridden state of English governments, and in 1768 had stolen an astute march on them. This was by acquiring Corsica from the decaying Republic of Genoa, which owned the island but could no longer control the perpetual revolts of its turbulent populace. By this he deprived the English of a possible naval base in the sheltered bay of San Fiorentino, and gave France a position that balanced that of England in Minorca. War nearly came in 1770 over the question of a claim to the Falkland Islands and, though it anticipated his planning somewhat, Choiseul was ready to wage it. But his appointment had been one of the few good ones that Mme de Pompadour had caused to be made, and she had died in 1764. Since then Choiseul had had to combat the intrigues that were endemic to the court of Louis XV, and a cabal backed by the new mistress, Mme du Barry, used this possibility of war to overthrow him. He and the duc de Praslin were incontinently dismissed at the end of the year. A few months before their fall they had seen their revived navy to a small extent successfully in action. The Dey of Tunis had objected to the French acquisition of Corsica and had tried to obtain English aid to oppose it. So a small force of two of the line, two frigates, and two bomb vessels under Captain de Broves (and joined by some Malta galleys) had bombarded Sousse and Bizerta so effectively that the Dey made a peace satisfactory to France.

Choiseul had aimed at a naval strength of eighty of the line. At his fall the number of such ships was sixty-four, plus fifty frigates and corvettes, and the officers available to man them were scientifically well trained and were doing their duties with a new heart. After a little, the Secretaryship of Marine passed into the hands of de Boynes, an administration lawyer. In 1772, he attempted to give the navy a rigid para-military organization, dividing ships, officers, and reservists for crews (the 'classes') into brigades attached permanently to different ports, even substituting army uniform for the naval blue and red. The scheme was unpopular, probably unworkable, and it was dropped by the next minister in 1774 for the service to revert to the *status quo*. But he did one thing that Choiseul had not found time and means to do; he established a 'squadron of evolution'. In other words, he sent some ships on manoeuvres. Perhaps Choiseul, like most of his predecessors, had failed to realize that no amount of lectures at a Marine Academy, or exercise in raising or taking in sail in harbour, can make a real deep-water sailorman: perhaps it was the question of expense in sending ships to sea that had restrained him. Certainly it was that that limited de Boynes, for the squadron on manoeuvres consisted only of three of the line, with six frigates and six lesser vessels. Even so it provided France with a solid body of sea-trained, gunnery-trained, and signal-trained officers and sailors who could be distributed throughout a fully commissioned fleet to give each ship a nucleus of first-class men in its crew. (A corps of gunners, given weekly training in harbour, had been created in 1767.)

When Louis XV died in 1774, there were sixty-six ships of the line, thirty-six frigates, and about fifty lesser craft. The liners contained three first-rates, and thirty second rates (which were powerful and heavily manned ships). The veteran was the DAUPHIN ROYAL of 1738, and a few others were nearing their thirtieth birthday; but thirty-nine of them were almost new, having been constructed during the Choiseul-Praslin administration, with various modern improvements in design. To officer them, or in training for that purpose, there were just over 1 200 gentlemen. The administrative officers, the former *plume*, now numbered only just over 250, including eight Intendants, of whom two were for the colonies. The constructor corps was just under forty strong. Weak in its lack of adequate sea training, and its shortage of able seamen, but otherwise a powerful force.

Nevertheless, both the planning for revenge, which had raised the spirits of the officers, and the drive in construction and in stocking the arsenals, that had been so prominent under Choiseul, had fallen astern under de Boynes. They were to be restored by his successor. Louis XVI, on coming to the throne, made an almost clean sweep of the ministers appointed by his grandfather. Though he only once visited any port, Cherbourg, he was unlike his predecessor in that he had at least a reasonable interest in his navy. His new Chief Minister was Maurepas, formerly Secretary of Marine and now brought out of his long retirement, and his new Naval Minister was Sartine, a popular and able police official. This gentleman was at pains to learn the business of his new office, and to restore the thrust and *élan* of Choiseul's day (which anyway was only recently passed), and he had the sense to seek, and act on, the advice of two able naval administrative officers, one of whom, Fleurieu, had progressed in active service from *garde* to captain before transferring to the administrative side.

De Boynes' unpopular regimentation of 1772 having been abandoned for a reversion to the *status quo*, a new ordinance was issued in 1776. By this the last relics of the *plume* disappeared; for the Administration Corps and the Constructor Corps were suppressed, and those of their members retained in service became commissaries and controllers of ports and arsenals. Their business was to see to the purchase and warehousing of supplies, to deal with payments, and to run hospitals and other port installations. Everything to do with ships, including their construction and their guns, was now in the hands of the active service, the seagoing officers. (They were however told that any who showed arrogance towards, or contempt for, the down-graded administrators would be dismissed from the service.) In theory, thanks to the Naval Academy and to their courses, these were well trained for the purpose. In practice, the change, though desirable, went a little too far, and the removal from shipboard of the administrative officers who had been borne by ships of the line and the larger frigates, and who had carried out the duties of the former paymasters in British service, was a mistake. For the active service officer proved, in the main, incompetent to carry out these functions.

A greater error was the suppression of the École de Marine that de Boynes had established at Le Havre two years before. Its place was taken by a body of *aspirants* at each port. They were aged fourteen to eighteen and they became *gardes*. Every *garde* had to serve for two and a half years at sea before getting further promotion, to *enseigne*. This was good, even though 'at sea' perforce meant largely 'on board in a roadstead'; for the amount of actual time spent by French ships on the high seas was absurdly little, so that for an officer to have been on an actual voyage was a thing to be noted. The great mistake about the suppression of the Le Havre school was that, whereas its entrants had had no need to produce any patent of nobility, now any entrant had to produce such patent – an odd requirement in a service that had known, *inter alios,* Jean Bart and DuGuay-Trouin. Not only did this increase the existing tension between the remaining *bleus* and the *noblesse,* but it was bad for discipline. For, being nobles, except when actually on watch or duty, the officers, regardless of rank and seniority, spoke and acted towards one another as equals. It was by no means unknown for a junior officer to argue with his captain or senior about the wisdom of an order that he had given, though at sea, in action, or even in tricky weather, prompt obedience to an order is all-important – even more so than in land forces. Promotion to flag rank, too, came much too slowly because of the amount of dead timber at the top of the list of officers. D'Orvilliers, du Chaffault, and Guichen, all thoroughly able and well tried, were well in their sixties before promotion to lieutenant-general.

Moreover, Sartine's eager and earnest administration concentrated far too much upon officers, and far too little upon crews. The system of *classes* had ceased to work. The benefits attached to it by Colbert at its creation had been whittled away or had fallen into desuetude. Pay was irregular, supplies of clothing so poor that ships' companies went half naked, hammocks often lacking, food bad. Living miserably aboard, the men had the added misery of knowing that their wives and families were starving ashore. As a result every effort was made to dodge the call-up and, since France had only a small merchant service, crews were completed with longshoremen, coastal militiamen, and soldiers. Ships' companies contained too small a proportion of real able seamen. That is why one often finds that a French ship, fleeing from greater numbers, is overtaken and brought to action, though superior in design and sailing qualities to its pursuers.

Sartine continued, and stepped up, the shipbuilding programmes, so that when he left the ministry in 1780 France had seventy-eight ships of the line and a large number of frigates and lesser vessels. He also continued the manoeuvres. Guichen took a 'squadron of evolution' to sea in 1775, though it consisted only of four frigates and eight smaller vessels. (A royal prince of the Orleans branch, the duc de Chartres, sailed in it as a volunteer.) In the next year du Chaffault took out a larger squadron, three of the line, seven frigates and some lesser ships. With one ship going

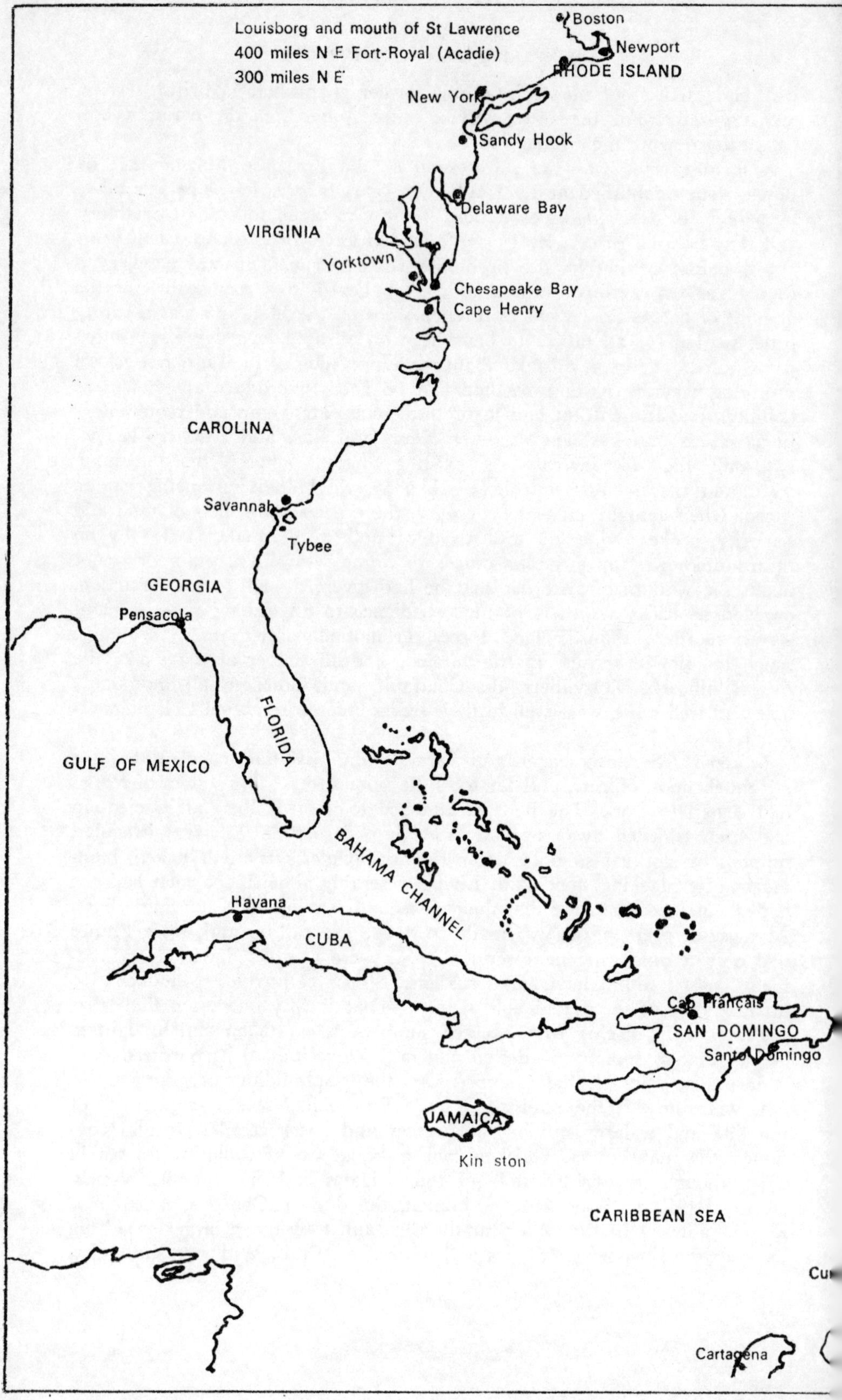
Louisborg and mouth of St Lawrence
400 miles N E Fort-Royal (Acadie)
300 miles N E
Boston
Newport
RHODE ISLAND
New Yor
Sandy Hook
Delaware Bay
VIRGINIA
Yorktown
Chesapeake Bay
Cape Henry
CAROLINA
Savannah
Tybee
GEORGIA
Pensacola
FLORIDA
GULF OF MEXICO
BAHAMA CHANNEL
Havana
CUBA
Cap Français
SAN DOMINGO
Santo Domingo
JAMAICA
Kin ston
CARIBBEAN SEA
Cartagena

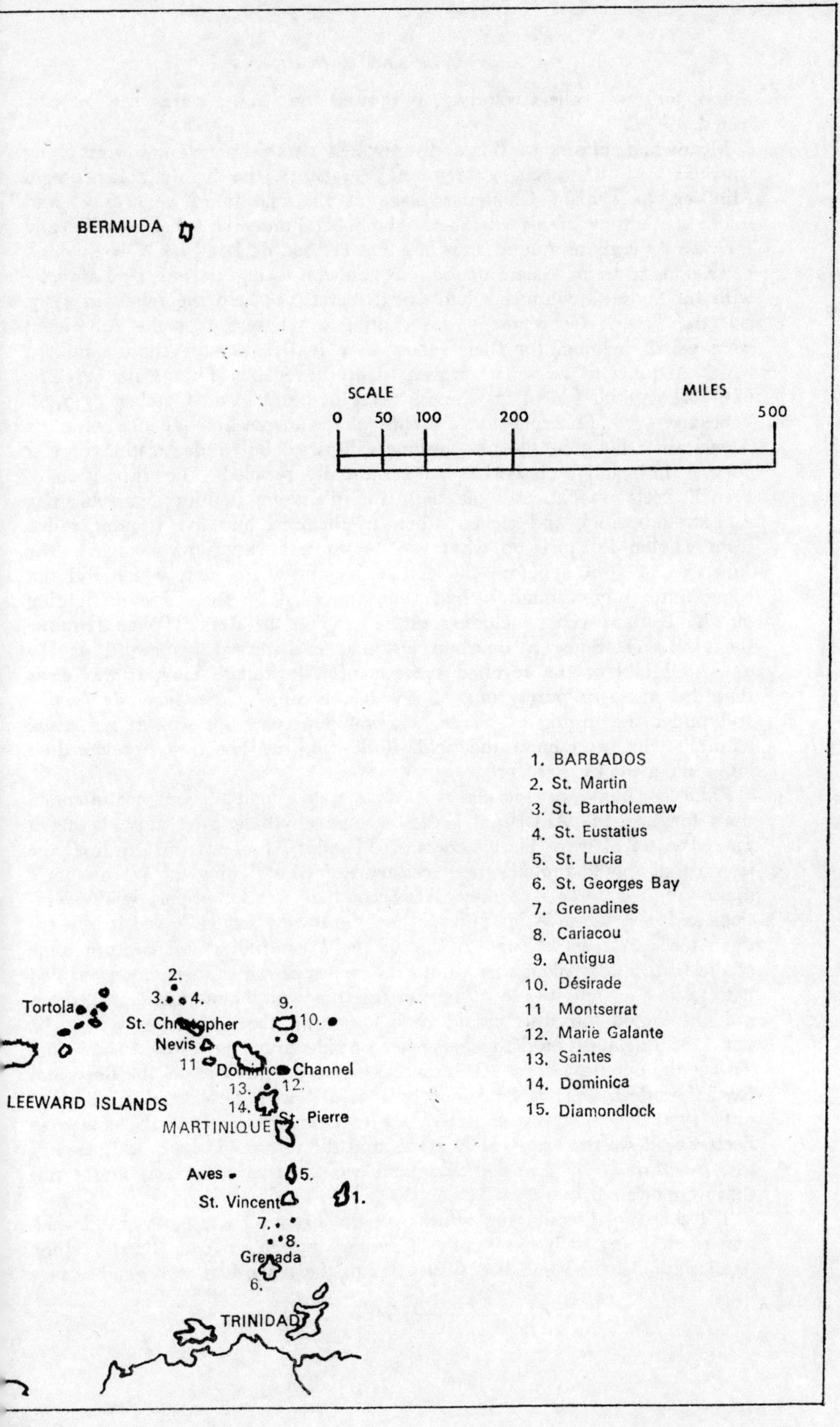
BERMUDA
SCALE
MILES
0
50
100
200
500
1. BARBADOS
2. St. Martin
3. St. Bartholemew
4. St. Eustatius
5. St. Lucia
6. St. Georges Bay
7. Grenadines
8. Cariacou
9. Antigua
10. Désirade
11 Montserrat
12. Marie Galante
13. Saintes
14. Dominica
15. Diamondlock
Tortola
St. Christopher
Nevis
Dominica Channel
LEEWARD ISLANDS
MARTINIQUE
St. Pierre
Aves
St. Vincent
Grenada
TRINIDAD

ashore and two others colliding, it showed that training was not yet perfected.

Meanwhile, affairs in England's revolted American colonies were being watched with the keenest attention. Vergennes, the highly able Foreign Minister, had taken Choiseul's place as the apostle of vengeance, and under the urging of his vitriolic anglophobia France was making all ready for war. As early as August 1775, George III had declared his New England possessions to be in a state of sedition, and the young marquis de Lafayette with his body of volunteers had sailed secretly to help the rebels in 1777. But the French Government was waiting to be sure that the Americans were worth helping, for their forces were really no more than a militia, which in previous wars had proved highly unreliable. The assurance came with the capitulation of an English army at Saratoga in October 1777. In February 1778, France made a treaty of commerce and alliance with the Americans which, in the circumstances, was all but a declaration of war on England, and ambassadors were mutually recalled. The disposition of French fleets was already planned, and d'Estaing (v-adm), with his flag in LANGUEDOC(80), and eleven others of the line, plus five frigates, sailed from Toulon in April on what was to prove a very slow voyage to the Americas. This was before the actual declaration of war, which did not come until July, though it had been preceded by some frigate fighting off the Brittany coast including that in which the BELLE POULE repulsed the *Arethusa.* Since a Bourbon autocrat had drawn his sword as the unnatural ally of the revolted subjects of a legitimate king, it was clear that the war was purely one of revenge. Spain, by the *pacte de famille* and under the urging of Vergennes, would prepare for war at her usual leisurely pace, and enter the field. Both countries hoped to recover their losses under the treaty of 1763.

France's grand strategic plan was, in a way, simple. It was not intended that, for her, the Americas should become anything even approaching a main theatre of war. The business of d'Estaing's fleet was first to land the troops that should provide the necessary trained stiffening for Washington's more irregular forces – as they did. Second, it was to support land operations as far as it was able. Third, when the stormy season forced it to leave the American coast, it was to sail to the West Indies and capture some of the British possessions there, until the return of the calmer season enabled it to resume its activities in support of the troops on the mainland. As France did not intend that this should be a long war, the main blow was to be struck at England herself, to bring her quickly to her knees by an invasion. To get the necessary force to obtain command of the Channel the Brest and Rochefort fleets were to be joined by that of Spain. This could not happen until 1779. Then, it was expected, the enormous strength of the combined fleets would see the army safely across and the war would be quickly ended; for, overthrown by invasion, England would have to accept any terms that were offered.

Till then no discouraging attacks on the French coast must be allowed, and to stop any such the comte d'Orvilliers put to sea from Brest in July. Aged sixty-eight, he had had fifty years in the navy, had seen good service in the previous war, had been early marked as an officer of considerable intellectual capacity, was known to be a skilful tactician and was generally liked and respected. With his flag in the BRETAGNE(110) he had thirty-two of the line, and a number of lesser vessels. Commanding the centre, under him, was the comte de Guichen (CE) in the VILLE DE PARIS(90): the van was led by du Chaffault (LG) in the COURONNE(80) with the vicomte de Rochechouart (CE) in the DUC DE BOURGOGNE(80) and the rear was led by the royal prince, the duc de Chartres (LG), with La Motte-Picquet (CE) in the SAINT-ESPRIT(84), having as second the comte de Grasse du Bar in the ROBUSTE(74). On taking over his command, d'Orvilliers had been sent a long letter by Sartine saying that the eyes of Europe would be on his fleet, that officers were to understand that the mistakes and misfortunes of the past were to be wiped out by dash in attack, that if any ship, forced on the defensive against odds, should strike before it 'reached the last extremity', the greatest royal displeasure would fall on its commanding officer, and (recognizing one of the great weaknesses of the service) that officers were to avoid disputation and obey the orders of their seniors instantly and unquestioningly. But now, on sailing, he was sent a shorter letter, instructing him to avoid risks. Though this doubtless had in view the need to keep the fleet intact for the great invasion campaign of the next year, it put d'Orvilliers somewhat in a heads-you-win-tails-I-lose position. The British Channel Fleet under Keppel was already at sea, twenty-one strong. Finding the French in such superior numbers, Keppel stood back for reinforcements, but was again off Ushant, thirty strong, before the end of July. When the fleets sighted one another there were a few days of foul weather and fog, during which the DUC DE BOURGOGNE and the ALEXANDRE got separated from the fleet and returned to Brest. So d'Orvillers lost his superiority of numbers, and, indeed, might be considered inferior in that three of his ships, being only fifty-gunners, were held to be not really stout enough to take their place in the line. In fact there was little real disparity between the fleets, because of the greater weight of metal, larger ship's company, and better design of a French ship compared with an English one of the same nominal strength.

The battle waged off Ushant was principally one of tactics. The wind was south-westerly, the French having the advantage of it, and both fleets were on the port tack. The French were in reverse order, that is, with the rear under the duc de Chartres leading, while Keppel's ships were beating against the wind in a general chase to force battle; but they came into line as they got into action. D'Orvilliers, thinking that Keppel intended a concentration on his rearmost ships, had made his fleet wear in succession so that the fleets cannonaded each other sailing on the same tack southwestwards. In this cannonade the French gunnery was the superior, Guichen's VILLE DE PARIS being notably effective. When he found himself drawing ahead of the English, who might thus concentrate on his rear, d'Orvilliers wore his ships again. The fleets were now passing and cannonading on opposite tacks, and as the division of the duc de Chartres drew clear d'Orvilliers signalled to him to wear yet again, to leeward of the English rear, so that, as the French centre and rear passed it, it would be caught between two fires. The signal was not obeyed. There was talk afterwards of cowardice on the part of the duc de Chartres and it is true that the prince was no great officer (nor, as he was the future 'Egalité', was he any great personality); but since he had La Motte-Picquet on board to advise him, and as he sent a boat (!) to make sure of the meaning of the signal, it is far more probable that, because it would have broken the sacred rigidity of the line of battle, he found the order incomprehensible. By the time he had made his inquiry the opportunity was lost.

Keppel had also attempted to bring his ships about to attack the French rear, but some of them had been too badly damaged in the rigging to make the complete alteration of course. This was particularly so in the case of the *Formidable*, flagship of Rear-admiral Palliser, and, because that admiral's division kept station on him, Keppel brought his other ships back to their original tack, and the fleets separated after some four hours' cannonading. Keppel stood away, and d'Orvilliers followed for a little; but by next morning the two fleets had lost touch. D'Orvilliers returned to Brest and Keppel to England where, by sneer and hint, he blamed Palliser for lack of victory. They were of opposite political parties, and there was a most rancorous and indecent squabble, for the English Parliament, like that of the 1960s, was one where it seemed that all were for the party, and none was for the state. Both sides claimed the victory. The battle had, in fact, been indecisive; but the French had had the better of it, for the damage to English ships and the number of English casualties were distinctly the greater.

D'Orvilliers soon put to sea again, and cruised mainly off the approaches of the Channel, but seemed to have an obsession against entering it, arguing that the prevailing westerly winds that made return after an action difficult, and the lack of any port that could receive even the large part of a fleet, made it unwise. It was hardly the right attitude for an admiral determined upon victory to take. He sighted no English fleet, and took only three inconsiderable prizes. With the approach of winter he came back to Brest.

Vice-admiral the comte d'Estaing had arrived on the American coast, at the mouth of Delaware Bay, in early July, having taken all but three months over this voyage from Toulon – a slowness which he attributed to the poor sailing qualities of some of his ships. Since France knew little of the Americas, or of the position there, his orders were vague: he was to do something positive to assist France's American allies, and then to go to the West Indies. This put all the responsibility on his shoulders, and he was not quite the man to bear it. He had been a dashing soldier and was a much better seaman than most of those who transferred from army to navy; and being also a thoughtful man, he realized that the Atlantic, with its varying winds and currents, offered not only a far larger but a far more difficult field for strategy than would have been presented by any land campaign. Also, coming into the service late, as a great noble and a pushful one, his rapid promotions, culminating in the creation of a third post of vice-admiral for his benefit, had caused him to be disliked. These things seem to have weighed on him so that he alternated short periods of dash with long periods of morose silence. When he came up against any frustration, such as the poor quality of American pilots, he would accept it rather than override it, wherefore his periods of dash usually petered away into caution. Since the Americans had no fleet, English naval forces on the coast were weak, just six ships of sixty-four guns and three of fifty. But his opponent, Lord Howe, who commanded them, was quicker and more enterprising.

D'Estaing had hoped to trap Howe's weak fleet in Delaware Bay, at whose head it was supporting the troops in Philadelphia; but the slowness of his passage defeated this scheme. For Howe, apprised of d'Estaing's sailing, had got his ships away just in time, to Sandy Hook, where he took up a skilful defensive position guarding the approach to the New York roadstead. D'Estaing followed him there but, though tides were high and Howe had taken in his lighter ships, the Frenchman accepted the word of his timorous American pilots that there was insufficient water, and made no attack. So he sailed further northward, to support an American attempt that was to be made on the harbour of Newport, in Rhode Island, which the British were holding. The entry to the roadstead was by channels between islands, and they were covered by batteries. D'Estaing forced the passage, four English frigates that were trapped were destroyed by their own captains, and the French prepared for a joint assault on the town.

Of course, the English Admiralty, at the news of d'Estaing's sailing, had sent out a reinforcing fleet, under Vice-admiral Byron. He was nicknamed 'Foul-Weather Jack', owing to his extraordinary ill fortune in running into storms, and he lived up to his tally, for his fleet was scattered, and some of its ships badly damaged, by a North Atlantic gale. One, however, struggled into Sandy Hook, whereon Howe at once went north after the French. Though he had now four more ships than formerly, including this seventy-four, Howe was still much the weaker. On his appearance off Newport, d'Estaing, who was in no mood to be attacked in harbour where fire-ships could be used, dropped the operation against Newport and stood out to meet him[1]. For three days Howe manoeuvred cleverly to keep the wind and avoid engagement by a much superior force; and then a gale scattered both fleets. The LANGUEDOC, MARSEILLAIS, and CÉSAR, the first two under jury rig and the third with its rudder carried away, had isolated engagements with weaker English ships which they met, and beat them off; and the battered fleet gradually reassembled off Newport. It was in no state to make, or resist, an attack, and so it stood away to Boston to make repairs[2]. Having spent two months in carrying out these d'Estaing sailed, in early November, for Martinique. He had taken, or driven ashore, several small English warships, but he had failed to execute the first part of his loosely worded orders – to strike some positive blow for the assistance of his American allies.

[1]The promptness of d'Estaing's decision, and his skill in getting his ships out of a tricky roadstead past batteries, were meritable; and it should also be said that his failure to take Newport was largely due to the tardy arrival of the United States militia, who were to co-operate.

[2]Nor were the disappointed Americans over-helpful in Boston – a French officer was killed by a mob.

In the West Indies, in the absence of any British force worth the name, the marquis de Bouillé, Governor of Martinique, had made a successful surprise attack on Dominica, which, lying between the French Islands of Guadeloupe and Martinique, was a nuisance in British hands. Rear-admiral Barrington, who had only two of the line, was merely waiting a reinforcement before making a *riposte*. That reinforcement was coming in the shape of five ships and some troop transports from New York under Hotham; and it is a poor reflection on French seamanship that he frankly outsailed d'Estaing, joining Barrington at Barbados. They moved at once to an attack on St Lucia, then a French possession, lying just south of Martinique. Its good and readily defensible roadstead was to prove an admirable fleet base. The British troops easily took the battery commanding it, and drove the small French garrison into the interior. Barrington then sent his transports to the back of the road and moored his squadron in a defensive position across its entry, throwing up a battery at each end. D'Estaing came bustling to the rescue. As he had twelve ships, of which only one was a fifty-gunner, while Barrington had only seven ships including three of only fifty guns, d'Estaing should have made an all-out assault on the British. But he did not like the look of the position and, after sailing past them twice and exchanging vigorous broadsides with them, he stood away to another part of the island and landed his troops. Their attack, which he led, failed expensively; and though one of his captains, Suffren, urged him to assail the ships again, rightly pointing out that nothing else mattered so much, he did not do so. The winds were not favourable, and he heard that Byron, having collected his ships and repaired their gale damage, was on his way from America. So d'Estaing returned to Martinique with another failure, and had a serious quarrel with the dashing and stormy Bouillé. St Lucia capitulated.

In Indian waters the French had only the BRILLANT(64), under Captain Tronjoly, and the POURVOYEUSE frigate(38), plus some armed merchantmen. At the outbreak of the war, English troops marched on the virtually defenceless Pondicherry, sending their artillery by sea in a small squadron of a fifty-gunner, a frigate, and some Indiamen. Tronjoly stood out to meet them. The two little squadrons, of which the French were rather the stronger, formed line and twice sailed past each other cannonading rather harmlessly at somewhat long range. Then the English, finding their road barred, stood away. When they returned, Tronjoly, ordered by the governor to put to sea, just sailed away to the Île de France, and Pondicherry fell in a few weeks.

The year had seen several fights between small ships, most of which went in favour of the French, who were usually the stronger. But, while the spirit of the service had been shown to be willing, things had somehow not gone quite right. This was disappointing.

The next year, 1779, was to prove more so. The main operation was to be the occupation of the English Channel by a grand combined French and Spanish fleet, under cover of which the troops were to be brought across to invade England; and France spent the early months of the year organizing this. Meanwhile, England was to be kept distracted by operations in the Antilles, and, if need be, the French fleet there was to give assistance to its American allies.

In the West Indies the year opened curiously quietly. Byron, after his usual battle with a gale, had arrived there from America in January, and had been mainly concerned with seeing to the safety of the trade (as any English admiral of those days had to be because of the strength of the 'sugar interest' in the Commons), and in bringing in safely a convoy escorted by Rowley. The arrival of Rowley's ships had made the English much superior in number, and d'Estaing had therefore remained quietly at Fort Royal, Martinique, limiting his activities to sending some of his small ships under Trolong du Rumain to take the little islands of St Martin and St Bartholomew. He was reinforced, first by five of the line under de Grasse (CE) in February, next, in April, by two of the line and some smaller ships under Captain the Marquis de Vaudreuil (who had brought out a convoy and, on the way, had sent his smaller ships to mop up the British trading posts on the Guinea Coast, from the Gambia round to the Volta), and finally, in June, by five of the line under La Motte-Picquet (CE). He had already, early in that month, seized St Vincent, which had been poorly defended, and on La Motte-Picquet's arrival, being now superior to Byron, he put to sea forthwith. (La Motte-Picquet gave himself high marks for being willing to sail without having a breather in harbour after his Atlantic crossing.) D'Estaing's objective was Barbados but, finding the winds contrary, he made for Grenada, landed his troops, and took it. Byron, then at St Christopher to the north, hearing of the island's danger, hurried along to save it, bringing a troop convoy with him, and not knowing, when he sighted the island, that it had already fallen.

D'Estaing had barely received the English surrender when he heard that Byron was only a day's sail away. He did not get to sea at once, for he feared to be carried to leeward away from his conquest; but he ordered his fleet to weigh during the middle watch. Getting a fleet of sailing ships out of harbour in pitch darkness can be a tricky business: the DAUPHIN-ROYAL(70) went aground and so took no real part in the action that was about to follow. When Byron and the French sighted each other, a little after dawn, the position was that, with an easterly wind, the French were heading roughly north-west, with their van and centre more or less in line, but their rear still in disorder. Byron, sailing almost due south on the port tack, was to windward of them and steering between them and the island. He had twenty-one of the line to the French twenty-four, though four of the French ships were weaker than any of the English. Seeing the disorder in the French rear the English admiral signalled for a general chase. The result was that Admiral Barrington with the three fastest sailers ran down on to the French rear far ahead of their fellows and thereby found themselves engaging a mass of eight ships, from whose superior weight of metal they suffered much damage. Three other fast sailers, getting too much to leeward, attacked the head of the French line, and, being on the opposite tack, passed down it, being hotly engaged and also suffering heavily in the process.

For the moment Byron did not mind this result of an over-precipitate

# GRENADA 1779

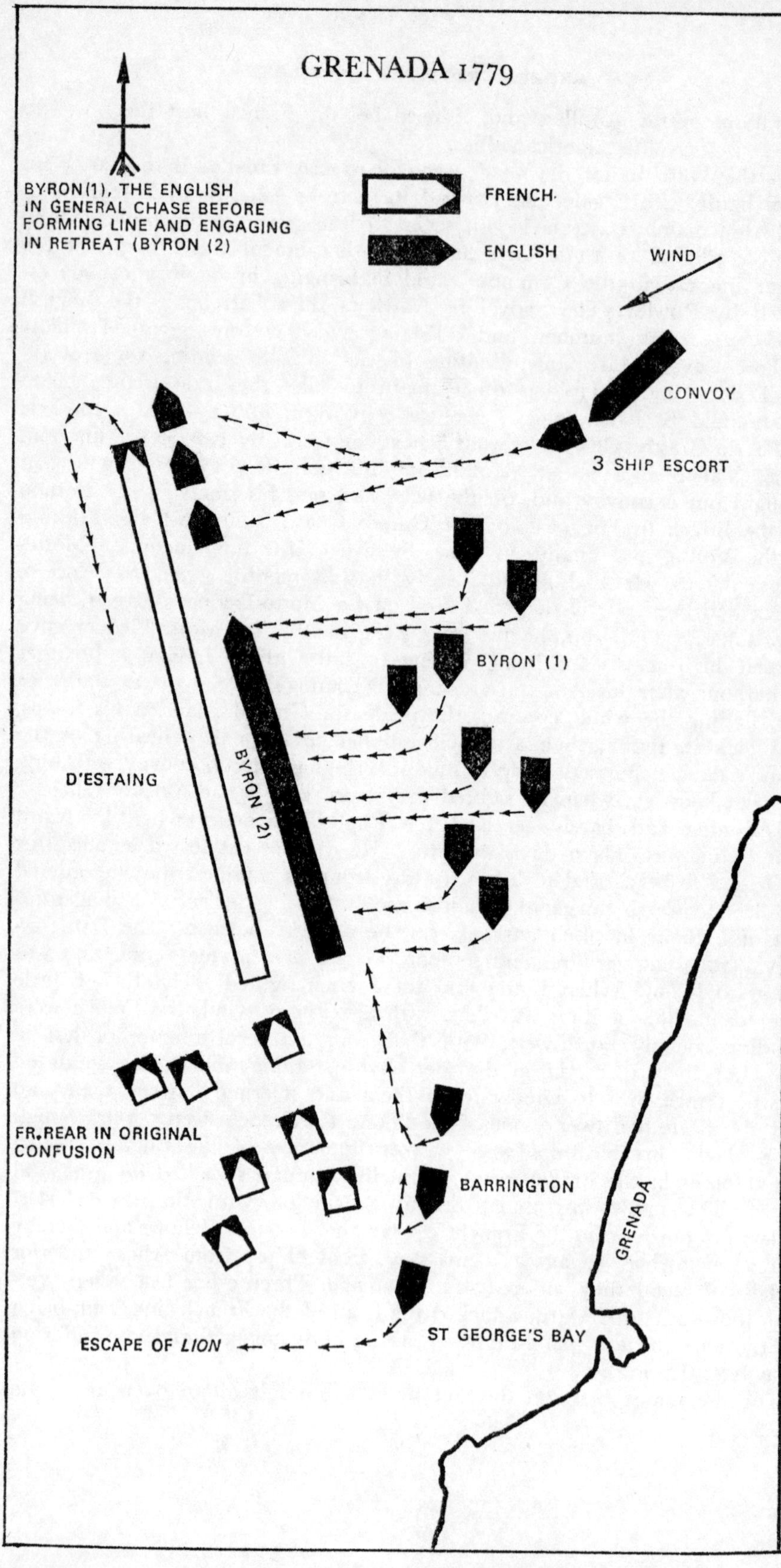

attack, for he believed that the English were still holding out in Grenada and that, by getting between the French and the roadstead, he would enable them to continue to do so while, by bringing in his convoy and landing its troops, he would capture the French forces on the island. But when he opened up St George's Bay and saw the French flags flying on the fortifications, he realized that he was all but in a trap between the guns on land and the fleet to his lee. He promptly signalled to go about and to form line on the starboard tack, while calling into line the three ships that he had left guarding the adjacent convoy. Two of these courageously took the shortest route and came in at the head of the English line, thereby preventing any doubling on it by the French, but coming under heavy fire from their leaders, Comte Barras de St Laurent (CE) in the ZÉLÉ(74) and Suffren in the FANTASQUE(64), whereby one of them, the little *Monmouth,* was virtually crippled. Both fleets were now heading northward, cannonading each other on the starboard tack; but several of Byron's ships, including the three rearmost, were partially dismasted and were dropping astern.

D'Estaing now brought his ships about together on the other tack but, to keep his line in order, formed it on his most leeward ship, whereby the fleets were mostly, for the moment, out of each other's range. Byron also went about, to succour his rear ships, but, finding some of his fleet too badly damaged to make the alteration of course, reverted to his northerly course on the opposite tack to his enemy. The game was now in d'Estaing's hands, if he cared to play it. The rearmost English ship, the *Lion*, was too crippled to have any hope of keeping in touch with its fleet, and was running westward before the wind under such sail as it could muster. To detach a single ship of equal strength in pursuit would have seen it easily overtaken and outfought; but no ship was detached : the doctrine of preserving the line in the presence of the enemy was sacred, and the *Lion* was allowed to escape. He could have borne up to cut off the damaged ships in the English rear, with the certainty of taking them unless Byron had come to the rescue by renewing the general action. This Byron would probably have done, but as, with so many ships damaged, he would have been fighting against odds of roughly two to three, a French victory, even at the cost of a few ships, was almost certain. D'Estaing thought it 'unwise' to take the chance; so he made back to St George's Bay, while Byron got his dismasted ships away northward – to his relief and rather to his surprise.

Probably d'Estaing was actuated by his sense of his 'mission', that strategical obsession of most French commanders in the days of sail : a fleet was really an adjunct to land forces : it had a mission, perhaps to escort a troop convoy and to support the troops in an attack on an island or a coastal place, perhaps to cut off an enemy convoy and so injure the trade by which the enemy lived, perhaps just to thwart an enemy objective. It never seems to have been realized that, if the enemy's fleet was utterly defeated, even at heavy cost, the remaining victorious ships would have such complete control over an ocean area that any 'mission', even if temporarily delayed, could be carried out with ease. So the battle of Grenada, while certainly a victory, was a singularly incomplete one. All that came from it was that a frigate was able to take one of Byron's convoy containing a company of soldiers, that the not over-important island of Grenada was secured, and that Suffren in the FANTASQUE and d'Albert de Rions in the SAGITTAIRE could be sent to take Cariacou and the little group of islets called the Grenadines without interference. As Suffren wrote after the action: 'If only M d'Estaing were as able as a naval officer as he is brave as a man....'

A month before this battle the main French fleet had left Brest to set in motion the great machinery of the invasion of England, but it will be convenient for the reader first to follow the fortunes of d'Estaing. La Motte-Picquet had brought him orders to leave de Grasse with eight of the line at Martinique, and La Motte-Picquet himself with three of the line at San Domingo, and to bring the rest of his ships, the original Toulon squadron, back to France, with all the trade that was ready for escort. But things had gone badly on the mainland. The English had overrun Georgia, had taken Savannah, and were threatening Carolina. Both the Americans and the French consul to them were appealing frantically to him for help. So d'Estaing rightly broke his orders and sailed for the mouth of the Savannah river. Here he was much handicapped by bad weather, lack of knowledge of local geography, and incompetent pilots. But he seized the key island of Tybee in the river mouth, eventually got his troops ashore, and himself, as was his custom, led the assault on the town. It was a failure and he was wounded. So he sent the troops back to Martinique with de Grasse and sailed for France with the original Toulon fleet as he had been commanded. A storm scattered the squadron and damaged several ships. The TONNANT got no further than Havana, the rest reached France in penny numbers and in various ports. D'Estaing's two years' vigorous campaigning[1] had resulted in the capture of a few of the less important West Indian islands, in the barren victory off Grenada, and in a failure to give any effective support to land operations in America.

We must now turn back a little to all that was happening in European waters. The plan for the invasion of England was comparatively simple. Two armies, each of twenty thousand, were to be assembled with their transports, one at St Malo, the other at Le Havre. D'Orvilliers was to take the main French fleet from Brest, join the Spanish fleet, and the combined force of over sixty of the line, which would give them odds of three to two over any force the English could put to sea, was to take command of the Channel. The troop transports were then to unite north of Cherbourg and be escorted to land the soldiers on the Isle of Wight and round Portsmouth, destroying the English naval base in the Channel preparatory to a march on London. This had been decided as the result of much planning over a long period by more than one group, to say nothing of numerous individual suggestions. The most notable of these groups was that led by the comte de Broglie which, in Louis XV's time, had corresponded with the

[1]To be precise, the Toulon squadron had been at sea for twenty months. Up till then this was the longest time that any French fleet had spent at sea without returning to base. In his formal report d'Estaing asked specially for some official recognition to be made of the ability and services of La Motte-Picquet, d'Albert de Rions, and Suffren. He seems at least to have been a good judge of men.

king behind the back of the minister, as was often Louis' custom (see p 145). Vergennes had combined the ideas of these groups but, though all of them had contained a naval officer or two (including a renegade British lieutenant, named Hamilton, who was given a French naval commission as captain), the attention given to the operations of the troops was greater and more careful than that given to the naval side[1]. Here the difficulties that could arise from the vagaries of wind and tide were largely ignored. So were the possible troubles of combined operations with allies, who seem to have been optimistically presumed to be at least nearly as efficient as the French. Yet, granted promptness of execution, and reasonable fortune with weather, the plan was perfectly feasible.

The whole machinery was to have been set in motion in May but, thanks to the haughty slowness of the Spaniards, arrangements took some time to concert, and it was not till June 3rd that d'Orvilliers set sail. Even so, he was hurried to sea with his ships somewhat short of supplies[2] and undermanned. For it was feared that the English, whose Channel fleet was stronger than that of the French alone, would establish a blockade of Brest. Their government was no longer that of a Pitt, and they made no attempt to do so. So d'Orvilliers sailed for the Sisargas, islets off the Spanish coast near Corunna, where the junction of the fleets could be made unobserved and undisturbed. As usual, the Spaniards were late. Indeed, they were extremely late and d'Orvilliers had to wait six weeks before they were fully mustered. Having arrived, they proved to know nothing of French squadron arrangements, which had been carefully drawn up for them months before, and a further precious week of summer weather had to be spent in exchanging signals and devising formulae whereby the proud Spaniards should be given orders by the French without being given orders by the French. For they disliked the French, felt that Gibraltar should be their first objective (despite their government's acceptance of the invasion scheme), and their admiral, Don Luis de Cordoba, who was no less than seventy-three years of age and a negligible person apart from his hauteur, was anything but co-operative.

The difficulty of prestige was overcome by forming a main fleet of three divisions, each of nine French ships and six Spaniards, a light division of three French and two Spanish fast sailers, and a so-called 'corps of observation', sixteen ships, all Spanish, and devised so as to preserve the pride and dignity of Don Luis by giving him what was apparently a separate command till battle should be joined[3]. Luc Urbain de Bouëxic, comte de Guichen (LG), who had a great reputation as a tactician, commanded the van of the main fleet, with his flag in the VILLE DE PARIS(100), d'Orvilliers, with his flag in the BRETAGNE(110), commanded the centre, and the rear was under Don Miguel Gaston in the *Rayo* (80), d'Arzac de

[1]Patterson, *The Second Armada* (Manchester University Press), gives the arrangements in detail.
[2]In particular and, as it proved, disastrously, they were short of doctors and of medicaments.
[3]It was also intended, if the British Channel Fleet should prove very weak, to detach these ships to block the Bristol Channel and destroy the trade there.

Ternay (CE) in the SAINT-ESPRIT(80) being the senior French officer: the light division was led by La Touche de Tréville in the COURONNE(80). With sixty-six of the line (thirty French and thirty-six Spanish) and its attendant frigates and lesser vessels, the whole armada was over a hundred sail. But during their seven weeks' wait at the Sisargas not only had the French cut heavily into their supplies of water and provisions, but the bad living conditions aboard had brought a serious outbreak of scurvy and smallpox, and of putrid fever. D'Orvilliers was heart-broken by the death in his arms of his only son, an officer in the flagship[1], and, at the age of sixty-nine and brought up in that school which had concentrated on repelling the enemy rather than destroying him, he was not quite the man to put the necessary drive into a fleet in the state in which were the allies as they neared England. For the powerfully built Spanish ships were, for that very reason, heavy sailers and, as they had poor crews, the passage across the Bay of Biscay had been a slow one.

So it was August before the great fleet entered the chops of the Channel. It had lost the favourable winds, and now met the August gales, with intervening calms. The scurvy and fever were now raging. So many had died or gone sick that there were barely the hands to work the ships, let alone to fight them. A shortage of water was seriously threatened, and one of provisions lay not far behind. The mourning d'Orvilliers, daily receiving increasingly distressing signals from his captains[2], in his turn sent by dispatch boats increasingly urgent and pitiful letters to the Minister, begging for supplies and, above all, for seamen to make good his appalling losses from illness. None came. He had got up as far as Plymouth, but was blown back again to the Atlantic. In mid August came a change of plan. He was to get possession, possibly of Plymouth, certainly of Falmouth, and bring the troops across to occupy Cornwall and winter there, to complete the invasion the next year. It was an absurd plan, and d'Orvilliers dared not attempt it with a probable English fleet on his flank, and his own ships in their parlous state. However, once again he struggled up to Plymouth only to be blown back. On August 25th he held a council of his flag-officers and it was decided to lie off the Scillies in the hope of engaging the English fleet but that, in any case, the provision and sickness position was such that the fleet must set sail for Brest on September 8th.

The British fleet, forty-two strong, did appear, under the veteran admiral Hardy. Britain was so riven by politics that Howe and other senior flag-officers had almost treasonably refused to serve under the existing government, and Hardy, who had not been to sea for many years, was brought from the command of Greenwich Hospital as likely to be acceptable to officers of all factions. This was true, but he was excessively cautious and pursued a policy of retiring before the French with his fleet in good order. D'Orvilliers tried to get up to him to engage, but, what with the clumsiness of the Spaniards and the short-handedness of the French, his fleet straggled badly, while Hardy withdrew even to Spithead before the odds against

[1]'Le Seigneur m'a ôté tout ce que j'avais dans ce monde,' he wrote.
[2]One or two French ships reached a stage when they dared not anchor because they had not enough fit men to weigh.

him, to the annoyance of his flag-captain, Kempenfelt, and some others. All the French could achieve was the capture of a lone ship of the line, the *Ardent*(64), which was skilfully out-manoeuvred and out-fought, mainly by two frigates.

On September 3rd, d'Orvilliers received orders to return to Brest, and he reached there on the 11th. He landed some 8 000 sick! Utterly downcast by the failure of the great campaign, and finding that he was reproached for it, he resigned from the Service, to the regret of his officers. He wrote that his fleet had 'made the proud English tremble'. *Ridiculus mus;* for the sole results of the great expedition had been to cause a lively, but temporary, panic and the capture of one small ship of the line and some merchantmen. This was no sort of return for a stupendously arranged campaign that was to have overthrown a nation. The queen, Marie Antoinette, had seen more clearly when she wrote to her mother: 'We shall have spent a lot of money to no purpose'.

With whom lies the blame for this great fiasco? A little, perhaps, with d'Orvilliers for not making some firmer attempt to seize the initiative; far, far more with the Spaniards whose dilatory arrival had caused the French ships to get into their desperate state of sickness and shortage of supplies and had made their arrival in the Channel so late that the fine weather had ended; most of all with the French planners and administrators. These had sacrificed the valuable assets of time and mobility in favour of a mass of force, and had made no arrangements for supplies and reliefs if things should not work out strictly to plan. The fleet was operating for a month within a few leagues of the French coast, yet no replacement of seamen or supply of food and water was brought to it, despite its agonized appeals. It was intended that the return of the fleet to Brest should be merely a temporary business of shipping stores and some fresh hands, and that then the invasion plan should be executed. It was found that the size of the sick list and the number of deaths made this impossible. So the idea was abandoned for good, though it was the one most likely to give France complete victory; instead, it was resolved to concentrate in Europe on the capture of Gibraltar, for which France had already pledged herself to Spain, and elsewhere on aid to the Americans combined with the expulsion of the English from the Antilles.

During the year an attempt to seize the Channel Islands had been foiled by bad weather, and there had been a number of single ship actions, usually ending in favour of the stronger vessel, in which the French had lost rather more than the English. But the capture of the *Ardent* by two frigates had been a good performance; so had been that of the frigate CONCORDE which, in a storm-battered condition that had caused her to jettison twelve of her thirty-two guns, yet fought off an attack by an English thirty-two-gunner. As against that, when two frigates and three smaller craft were chased into the little bay of Cancale by superior English forces their crews seized the boats and abandoned their ships against the orders of their officers, who could not even fire their vessels to keep them from capture because they had no means of getting away their sick and wounded. There was, too, an instance of the chivalry of war in a fight between the

SURVEILLANTE(32) and the EXPEDITION(10), and the thirty-six-gun *Quebec* and the ten-gun *Rambler,* off Brest. The two frigates had so fought each other to a standstill that the whole of their masts came down, almost simultaneously. As the *Quebec* had the wind her wreckage fell between her and her opponent whose shot set it afire. With all her officers but an ensign *hors de combat*, the SURVEILLANTE had the greatest job to get clear before the *Quebec* blew up. At this both cutters ceased their battle and came to the rescue of survivors. Between them, and such help as the battered SURVEILLANTE could give, they picked up eighty-three men.

At the very end of the year a somewhat unusual action took place almost in the roadstead itself of Fort Royal. La Motte-Picquet, who had brought three of the line to Martinique from the failure of Savannah, was really caught napping through having put out no vedettes; and found a convoy from France, with only frigate escort, being chased into the very roadstead by Hyde Parker's squadron. However, the French CE got the canvas on to his ships with commendable speed and, standing boldly out to engage his superior enemy, succeeded, in a fight of four hours, in getting nearly half the convoy into safety. It is not often that a naval action is watched by a cheering population! The skilful gallantry of the French ships was admirable, and conveniently obscured the fact that they had been so casually unalert that half the convoy was lost: nor did La Motte-Picquet, who was as boastful as he was vigorously active, call attention to that in his report.

Early the next year, off the coast of Portugal, the PROTÉE(64) was taken, sacrificing herself to save her convoy; and a few months later France got ample revenge when the ships of de Beausset (CE), in conjunction with some Spaniards, seized the whole of a large but inadequately escorted British convoy. De Beausset was assisting the Spaniards in their rather ineffective blockade of Gibraltar, but it is not there but in the West Indies that interest chiefly lies for most of the year. La Motte-Picquet, who had sailed to Cap Français (San Domingo) in March to pick up a convoy, with his flag in the ANNIBAL(74) and three others of the line, sighted and pursued three weaker liners under Commodore Cornwallis. Only his own ship was sufficiently well sailed to come up with them, and it was therefore roughly handled till, the others approaching, Cornwallis stood away again. The ordinary ship-mastery of the French captains clearly left much to be desired. The other French ships that had been left in the Antilles remained at Martinique, ready to attack St Lucia, but awaiting the arrival of Guichen (LG) who was bringing sixteen of the line from Brest, and was to take command. Arriving, Guichen landed his sick in Martinique, and then sailed forthwith to take St Lucia with twenty-two of the line. He found that Hyde Parker had drawn up his sixteen liners across the entry to its bay in a strongly defensive position, judged it unwise to attack, and returned to Martinique. He would probably have gained a victory, but only at a cost that would have prevented him from carrying out any further major operations.

Hyde Parker was awaiting the arrival of his superior, Rodney, who came in a few days later, having relieved Gibraltar on his way. Of all the British admirals of his day Rodney was by far the most able tactician – only Howe even approaching him – and Guichen occupied the same position among French flag officers. The difference between them was that Guichen's tactics were based mainly on the defensive, those of Rodney being designed to obtain a favourable position for attack. Some complicated manoeuvring was therefore to be expected, and indeed took place. The fleets met in April, nearly thirty miles west of the Dominica Channel, when Guichen had sailed to cover a convoy going to San Domingo. Guichen had twenty-two of the line, his flag being in the COURONNE(80), his van commanded by the comte de Grasse (CE) in the ROBUSTE(74), and his rear by the comte de Sades (CE) in the TRIOMPHANT(80). Rodney in the *Sandwich*(90) had twenty-one of the line, and under him were Rear-admirals Hyde Parker, in *Princess Royal*(90), and Rowley in *Montagu*(74). After some manoeuvring, he obtained the advantage of the wind, which Guichen probably yielded to him without regret since French tactics were based on defensive fighting from leeward, and the fleets found themselves sailing northward, both on the starboard tack with the usual easterly wind of the neighbourhood. The French drew somewhat ahead, whereon Rodney signalled to alter course together eight points to port and engage. This meant that he was about to effect a concentration on the rear half of the enemy fleet. Guichen promptly saw the danger, and wore his ships all together, which would bring them to a position whereby, by the time the English were in medium to close range, their attack would fall on the whole line. At this Rodney, thwarted, altered a further eight points to port, so that both fleets were now standing southward on the same (port) tack.

The French line began again to stretch ahead, with rather large gaps between the ships. Rodney signalled again to bear down and attack, whereby he would have effected his desired concentration, this time on the French centre and rear, by cutting through astern of the French van. Guichen gave himself up for lost, and, like the chivalrous gentleman he was, wrote to Rodney after the action to say so. He was saved because the captains of the leading British ships, devoted to the line of battle, misunderstood the signal and made sail to stretch ahead and engage the French van. The result was that close fighting was limited to the rear and, especially, the centre, where the *Sandwich* made a determined effort to break the French line and was so damaged in the process that Rodney had to shift his flag. Thinking his rear was now going to be cut off Guichen had again signalled to wear together, but had then cancelled the order as the English failed to break through. Some ships had executed it, and thereby had gone to leeward, whereat Guichen signalled to re-form line on them. This broke off the action and the fleets separated, each to make for harbour and repair damages, which were rather the heavier on the English side because they had been exposed to broadsides as they ran down. Rodney reproached his van captains for their lack of intelligence, while Guichen realized that he had met a foeman fully worthy of his tactical skill, and that, despite common French practice, he would be wise to deny him the wind in any future engagements, so as to limit his chances of attack.

There were to be two such, within a few days of one another, in May. Both fleets had got quickly to sea after repairing damages and landing wounded, and each had been reinforced by a ship of the line, so that the French retained their slight superiority of numbers. Guichen had the wind when they met, and Rodney, in several days of tacking and retacking, could not manoeuvre him out of it; but he always avoided being brought unfavourably to battle. On the 15th, each admiral thought that he saw a favourable opportunity and they closed. But some shift of wind altered the position and brought a certain amount of confusion into both lines, so that only the two van divisions were at all seriously engaged, and that for no more than an hour, at which Guichen, fearing that a further shift of wind might expose his rear ships, signalled to go about in succession. This brought the fleets on to opposite tacks and ended the firing.

For four days the fleets continued to try to get the better of each other, but Guichen succeeded in keeping the advantage of the wind. By the 19th, they were 120 miles west of the Windward Islands, when their twisting and turning brought them so close that action could hardly be avoided, and Guichen signalled to go about and attack the English van. The French found their enemy bearing up against the wind with a threat to cut off their leading ships, and so paid off on to a parallel but opposite course to him. The three-hour battle therefore consisted of the fleets passing on opposite courses and cannonading vigorously. At one stage, Rodney threatened a concentration on the French rear, whereat Guichen came about with his van and centre to support it. The English resumed their former course and the fleets again separated. Both had suffered a good deal and, as the *Cornwall*(74) could not be kept afloat after reaching St Lucia, the French could claim to have had the better of it. In fact all three battles had been indecisive, and had settled nothing.

A fleet of twelve Spanish ships, under Solano, was expected, and the fleets again put to sea, the one to intercept it, the other to effect a junction with it. As it pursued a much more northerly course than usual the French, who knew of this, were able to meet it without interference. But this served no purpose, for Solano was under orders to go to Cuba, and in any case the Spanish ships were in so poor a state, and had so heavy a sick-list, that they were in no sort of fighting trim. So Guichen had to abandon his intentions to use what would have been a great superiority of numbers to attack and take St Lucia and Barbados as he had wished. He sailed with the Spaniards as far as San Domingo. There he put in to Cap Français, collected La Motte-Picquet who lay there, and, being under orders to return to Europe before the hurricane season, which was nearly due, set sail with all but ten of the ships of the line. As de Grasse, who should have remained in command of them, pleaded ill-health, de Monteil (CE) was left in his place. He spent the remainder of the year doing nothing. Rodney also left ten ships behind under Rowley and took the rest to New York. Guichen's handling of his fleet in the season's campaigning had been highly skilful, and he was proud of it; but it had also been almost completely defensive, and therefore negative. He had gained no victory: he had made no capture of territory: he had merely denied victory and capture of territory to Rodney. It is not by such tactics that wars are won.

A little before he had sailed for Europe there had been an example, to the north of him, of a similar failure to turn defence into advantageous attack. The inadequate quality of American troops had shown that such French forces as were helping them on land needed to be strongly reinforced if victories in America were to be won, and with some difficulty transports had been found to take out an army of 6 000 under Rochambeau. They sailed from Brest early in May under the escort of eight of the line (one armed only *en flûte*) and some frigates. In command was the chevalier d'Arzac de Ternay (CE), who had a considerable reputation because of his success in getting the stranded ships out of the Vilaine after Conflans' defeat at Quiberon. His flag was in the DUC-DE-BOURGOGNE(80). Their crossing was slow because of the difficulty of keeping the transports together, and late in June, some way off Bermuda, they ran into a division of four of the line and a fifty-gunner under Commodore Cornwallis. Both squadrons formed line of battle, running before an easterly wind on the port tack. They opened fire at 5.30 pm and, as the leading French ship drew ahead, the French went about in succession so that they now passed and cannonaded the British on the opposite tack till 7.00 pm. Though the British were much the inferior, and one of their ships, the *Ruby*(64), had dropped astern of the rest so that she, at least, could readily have been overwhelmed, de Ternay discontinued the action. Getting his convoy safely to harbour was, he felt, more his concern than the destruction of an enemy squadron. He took it into Newport (Rhode Island), where he put his ships in a strongly defensive position and refused to budge without reinforcements.

Meanwhile, the siege of Gibraltar had been continuing, vigorously on the part of the defence, rather lackadaisically on the part of the allies, though de Bausset's French ships had been reinforced. As it dragged on the Spaniards agreed that d'Estaing should come and take overall command. He arrived in October. So, at Cadiz, did Guichen, La Motte-Picquet, and de Grasse, directed thither with the fleet that they were bringing home from the Antilles together with their convoy of the West Indian trade. This was sent on, under Suffren's escort, into the Mediterranean. D'Estaing now had forty ships under his command. But he rightly considered a purely, or mainly, fleet assault on the Rock to be impossible, everything being in favour of the defence. Many of the ships from the Caribbean, too, were in poor shape, for the new copper sheathing was still fitted only to a proportion even of the English ships, and as yet to but few French ones, so that several ships' bottoms were in faulty state indeed. He was therefore ordered to bring the whole lot home, and reached Brest, without sighting any enemy, at the turn of the year.

He put in a report unfavourably comparing the manning of French ships, and the supply of ammunition to them, and their care of their small arms, with those of the British. This went to a new minister, for Sartine had been dismissed in October. Necker, a Genevan-born banker, had been made Director-General of French finances in 1777 and was endeavouring to get France's hopelessly chaotic and desperately overstrained fiscal system into some state of order and soundness. He and Sartine happened cordially to dislike one another, so that when Necker discovered that Sartine had pledged the Treasury's credit for some heavy naval expenditure unknown to him and contrary to a decree that he had issued his anger was extreme, and Sartine had to go. He had been a good Minister of Marine, though not a faultless one, for it was somewhere in the administration – it has never been discovered precisely where – that the breakdown occurred which had left d'Orvilliers in the Channel without supplies or replacements for his depleted crews in the previous year. Yet he had done the Service well, and it was sorry to see him fall. The opportunity was not taken to fill his place, at last, with a suitable naval officer, and the post was given to a M de Castries, a soldier. Nevertheless he was to prove an able and forceful administrator; and he received his marshal's baton at the end of the war.

Of course a year containing so much hither and thither upon the sea had seen a number of isolated combats, mostly between small ships, and mostly ending in favour of the stronger. This usually happened to be the enemy, but the English had lost a number of lesser vessels by wreck and accident.

In the next year, 1781, events pursued a more definite course, with much more important results. When the campaigning season started the French dispositions were as follows: Holland had entered the war against England, and a squadron of five of the line was shortly to leave Brest, first to reinforce the Dutch in the Cape of Good Hope, and then to continue to India. It was under Captain the commandeur de Suffren St Tropez. (Though commonly known as 'le bailli', of which title he was proud, Suffren had not then been promoted to that higher rank in the Knights of St John.) As the Indian Ocean forms a quite separate field of operations, we will follow his fortunes later. De Monteil (CE) was still in the West Indies with his small squadron, and it may be said, now, that he did nothing with it except, in May, to support a successful attack by the Spaniards on Pensacola, then the key point in Florida. In North America, at Newport, Rhode Island, were the seven of the line which had brought out Rochambeau's army. They were under Captain Destouches, who had succeeded to their command on the death of de Ternay (CE) at the close of 1780. A further liner was to come to them from Brest, and de Barras St Laurent (CE) was to take the command. The comte de Grasse (LG) was shortly to leave Brest with twenty of the line for the West Indies and, when his ships had cleared, nineteen others were to be fitted out under Guichen to support the Spaniards at Gibraltar. As to the English, they had in North America a squadron at New York under Arbuthnot of about the same strength as that of Destouches, and Rodney had already reached the West Indies to be followed and joined by Hood. The Channel fleet, under Darby, was preparing to carry out a further relief of Gibraltar. It will be noticed that both sides had dropped any idea of settling the naval war by a decisive action near home though the allies were to toy with it, in a half-hearted way, later in the year.

The first activities were on the North American station. Destouches had sent a sixty-four and some smaller vessels to deal with English commerce raiders off the Chesapeake and they had captured the forty-four-gun *Romulus*. In March, at the request of Washington, he sailed with his whole squadron from Newport for the Chespeake, carrying troops who were to stop the English depredations in Virginia. Arbuthnot followed him from New York and, outsailing him, cut him off as he approached his destination. The two squadrons were of roughly equal strength, seven of the line with some lesser vessels, and their battle followed the usual course. With both heading east, away from the shore, Arbuthnot attacked from windward. As he did so, Destouches brought his line on to the opposite tack, heading for the bay. Arbuthnot followed suit. After a hot cannonade Destouches went about again, heading out to sea and neatly giving the leading English ships a concentration of fire as his line passed them. Three English vessels suffered much damage, but so did two of the French, and Arbuthnot was still between them and the Chesapeake. So Destouches gave up the attempt to get into the bay and returned to Newport. At Versailles he was considered to have done poorly, and it is typical of the French service of that day that some of his officers complained vigorously when no rewards and commendations were issued for the action.

Holland's entry into the war aided the French in an unexpected way; for Rodney, when he heard of it in February, swooped on the little Dutch West Indian island of St Eustatius. During the war this had become a storehouse of valuable (and often contraband) goods. The English admiral, whose private fortunes were at a low ebb, put the replenishing of his pocket from this vast prize booty above his duty to his country. Keeping several of his ships with him while he supervised its loading into merchantmen and made arrangements for their convoy, he sent only the greater part of his fleet under Hood to intercept de Grasse. Moreover, he lost his normal tactical grip by insisting that they should blockade Martinique, where the offshore winds are tricky, instead of allowing Hood to keep east of the island where the reliable trades would have enabled him to lie athwart de Grasse's path. That officer, who, handling his fleet and convoy vigorously, had made an expeditious crossing, was therefore able to get his convoy into Fort Royal past an almost immobilized enemy. There was an engagement, for de Grasse stood out to sea to cover the convoy's entry, but it was one in which the fleets cannonaded at almost maximum range on opposite tacks and was largely a waste of powder and shot. As the *Russell* sustained bad damage and de Grasse, joined by the four of the line that had been under the guns of Fort Royal, was markedly his superior in numbers, Hood sailed to rejoin Rodney. So de Grasse had safely brought in his convoy and picked up the ships on the station, and the seas were open to him for further operations. He made no great use of his opportunities, merely joining with the energetic commander of the troops, Bouillé, in an abortive attack on St Lucia and a successful one that took Tobago, which was weakly held. He deliberately avoided action with Rodney, who had hastened to Barbados on hearing of his arrival.

In European waters, Vice-admiral Darby, in April, had carried out a relief of Gibraltar, without interference thanks to Cordoba's passivity and the poor state of the Spanish fleet, and had sent on reinforcements to Minorca. For England, fighting the three other maritime powers, was strained to her limit to find ships to contain them, and could not attempt to control the Mediterranean, but merely to hold her posts there. The French had one great piece of fortune, however, for in May La Motte-Picquet, who had been sent for the purpose with six of the line and some frigates, succeeded in cutting off the convoy that was bringing home Rodney's booty from St Eustatius. Its escort of two liners and two frigates under Commodore Hotham got away with some of the ships, but La Motte-Picquet took eighteen of them and brought them safely into Brest, despite an attempt to intercept them by Darby, who had heard of the disaster while returning from Gibraltar. (This did lead to a fierce, and indecisive, fight between two outlying vessels of each force. ACTIF and *Nonsuch.*) The recovered booty amounted to almost five million livres!

In June, Guichen sailed from Brest for Cadiz with twenty-four of the line. His flag was in the BRETAGNE(110) and La Motte-Picquet was his second-in-command in the TERRIBLE(110). With extraordinary forbearance to a dilatory ally, France had agreed that he should be under the command of the very elderly and extremely inept Cordoba. (This was possibly because Spain had formally complained that the French fleets had not stopped the relief of Gibraltar – though she said nothing of the inactivity of her own fleet, which had been almost on the spot at Cadiz.) The combined fleet first saw an expedition against Minorca safely into the Mediterranean. Though this had a French commander, the duc de Crillon, the ships and troops were nearly all Spanish. In the absence of British ships from the Mediterranean Crillon had no difficulty in landing his army, but it was not till more than six months later that the British garrison surrendered. The huge allied fleet, forty-nine sail of the line, had made for the English Channel. It is hard to say what was the positive object of this great effort; for, as at Beachy Head ninety years before, the huge fleet might take command of the Channel, but there were no forces ready to follow it with an invasion. Perhaps it was hoped to gain such a considerable naval victory that invasion the next year would be easy, or that the threat of it would bring an advantageous peace. The much strained British had found it hard to collect a fleet of thirty of the line, which lay in Torbay under Darby, waiting upon events. Though Torbay is a wide roadstead, giving ample room for manoeuvre even by a large fleet, and though the allies were in the strength of five to three, Cordoba refused to consider an attack. He seems to have thought that by appearing in the chops of the Channel and causing a lively alarm in England he had fulfilled his task. So the allies separated, Guichen, much discouraged, taking the French ships back to Brest and describing the campaign as physically wearing and utterly pointless. It was certainly an absurd waste of a powerful force.

On the other side of the Atlantic events of infinitely greater importance were taking place. As the hurricane season in the West Indies drew near, de Grasse left Martinique in July with a big convoy for Cap Français in San Domingo. Here the rest of the trade had been collected under the slightly soporific eye of Monteil's squadron, and the whole lot were then to be sent under escort to France. On the American mainland, however, what was to prove the crisis – and, indeed, the absolute turning point of the war – was approaching. The English had two main armies: one was holding the New York area, the other, under Cornwallis, had overrun the southern states and invaded Virginia. Washington and Rochambeau were in a position to make a large concentration of troops against either of them, but this would be useless unless de Grasse could come from the Antilles and establish a blockade that would cut off the attacked army from any support or relief by sea. A frigate arrived at Cap Français giving de Grasse this news, and asking for a supply of money for the French troops. The French admiral acted with admirable decision and promptitude. He sent the frigate back with the news that he would sail at once for the Chesapeake (which meant that the allied army must move against Cornwallis), ordered his merchantmen, regardless of their owners, to stay put at Cap Français so that he could use every available man-of-war, arranged for Spanish ships to guard them, borrowed the needed money from Havana, took on board every soldier that the governor of San Domingo could spare from his garrison, and sailed via the Bahama Channel, where he was unlikely to be sighted, with twenty-eight of the line.

Rodney, whose health was in a deplorable state – and this largely explains why he had not forced a trial of strength with de Grasse during the summer – left for England with four ships and a convoy and, guessing that the French would make for the American coast, dispatched Hood thither with fourteen of the line. Making all sail, Hood reached the Chesapeake ahead of de Grasse and, finding that though Cornwallis was falling back on Yorktown on the western shore of that vast inlet all was otherwise quiet, went on to put himself under the orders of Rear-admiral Graves, who had relieved Arbuthnot at New York. He was not many miles below the horizon when the French fleet arrived. De Grasse sent some of his ships up the Chesapeake to land the troops, facilitate the movements of Washington and Rochambeau, and blockade Yorktown, and kept twenty-four with him in Lynnhaven Bay, which is the outer roadstead of the Great Chesapeake inlet, just west of Cape Henry, its southern point. It was now the start of September.

Hood's arrival having brought his strength to twenty-one of the line, Graves sailed for the Chesapeake. He had no knowledge whatever of the French dispositions, for the English frigates watching the bay had anchored instead of cruising outside it, and so had been caught napping by de Grasse's arrival and had fled up the inlet, being unable to beat out to sea. He thought that probably he would have to deal only with the small squadron of de Barras St Laurent from Newport, and was surprised to find himself faced with a fleet of greater strength than his own. He might have been wise to stand out to sea while considering his best move, and in that case he might have cut off and taken de Barras' ships that were bringing the army's siege train and had steered a course that was a long loop out to sea in order to avoid him[1]. As it was, having the advantage of the wind, he came in to the attack at once. De Grasse, when the English arrival was signalled, slipped his cables and stood out to meet them in the following order:

| Ship | Guns | Captain |
|---|---|---|
| PLUTON | 74 | d'Albert de Rions |
| BOURGOGNE | 74 | chevalier de Charitte |
| MARSEILLAIS | 74 | de Castellane-Majastres |
| DIADÈME | 74 | chevalier de Montecler |
| RÉFLÉCHI | 74 | Cillart de Suville |
| AUGUSTE | 80 | de Bougainville (CE) |
| SAINT-ESPRIT | 80 | marquis de Chabert Cagolin |
| CATON | 74 | comte de Framond |
| CÉSAR | 74 | chevalier Coriolis d'Espinouse |
| DESTIN | 74 | Dumaitz de Goimpy |
| VILLE DE PARIS | 100 | comte de Grasse (LG) |
| VICTOIRE | 74 | d'Albert-Saint-Hyppolite |
| SCEPTRE | 80 | comte de Vaudreuil |
| NORTHUMBERLAND | 74 | marquis de Briqueville |
| PALMIER | 74 | baron d'Arros d'Argelos |
| SOLITAIRE | 64 | comte de Cicé-Champion |
| CITOYEN | 74 | comte de Thy |
| SCIPION | 74 | comte de Clavel |
| MAGNANIME | 74 | comte de Bègue |
| HERCULE | 74 | vicomte de Turpin de Breuil |
| LANGUEDOC | 80 | chevalier de Monteil (CE) |
| ZÉLÉ | 74 | chevalier de Gras-Préville |
| HECTOR | 74 | Renaud d'Aleins |
| SOUVERAIN | 74 | chevalier de Glandevez |

Rear-admiral Graves had his flag in the *London*(98), with Rear-admiral Drake in the *Princess*(70) leading his van, and Rear-admiral Sir Samuel Hood in the *Barfleur*(98) in command of the rear. Graves was typical of many English admirals of the time: thoroughly game to fight, but not over-intelligent, and wedded to the preservation of the line of battle in accordance with the fighting instructions. He refused to allow Hood to concentrate on the French rear to cut off a few ships, nor would he attack while the French were forming line because his own line was not formed. Nevertheless, he came down in quarter-line, that is to say with each ship keeping station on the next ahead's quarter, approaching the enemy also on a diagonal, so as to spare his ships exposure to broadsides while they were

[1]One or two historians have given the opinion that this loss of the French siege train would have saved Cornwallis. It is improbable; for so long as de Grasse remained in the Chesapeake, Cornwallis could have been reduced by famine. But it would certainly have enabled him to prolong his resistance, and in that case the incidence of foul weather might just possibly have enabled the English to relieve him.

closing the foe, without being able to reply. As he kept the signal flying for maintaining the line (ie the quarter-line), this, plus some difficulty of navigation owing to the large Middle-ground bank, meant that his van came into close action, his centre into fairly close action, but that his rear fought only at long range. He seems to have hoped to knock his enemy out of the way by superiority in a gun-fight. As the French corps of gunners was able, and de Grasse could keep carefully on the defensive – quite rightly, for his business was simply to keep the English out of the Chesapeake – he failed. Five of his leading ships suffered severely, and with the dark the fleets drifted apart.

In this rather indecisive-looking action, though it was not then realized, one of the world's decisive battles had been fought. The United States had been established. Graves manoeuvred for four days in sight of the French, but he would not risk Hood's proposal to take advantage of the wind to run into the anchorage of the French and leave them to turn him out, and with de Grasse making no mistake, with de Barras St-Laurent having slipped in to add to the French superiority, and with five of his own ships in poor shape (the *Terrible* had later to be destroyed), he stood away for New York. Reinforced, he returned in the third week of October; but Cornwallis had been forced to surrender at Yorktown five days before and, as de Grasse was still in superior strength, Graves sent some ships back to Sandy Hook and took the rest to the West Indies. Though pressed by the soldiers to remain and aid other operations, de Grasse soon followed him; for the stormy season was approaching, the PLUTON had been hard hit in the battle, others of his ships needed a partial refit, some, long at sea, must go back to France, and there was the trade convoy to be got away from San Domingo.

France had had the revenge for which, under Choiseul and his successors, she had been preparing so long. Though the war in America was to drag on for more than a year, and the British were to hold New York to the end, it was clear that, with the surrender at Yorktown, their American colonies had gone. Britain had no more resources of soldiers with which to recover them. The French troops, and particularly the French fleet, had taken from her the most valuable part of her Empire. But France had not taken them for herself; they were to be a new nation. Could she now add to her successes such further victories as would enable her to recover her own vast losses of the Seven Years War and carry out her promise to regain Gibraltar for her ally? The next year was to show that.

Nor did 1781 end on too happy a note. In early December Guichen (LG) left Brest with nineteen of the line and a large convoy carrying troops, munitions, and supplies. He was to detach two ships and some merchantmen for India, seven ships and the rest of the convoy, under Vaudreuil (CE), for the Antilles, and take the remainder to Cadiz. When he was some 150 miles south of Ushant, Guichen – Guichen the tactician, of all people – allowed himself with his escort to get ahead of the convoy, which, as they were running before the wind, put him to leeward of it. The weather was foggy and, as the fog lifted, Rear-admiral Kempenfelt appeared out of the mist with twelve of the line and a fifty-gunner. Though he was much inferior in strength he saw his opportunity and, detaching a few ships to hold off Guichen should that admiral be able to beat back, he sprang on the convoy. Guichen tried desperately to get to the assistance of his charges. But Kempenfelt had taken twenty-four of them (these ships were carrying a thousand soldiers) and Guichen, unable to beat up to the rescue in time, could only signal to the remainder to scatter. As the English stood away with their prizes, the Frenchman collected his escort and some of the convoy as best as he could, and then set off in pursuit. Kempenfelt had too good a start, and only Vaudreuil's TRIOMPHANT(80) and the fast-sailing ACTIF (64) were able to come up with the *Edgar*(82). TRIOMPHANT had her main top-mast and main-yard shot away and ACTIF could not alone hold her more powerful opponent. Worst of all, while Guichen was endeavouring later to reassemble some of the convoy, a violent storm arose, scattered the ships again, and did much damage to the flagship (BRETAGNE) and to La Motte-Picquet's COURONNE. Nothing better could be done than to return to Brest. Only Vaudreuil with the TRIOMPHANT and the BRAVE and five ships out of 150 in the convoy reached the Antilles.

Of the encounters between separate or smaller ships during the year, one was unusual. The thirty-two-gun frigate FRIPONNE, Captain Macnémara, off Cadiz, sighted three frigates flying English colours, and in so obviously poor a state that he closed to attack, despite the apparent odds. They were the *Flora*(44), the *Crescent*(34), and the *Castor*. This last was a prize taken from the Dutch in a fight so fierce as to have disabled the victor. The *Castor*, with only a prize crew aboard, struck at once, as did the *Crescent* soon after, for she had only her foremast standing; and the Frenchman was able to bring in two prizes.

The triumphal year, 1781, had ended badly with the convoy disaster: nor did 1782 begin well. Returning from the American coast and the success of the Chesapeake to Fort Royal near the close of 1781, de Grasse found that Bouillé had taken advantage of the absence of both fleets to recapture St Eustatius by a brilliant *coup de main*. The two commanders now decided to attack Barbados. De Grasse's orders for 1782 were to break British power in the Antilles by taking Jamaica in a joint expedition with the Spaniards from Havana; but for that he was awaiting the reinforcements and supply convoy that were coming from France. Barbados would be a good interim objective, and Hood, who lay there with the ships that he also had brought back from America, was in inferior force. So 6 000 soldiers under Bouillé were taken on board and the French set sail at the turn of the year. The winds were persistently, and strongly, adverse, however, and after struggling against them the expedition returned to Martinique. The objective was then changed to St Christopher (St Kitts). That island was reached on January 11th, the troops were landed, and the English garrison retired to the fortified position of Brimstone Hill, which would have to be carried by a regular siege operation. The fleet moored a few miles along the coast to the east.

Hood, embarking what few troops were available in Barbados and Antigua, came along to the rescue, though he had only twenty-two ships to de Grasse's twenty-six. Thinking that Hood would make for an anchorage under Brimstone Hill where his men would reinforce the garrison and his fire would enfilade the attackers, de Grasse stood out southward to stop him. Hood, in fact, had quite another scheme in his mind. He was skirting the west coast of the little island of Nevis, just south of St Christopher, with his line in good order. As the French came out he altered course a little to the northward, aiming to skirt the southern coast of St Christopher, getting between the French and the land. Now de Grasse's fault as a tactician was to make over-many and over-hasty signals. That happened on this occasion, with the result that, having some captains who were not over quick in their interpretation of signals, and not over-facile in manoeuvre, he failed to get the whole of his fleet into action. The ships passed one another cannonading on opposite tacks, and de Grasse, going about, threw the weight of the ships that were around him on to the English rear. This suffered somewhat, but was able to hold off the French and keep station, and followed its admiral till he brought all his ships to an anchor not very far west of the previous French moorings. In other words, Hood had neatly got the Frenchman out of his anchorage and had occupied it, now being between him and his troops on the island.

Hood quickly got his ships into a strong position. The curvature of the coast protected him to the east and north, he had a few ships lined from north to south which could be assailed only from the west, a side virtually banned to any attacker because of the direction of the prevailing winds, and the bulk of his ships lay in a line from east to west, broadsides to the south, and with springs out so that they could be moved a little at their moorings to use those broadsides to the best effect. The next day de Grasse launched his attack, his ships coming in line ahead, cannonading along the English line, then standing seaward and circling round to do the same thing again. It had no effect, and some French ships, especially the PLUTON under d'Albert de Rions, which had led and so had received concentrated fire, were damaged. After a less earnest second attempt de Grasse gave it up, and waited for his convoy. Only a small part of it, with Vaudreuil's two of the line, arrived; for, as we have seen, Kempenfelt had successfully cut it up soon after it left France.

Probably de Grasse did not greatly mind his failure to drive out or overwhelm Hood for, as the latter had too few troops to effect a relief of Brimstone Hill, that fort was bound to fall, and then the English ships would be trapped between the stronger French fleet at sea and the French guns ashore. Hood, however, knew a trick worth two of that, which reads rather like a schoolboy adventure story. The fort having fallen, riding lights were fixed at dark, not on his ships, but on buoys, so that the fleet appeared to the French to be at anchor all night; but, at a carefully synchronized hour, all cables had been cut and it had stood quietly out to sea. By daylight it was beyond pursuit. De Grasse had been outwitted. He returned to Martinique, taking Montserrat on the way, where he was joined by three of the line with some supply ships from France, and got the good news that a little division of small ships sent from France under Captain Kersaint de Coëtrempren had retaken Dutch Guiana, which had been seized by the English. Hood returned to Barbados, where he was joined by Rodney with twelve of the line. This gave the British superiority of numbers.

Owing to the greater weight of the French metal there was not – at the moment – much in it. When one or two outlying vessels had come in, Rodney had thirty-six effective ships of the line and de Grasse thirty-three. The latter had also two fifty-gunners; but he did not intend to put these into the line. The latest arrivals from France had brought him dispatches from Castries which stressed that above anything else the object of the year's campaigning was to be the attack on Jamaica by combined French and Spanish forces. For this it was necessary for de Grasse to meet the Spaniards, who would have come from Havana, off Cap Français. As he had also a large convoy to take to that port, where it was to join other merchantmen for the homeward voyage, de Grasse did not sail direct for San Domingo, but decided on a northerly course, keeping close to the islands. These were almost all in French or Spanish hands: he could see that Rodney's frigates were observing and signalling his every movement: pursuit was certain and by keeping near to the islands he would be able, in case of need, to refuge his convoy into a friendly harbour. In doing this, de Grasse was taking something of a gamble with the winds. Though, of course, they vary in strength, the trades blow consistently from the east, or a bit south of it. The mountainous islands check them, so that in the gaps between the islands they blow freely, but in the lee of the islands, which stretches a long way out to sea, there are liable to be areas of dead calm varied only by occasional gusts down from the mountains or occasional lights airs. De Grasse sailed on April 8th: this was quickly signalled to Rodney, who was all prepared and weighed at once; by sunset, the fleets were distantly in sight of each other.

The tactics of the French during the next four days were bedevilled by two things. One was that deeply ingrained principle of French strategy that the carrying out of 'the mission' was more important than a successful battle. The other was the incompetence of the captain of the ZÉLÉ(74), a certain Gras-Préville, who had been promoted from lieutenant by virtue sheerly of seniority, and given a command though official reports on him specially recorded his almost complete inexperience in handling a ship. At daybreak the next day, the 9th, the position was that the bulk of the French fleet and its huge convoy was clear to the north of Dominica, though in no great order, and was getting the winds blowing through the Dominica Channel, while the bulk of the English fleet was becalmed under the lee of the island. Its van, under Hood, however, was beginning to feel the winds and was making for two French ships, the AUGUSTE and ZÉLÉ, that had let themselves be carried some way to leeward of their colleagues. Seeing that they were in danger de Grasse ordered Vaudreuil (CE) to sail down before the wind to attack Hood with fifteen of the line: at the same time, seeing that a fleet action might be imminent, he ordered the convoy, under the escort of his two fifty-gunners, into Basse-Terre roadstead in Guadeloupe, a few miles to the north.

Hood, seeing himself about to be attacked by almost double the number of his own ships, hove to in order not to get any further away from the British centre and rear. Instead of making an all-out attack on Hood by a general chase and using his superior numbers to assail him from both port and starboard, Vaudreuil formed his line with almost excessive care, and brought it along Hood's line, from south to north, cannonading as he came. As each ship reached the head of the British line it tacked and, sailing round behind its fellows, came in to cannonade again from the south – not unlike sixteenth-century reiters *faisant la caracole*. This went on for nearly four hours. By then Rodney and the English centre, getting some light airs, were working up so as to threaten to come on to Vaudreuil's starboard (or windward) quarter and thus to catch him between two fires. At this de Grasse signalled his whole fleet to form line, and so, being to windward, broke off the action. He could, earlier, have brought down nearly all his ships on to Hood and overwhelmed him, thus destroying about a quarter of Rodney's force. The cause of his missing so great an opportunity was that he considered, rightly, that Vaudreuil was strong enough to deal with Hood, and also that he must not risk more of his fleet for fear of imperilling his ultimate and main 'mission', the junction with the Spaniards and the attack on Jamaica. And it is interesting to note that the court which later inquired into the whole campaign considered his decision to have been a prudent one. Maybe; but he had missed a great opportunity. Of course, Hood's ships, having fought at odds, had suffered more damage than the French, but they repaired this at sea, while one French ship, the CATON(64), considered herself too injured for further action and stood off, without permission, to Guadeloupe.

Grasse's strategy was, of course, designed to draw the English away to leeward through the Dominica, or Saints, Channel, and, using the start that he had of them, to outsail them till he effected his junction with the Spaniards, which would put him in greatly superior force. Meanwhile the convoy, escorted by the two fifty-gunners in case of attack by frigates or privateers, would be free to steer direct for Cap Français. The channel between the north of Dominica and the southern point of Guadeloupe is roughly twenty-three miles wide; but for practical purposes it is reduced to about thirteen miles by the presence of the rocky islets called the Saints, which lie some nine miles south-eastward of Guadeloupe. Into this channel the French were now beating; but on the night of the 10th they lost the services of a second ship, for the ZÉLÉ collided with the JASON(64), and the latter was so damaged that, like the CATON, she had to make for Guadeloupe. On the 11th Grasse had gained considerably on the English; but that night the ZÉLÉ ran into the flagship, broke her own bowsprit and brought down her foremast. La Pérouse in the ASTRÉE frigate was ordered to take her in tow for Guadeloupe. It took rather a long time to get the tow fixed, no doubt because of the breakage forward in ZÉLÉ, and she was still in sight of both fleets at daybreak on the twelfth. Rodney signalled to Hood in the van to detach his fastest sailers after her.

De Grasse would have been wiser to let her go, and chance whether her tow would get her to her refuge in time or whether she would be taken. As it was he decided to save her, came about, and signalled his ships to form line of battle and to clear for action. By so doing he was not merely sacrificing the windward distance he had gained on the English, but was

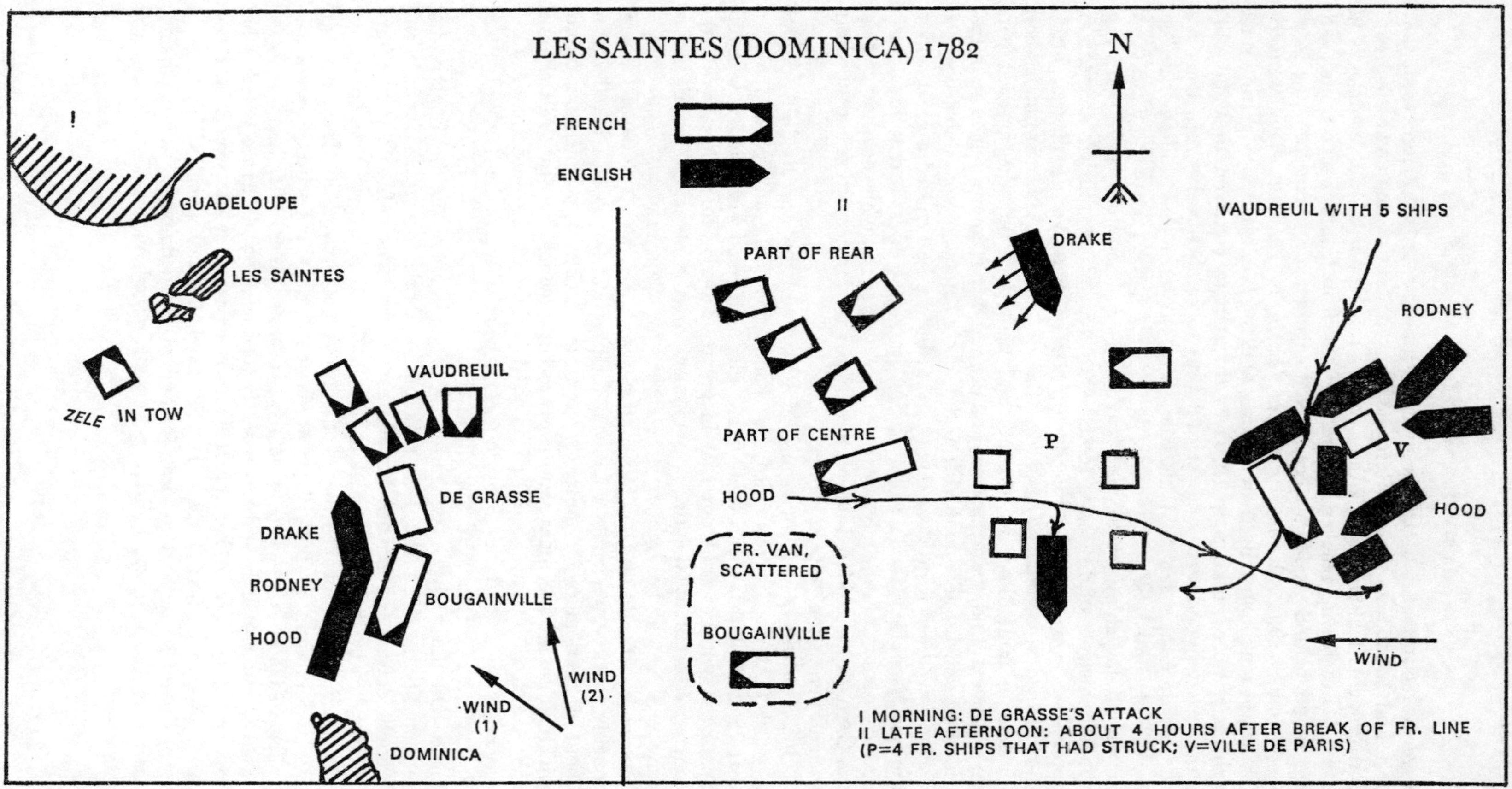

LES SAINTES (DOMINICA) 1782
N
FRENCH
ENGLISH
I
GUADELOUPE
LES SAINTES
ZELE IN TOW
VAUDREUIL
DE GRASSE
DRAKE
RODNEY
HOOD
BOUGAINVILLE
WIND (1)
WIND (2)
DOMINICA
II
PART OF REAR
PART OF CENTRE
HOOD
FR. VAN, SCATTERED
BOUGAINVILLE
DRAKE
P
VAUDREUIL WITH 5 SHIPS
RODNEY
V
HOOD
WIND
I MORNING: DE GRASSE'S ATTACK
II LATE AFTERNOON: ABOUT 4 HOURS AFTER BREAK OF FR. LINE
(P=4 FR. SHIPS THAT HAD STRUCK; V=VILLE DE PARIS)

committing himself to a battle with a fleet now superior by six in number to his own, much superior therefore in weight of metal, and superior in the quality of its ships' companies – for some of the French officers who had gained experience in previous battles had gone home, and the wastage of some years of war had reduced the efficiency of his gunners, since France had less good reserves with which to replace such wastage than had the more naturally seafaring English. (In this action, for the first time in this war, we hear of French gun crews running from the guns at the threat of an enemy broadside.) The French fleet, in order of battle, was as follows:

| Ship | Guns | Captain |
|---|---|---|
| HERCULE | 74 | Chadeau de Laclocheterie |
| NEPTUNE | 74 | Renaud d'Aleins |
| SOUVERAIN | 74 | chevalier de Glandevez |
| PALMIER | 74 | de Martelly-Chautard |
| NORTHUMBERLAND | 74 | de Saint-Césaire |
| AUGUSTE | 80 | de Bougainville (CE) |
| ARDENT | 64 | de Gouzillon |
| SCIPION | 74 | Clavel |
| BRAVE | 74 | comte d'Amblimont |
| CITOYEN | 74 | comte de Thy |
| HECTOR | 74 | de la Vicomté |
| CÉSAR | 74 | vicomte Bernard de Marigny |
| DAUPHIN-ROYAL | 70 | comte de Roquefeuil-Montpéroux |
| LANGUEDOC | 80 | Baron d'Arros d'Argelos |
| VILLE DE PARIS | 104 | comte de Grasse (LG) |
| COURONNE | 80 | Mithon de Genouilly |
| EVEILLÉ | 64 | Le Gardeur de Tilly |
| SCEPTRE | 74 | comte de Vaudreuil |
| GLORIEUX | 74 | vicomte Des Cars |
| DIADÉME | 74 | chevalier de Montecler |
| DESTIN | 74 | Dumaitz de Goimpy |
| MAGNANIME | 74 | comte le Bègue |
| RÉFLÉCHI | 64 | chevalier de Médine |
| CONQUÉRANT | 74 | de Lagrandière |
| MAGNIFIQUE | 74 | MacCarty Macteague |
| TRIOMPHANT | 80 | marquis de Vaudreuil (CE) |
| BOURGOGNE | 74 | chevalier de Charitte |
| DUC DE BOURGOGNE | 80 | chevalier Coriolis d'Espinouse (CE) |
| MARSEILLAIS | 74 | de Castellane Majastres |
| PLUTON | 74 | d'Albert de Rions |

Seeing de Grasse come about, flying his signals, Rodney recalled the ships pursuing the ZÉLÉ and, to give them time to take their places in the line, ordered the rear division, under Rear-admiral Drake, to make sail ahead and lead the line into action. The centre followed, and Hood's van division, its returned ships now in place, formed the rear. The line was regular and compact and, though more numerous, stretched no more than two thirds of the distance, from head to tail, that was covered by the French line. Nor was this line regular, for the fleet had been in some disorder when de Grasse hoisted his signals. The French were so stretched as to threaten to leave gaps, nor was each ship always properly in line astern of its next ahead. Indeed, Vaudreuil's rear division was still getting into formation when it came under fire.

The wind was easterly, and de Grasse was coming down on the port tack, aiming to cross the head of the English line and so get to windward of it. Rodney, approaching on the other tack, was more willing to fight to leeward than most English admirals, and left the French to win the race, but signalled for close action. The French crossed the head of the English line at a fine angle, then altered course a little to starboard to range along it. The leading English ship, *Marlborough*, came on to the French line somewhere between the fourth and ninth ship, and the latter, BRAVE, opened the battle with a broadside at about 7.45 am. At 8.30 de Grasse signalled his ships to wear together. It was a bad signal, for the fleets were too close together – well within musket range – for it to be obeyed. He then signalled to wear in succession. This was not seen, or properly repeated, through the smoke which was now very heavy, and was not executed. So far the battle had followed the routine course of the age – fleets passing on opposite tacks and cannonading hard – in which, for once, the French were having distinctly the worse of things. For the English guns had been supplied with locks that increased the rate of fire, and with improved carriages that widened its sector. The French van ships, in particular, were much cut up in sails and rigging.

Shortly after nine the wind shifted southerly, sending the bows of the French a little to starboard. As their line was already stretched, the ships being rather too far apart, this left gaps; and through one of these, astern of the GLORIEUX, Rodney led with his flagship, the hundred-gun *Formidable*, blasting GLORIEUX as he passed, and giving DIADÈME broadsides that caused her to pay off into irons and thereby to halt the ships behind her. Astern of Rodney, the *Bedford* came through a gap between CÉSAR and DAUPHIN-ROYAL, and confusion of the French line in the smoke became complete. A flat calm then fell, but a breeze a little after midday cleared the smoke and showed the French fleet to have been cut into three parts, each some distance from the others and the ships in each being somewhat scattered. The van was well ahead, for it had kept its wind after passing the tail of the English line, and had made no attempt to come about to the support of its centre. The cut-off ships of the centre, and the rear division some distance away on a different tack, were both to leeward of the English, while CÉSAR and GLORIEUX dismasted lay between the two fleets, as did the damaged HECTOR and CITOYEN.

Drake, with the English van, had got ahead of the French rear and his ships were, for the time, out of action; but Rodney in the centre, and Hood with the rear, were in a position to attack as they pleased, and Rodney hauled down the signal for the line. The English then ran in upon the nearest foes, especially upon the centre and upon the ships that lay damaged between the fleets. De Grasse made signals to close the admiral and to form line. They were largely not obeyed. Bougainville flew a signal to his van division to support ships in trouble, and signified that he was out of action. He had shown himself a clever and intrepid navigator, but he had little battle experience and now, concerning himself mainly with the damage to his own vessel, he thereby lost control of his squadron. Only the ARDENT made anything of a determined effort to reach the admiral. Some of de Grasse's centre division, moving to join him under heavy fire, failed to take in sail in time and so overran him and went largely out of action. With the breaking of their line the spirit seems to have gone out of the French, though some did their duty and strove to close the flagship. Vaudreuil tried to effect a rally by leading in TRIOMPHANT, MAGNIFIQUE, BOURGOYNE, MARSEILLAIS, and PLUTON (which offered the admiral a tow). LANGUEDOC, after first paying off to leeward, also closed the flag. COURONNE, SCEPTRE, DAUPHIN-ROYAL, and the much damaged CITOYEN also kept in action for a while before standing to leeward. But as the VILLE DE PARIS was to windward of them all, they could not form line both ahead and astern of her and so, heavily engaged as they were by some of the English, could not take off any of the weight of the attack made on her by a number of enemy ships. Nor did her frequent signals, rendered useless by her frequent changes of course to get her guns to bear, make their task any easier.

Hood, largely ignoring the scattered French van, led his ships first against HECTOR, GLORIEUX, CÉSAR, and ARDENT. The first three fought desperately against odds before striking: the last, unable to join her colleagues, struck with little resistance[1]. Hood then, at about 4.30, moved in upon Vaudreuil and what was left of the French centre, already heavily attacked by Rodney's ships. Some of the French van, having repaired their rigging, were now coming near the scene, but so much to leeward as to be virtually out of action – as were the ships that had run past the flagship. That great French three-decker made a tremendous resistance, but the slaughter aboard her was appalling (her casualties were more than the total of those in the English fleet), and a little before six she had become unmanageable, while the survivors of her guns' crews had used all cartridges and were reduced to scooping powder out of barrels hoisted from the hold. She was plainly doomed, and Vaudreuil led away the ships that had tried to support her, collecting the others as he went, and made for distant Cap Français. VILLE DE PARIS was now under the fire of no less than eight enemies, and de Grasse could only yield. He was one of only three left unwounded on her upper deck. As darkness fell the violent explosion of the surrendered CÉSAR provided the last tragic incident of a disastrous day[2].

The French were lucky that there was no pursuit, for DIADÈME, DESTIN, MAGNANIME, and CITOYEN were much damaged, PALMIER and AUGUSTE had been in collision (as had DUC DE BOURGOGNE, but not seriously enough to excuse her CE's shameful inaction), and RÉFLÉCHI was a notoriously

[1]She had been an English ship, whose capture by d'Orvilliers had been the one trophy of the combined operation in the Channel in 1779.
[2]The battle is known variously as that of Dominica and of the Saintes.

bad sailer. The battle did, however, have an aftermath. Hood went to blockade the ships that had put in to Guadeloupe – CATON, JASON, and the frigates[1]. Under the comte de Framond, who was the senior officer, they slipped out one dark night; but they were pursued, and all but one frigate were taken. There was also another, and unpleasant, aftermath with which we may as well deal now, though it occurred two years later.

De Grasse might be considered unfortunate. His tactics until the last day had been shrewd and it was the accidents of collisions, due to the lack of skill of some captains, that had largely negatived them and had reduced him to inferiority of strength. Bearing down to save the ZÉLÉ was possibly a mistake, but it was a brave and honourable one, and he had fought his flagship with the greatest courage. Had he kept quiet about his defeat he would have been received with honour in France. But he did not keep quiet, for he attributed his defeat to the failure of his captains to support him. From England (where he was almost fêted) he sent a report which led to the unjust imprisonment of two of his captains; and when he got back to France he wrote such libellous attacks upon the honour of a number of them that a court martial was held. Bougainville and Coriolis d'Espinouse were publicly admonished for having failed to set a proper example as CEs once the line had been broken : less public admonitions were given to the captains of NEPTUNE and JASON : the captains of ARDENT and ZÉLÉ were dismissed their ships : the comte de Framond of the CATON was sent to life imprisonment : the captain of MAGNANIME was warned to keep his mouth shut in future; and the captain of DAUPHIN-ROYAL was given a hint that he might have done more. But all others were honourably acquitted, and in some cases positively praised, and it was specially ordered that all de Grasse's written attacks upon them should be suppressed. His vital victory of the Chesapeake was forgotten, the king refused to receive him in future, and he was advised (ie virtually ordered) to retire to his estates. There he died in 1788. Clearly the haughty quarrelsomeness that had often done harm had not yet been eradicated from the Service.

Arriving at Cap Français, Vaudreuil found the convoy and its two fifty-gun escorts safely at anchor, plus the Spanish squadron, and he was gradually joined by all the survivors of the battle[2]. But though this assembled and combined fleet was over forty strong, the attack on Jamaica was clearly impossible. Some of the French ships had been hard hit, seven of their captains had been killed and other casualties were heavy, and most of the siege train had been lost in the ships that had been taken. So, as the hurricane season approached, Vaudreuil left two of the line with the Spaniards at San Domingo, sent La Pérouse in the SCEPTRE to beat up the English settlements in Hudson Bay (which he did effectively), and took the rest to Boston. Though only seven of the line had been lost (five in the battle and two later) the battle of the Saintes had completely reversed

[1]ZÉLÉ, was also there, but was too damaged to move.

[2]In his report on the battle Vaudreuil stressed the advantage given to the British by their introduction of the carronade. This was a light-weight gun with a heavy charge, of use only at really short range. But at such range it was dreadfully destructive and was known to English sailors as 'the smasher'.

the position in the Antilles. The captured islands were now exposed to recapture by the British.

In Europe, the great attack on Gibraltar was to be the main event. The comte de Guichen (LG) had left Brest in February with his flag in TERRIBLE (110), taking with him reinforcements and convoys for both the West Indies and the Indian Ocean. Having seen them on their way he took the remaining five of the line to Cadiz, to put himself once more under the command of the very elderly Cordoba. There were two other French flag officers – vicomte de Rochechouart (LG) in MAJESTUEUX, and de Beausset (CE) in the ROYAL-LOUIS (110) – in the combined fleet, which was preponderantly Spanish. The preparations for the great attack not yet being complete, the fleet sailed in June for the approaches to the English Channel in order to block the trade, and succeeded in cutting up a convoy. It was joined by La Motte-Picquet (LG), with his flag in INVINCIBLE (110), and eight of the line, and sighted the inferior British Channel fleet under Howe. That skilful admiral would not let himself be brought to action, so the allies returned to Cadiz in July to be ready for the assault on the Rock.

This took place in September and, except that the fleet acted as a screen to prevent any assistance from arriving from outside, was not really a French naval affair. The main attack was by ten Spanish warships which had been specially converted into formidable floating batteries by a French engineer officer, d'Arcon, and were manned by soldiers. There were to be diversionary attacks along the isthmus at the north, and by landing troops under cover of fire from some ships at Europa point, the southern extremity of the peninsula. This last did not take place and, for the rest, General Elliott's defence was as stout as ever. A deluge of red-hot shot set nine of the floating batteries afire and they blew up, with appalling loss of life, for no boats had been provided to take off the soldiers in case of misadventure; the English came out in their boats at night and destroyed the remaining one. The great attack had failed, and the Spaniards sullenly withdrew to wait for famine to do their work for them.

It was known that an attempt at relief was coming, and the combined fleet of forty-nine ships, of which fourteen were French, remained at Algeciras, just across the bay from Gibraltar, to intercept it. In October, Howe was signalled off Cape St Vincent. He had thirty-four of the line, and a large convoy. As he approached the Straits, so did a violent gale. This drove one Spanish liner ashore on Gibraltar (where it was seized), damaged some others of the combined fleet, forced a number out to sea, and generally scattered it. With fine seamanship Howe's fleet and convoy had kept well together, though only a few of the merchant captains were sufficiently skilful to make Gibraltar on that tack. The rest, with the escorting fleet, were driven by the strong winds into the Mediterranean. The allies gathered to the east of them, and nearly a week was spent with Howe manoeuvring to get his charges into Gibraltar, and the combined fleet endeavouring to get at them. Eventually the convoy was seen into its destination, and Howe stood away through the Straits with his fleet. He always held this relief to be the finest of his achievements.

The allies followed in pursuit, but straggling. La Motte-Picquet in the copper-bottomed fast-sailing INVINCIBLE was in the lead and some thirty others were able more or less to keep up with him; the remainder, including Guichen's TERRIBLE, being slower sailers, pounded along astern. The leaders, in something roughly resembling a line, came up with the British off Cape Spartel and opened fire; but the engagement never became anything like a battle. For Howe had no intention of fighting a full-scale action which would have allowed the lagging ships of the allies to come up so that they could exercise their considerable superiority of numbers. His ships continually gave way and, at dusk, went off under full sail. The allies returned to Cadiz: Gibraltar had been relieved, and their failure there was complete. D'Estaing came to Cadiz to take command and prepare for a joint expedition against Jamaica (which never took place) and Guichen left for home.

This was his last piece of service at sea of the least importance. He was then seventy, had had over fifty-two years in the navy, and typified all that was best in French naval officers. He was a chivalrous man, far less addicted to pushing himself and blowing his own trumpet than many other officers, and was utterly loyal and devoted to his Service, never standing on his dignity whatever position he was asked to fill or whatever orders were given to him. He had shown himself in the battle of Ushant to be a brave and skilful fighter and, despite his one grave error with a convoy, he was the most able tactician of the navy when in chief command. But if he typified what was best in French naval officers he also typified their weakness: his strategy and his tactics, clear-headed, prompt, and clever though they were, yet were almost entirely defensive. A Fabius rather than a Scipio. He died at Morlaix early in 1790.

A good illustration of the relative power of French and English liners was provided during the year by a combat between the SCIPION(74) and the hundred-gun *London*, a perfectly efficient ship. The Frenchman, superior in design and carrying some heavier pieces, fairly beat off his nominally stronger opponent. There was also an unusual incident when, during a frigate action, the French captain was killed and his second in command was an Italian: foreigners were not allowed to command in wartime and the French frigate stopped fighting while he and the next senior officer argued it out who should give orders! It is also of interest that none of the ships taken at the Saints ever reached England: they were either condemned or were overwhelmed and sunk by a North Atlantic gale. The PALMIER and SCIPION also sank during the year.

What with the defeat off Dominica and the failure at Gibraltar, both sides of the Atlantic presented France with a disappointing naval picture. But there was one area, the Indian Ocean, where things went differently. We put it aside for future consideration, and to it we must now turn.

When de Grasse had taken his fleet out of Brest in the spring of 1781, five of the line had been detached for the Cape of Good Hope and thence for the Indian Ocean under what one French historian calls their 'young' captain. This was the Chevalier-commandeur Pierre-André de Suffren St Tropez: he was the third son of a minor Provençal noble, and, as he was just on fifty-two years of age, the adjective 'young' seems hardly appropriate – the more so as he had put in thirty-eight years' strenuous service. As a garde he had fought in the battle off Toulon, had served under the duc d'Anville at Louisbourg in 1746, and in MONARQUE in l'Étanduère's brave fight against Hawke, when he had been taken prisoner. He had been a lieutenant in ORPHÉE with La Galissonière at the battle of Minorca, and with La Clue in OCÉAN in the disaster at Lagos, when he had become a captive of the English for the second time. He had been with du Chaffault's somewhat unsuccessful expedition to Morocco in 1765, and had commanded FANTASQUE under d'Estaing on the American coast and in the Antilles at the battle of Grenada. Between the wars he had served with the Knights of Malta, from whom he had his title. He was short, very fat, full of *élan* and initiative and highly critical of his country's naval operations in the present war. Ruyter was his fighting model. Schooled in adversity, he knew his ships and his profession from truck to keelson.

He was in HÉROS (74) and had with him ANNIBAL (74), ARTÉSIEN, VENGEUR, and SPHINX (all 64). Since Holland had entered the war, and because she was now only a minor power, her colonies were exposed to English attack. Suffren's first business was to take troops to reinforce the Dutch at the Cape of Good Hope, to conquer which Commodore Johnstone with five ships (three of them fifty-gunners) and a convoy had already sailed ahead of him. As Suffren's squadron had been somewhat hastily fitted out, some ships were short of water, wherefore he made for Porto Praya of Santiago Island in the Cape Verde group. The fast-sailing ARTÉSIEN, scouting ahead, reported war ships and a convoy in the roadstead, which could only be those of Johnstone. Two courses were open to the Frenchman: one, the normal French tactic of seizing the opportunity to out-sail the enemy to the Cape and thus to fulfil 'the mission'; the other to deal with him there and then so that he might, perhaps, not be able to get to the Cape at all. Suffren, with his disapproval of the previous negative French tactics, departed from custom by choosing the latter. Ordering his troopships to heave to, he signalled to clear for action and prepare to anchor, and made straight for the enemy. The Praya roadstead was a wide one of only slight depth and facing south so that, as the wind was north-east, the French, approaching it from the eastward, could enter easily on the starboard tack to get at the enemy lying at its west end.

Of course, in sailing-ship days, with a best speed of not many knots and with guns effective only at rather close range, complete surprise was impossible, but Suffren did achieve partial surprise. For Johnstone, in a careless and unseamanlike way, had allowed his ships to anchor anyhow, and though he had time to clear for action, he did not have time to move them to a proper fighting line, so that some at least partly masked the fire of others. While the number of fighting ships on each side was equal, the French carried distinctly the greater number of guns, though this was somewhat balanced by the number of well-armed Indiamen with Johnstone;

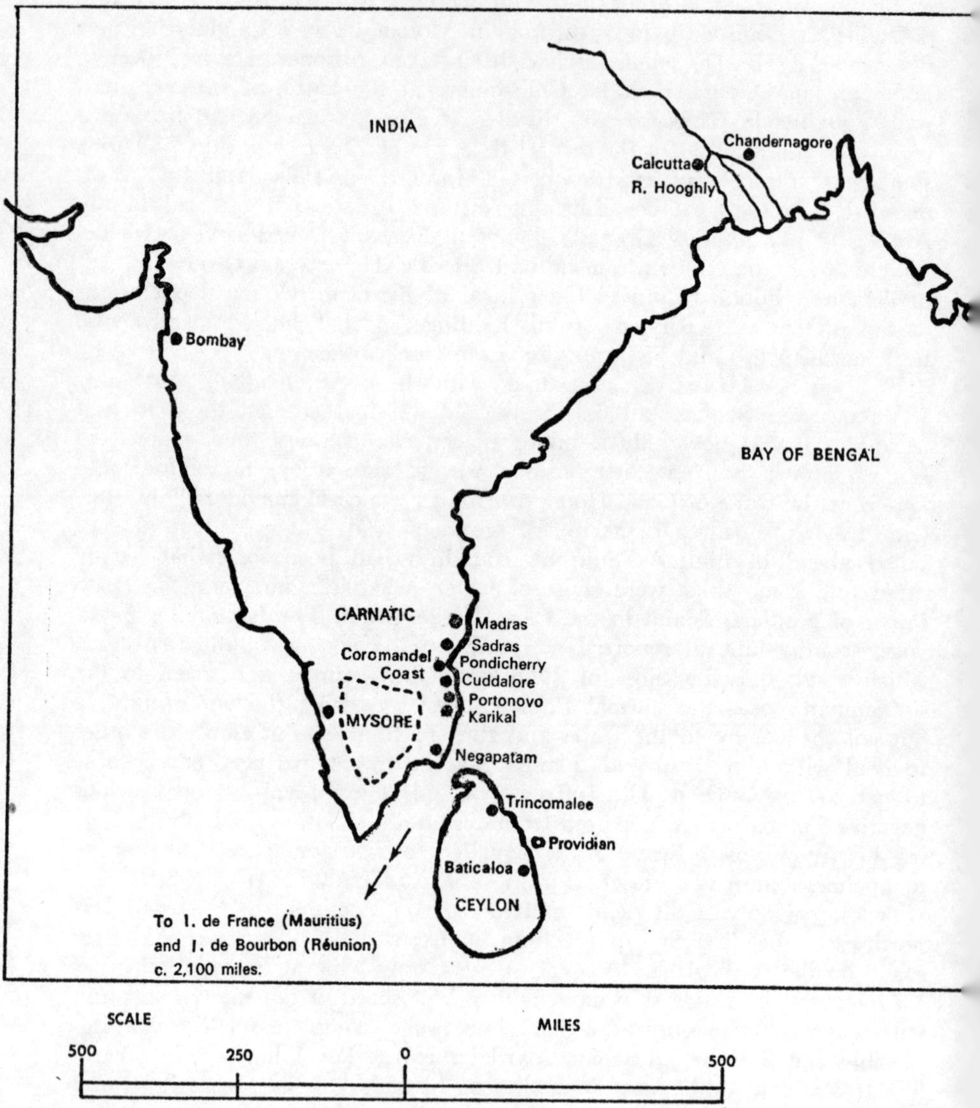
INDIA
Chandernagore
Calcutta
R. Hooghly
Bombay
BAY OF BENGAL
CARNATIC
Madras
Sadras
Coromandel
Coast
Pondicherry
Cuddalore
Portonovo
Karikal
MYSORE
Negapatam
Trincomalee
Providian
Baticaloa
CEYLON
To I. de France (Mauritius)
and I. de Bourbon (Réunion)
c. 2,100 miles.
SCALE
MILES
500
250
0
500

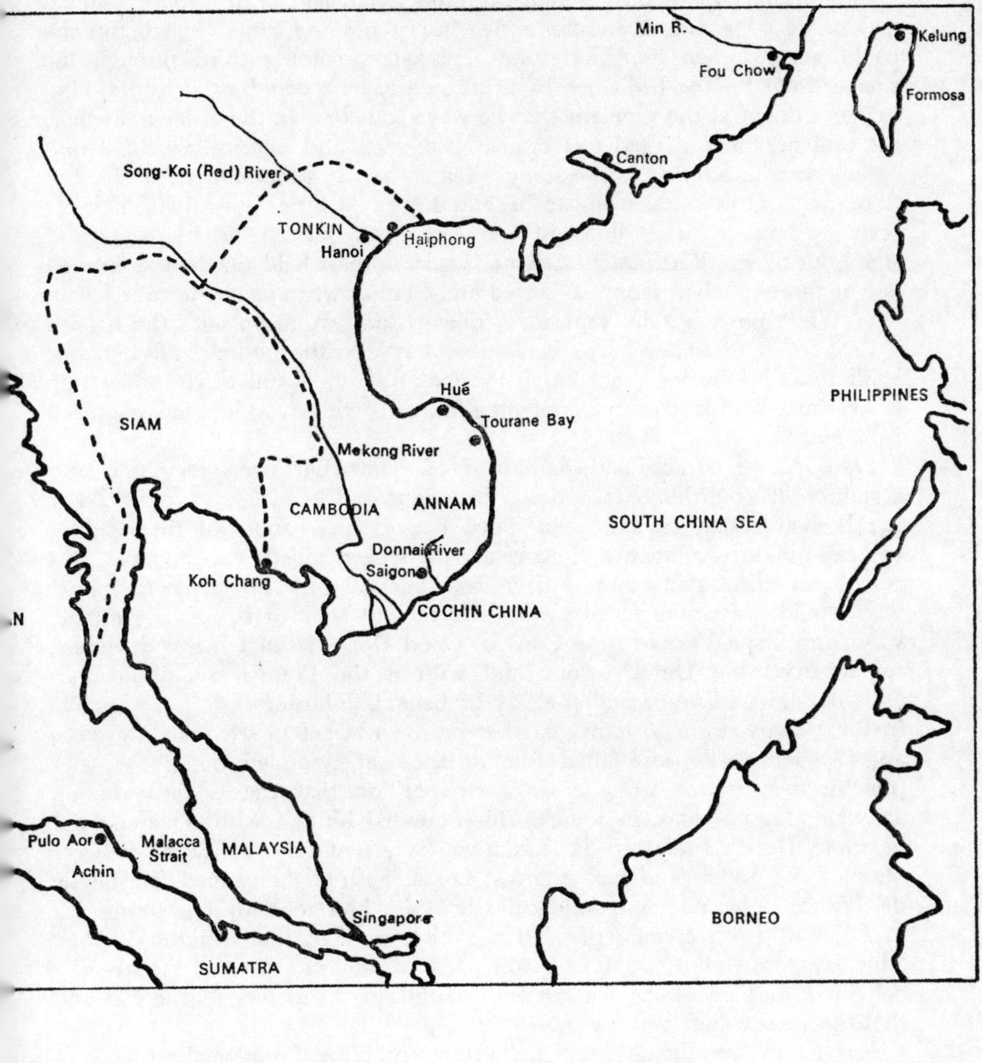
Min R.
Fou Chow
Kelung
Formosa
Canton
Song-Koi (Red) River
TONKIN
Hanoi
Haiphong
Hué
Tourane Bay
PHILIPPINES
SIAM
Mekong River
CAMBODIA
ANNAM
SOUTH CHINA SEA
Donnai River
Saigon
Koh Chang
COCHIN CHINA
Pulo Aor
Achin
Malacca Strait
MALAYSIA
Singapore
SUMATRA
BORNEO

it was also to be more than balanced by the poor showing made by two of the French captains.

Suffren went slap among the warships in the HÉROS, dropped anchor, and opened with both broadsides. ANNIBAL followed him closely; but her captain had not properly cleared for action, thinking that, as the Portuguese harbour was neutral water, Suffren's signal had been a mere nominal precaution! He was therefore under heavy fire for some time before he could get his own guns into play. ARTÉSIEN, coming third through the smoke, went for the Indiamen by mistake, and took one, but her captain being shot down at the moment that he was about to give the order to anchor, she and her prize drifted to leeward[1]. VENGEUR and SPHINX never got into action at all. So the two leading French vessels were under all the fire that the English could bring to bear and, after a little more than an hour, ANNIBAL having lost both main and foremasts, Suffren cut his cable and stood out to sea. Fortunately ANNIBAL's mizzen mast held till she was headed out of the bay when it too collapsed and SPHINX was ordered to take her in tow. When he was able, Johnstone followed and came up with the French by the evening. They were in line of battle with SPHINX and her tow leading, and Johnstone, not liking the look of things, and fearful of getting to leeward of his convoy, contented himself with retaking the captured Indiaman.

Frankly, the attack had perhaps been somewhat over-precipitate and certainly ill co-ordinated; and at first sight it looked like a defeat. For the French casualties had been much heavier than those of the English, the captains of ANNIBAL and ARTÉSIEN had been killed, and ANNIBAL was completely dismasted – and Suffren was one of very few French officers who would have thought of towing her in that state or in jury rig all the way from Cape Verde to the Cape of Good Hope. In fact it was a success, for it saved the Dutch colony and with it the French communications with the Île de France and the Bay of Bengal. Johnstone's ships, especially in the convoy, were so damaged that he did not get to sea for some days, and Suffren easily forestalled him at the Cape, landed his troops, and put his fleet in so strong a defensive position that the Englishman, on arriving, did not dare to attack. He contented himself with the rich haul of some Dutch Indiamen in Saldanha Bay, sent on such ships as were destined for India, and took the rest home. Suffren then sailed for the Île de France. The ANNIBAL followed when she had refitted. His promotion to CE and from commandeur to the high rank of bailli in the Maltese Chivalry also followed; for, despite his breach of custom, France was delighted that her ships had roundly launched themselves at the foe and that the Cape colony was saved.

Arrived at the Île de France, Suffren put himself under the orders of the comte d'Orves, who had local rank equivalent to commodore. He was an officer at once mediocre and ailing, and he had missed a great oppor-

[1]There was, at this time, much discussion as to whether in action the second in command was best placed in the battery, as Suffren believed, and as was so in this case, or forward on the upper deck, where he served no purpose save to give prompt orders if the captain was killed.

tunity. When France entered the war, the rule of Britain in India was under severe pressure. She was at war with the powerful Mahratta Confederacy, and her seizure of the French posts in the subcontinent had angered Hyder Ali, the warrior head of the usurping dynasty in Mysore. No help could come from an overstrained England and it was due to the firmness of a great Governor-General, Warren Hastings, and the skill of a veteran soldier, Eyre Coote, that that rule was by the narrowest of margins maintained. Even they could hardly have saved the situation had d'Orves had the least grasp of its possibilities and his duty. It was not until early in 1781 that he arrived off the coast of the Carnatic. Cuddalore was then closely besieged, Madras was all but surrounded, and the French squadron was rather stronger than that of Rear-admiral Edward Hughes which lay off that town. D'Orves declined to attack it: he declined to land some men to give moral support to Hyder's army: he declined to remain on the coast to cut off English supplies for Cuddalore, which would have fallen. He spoke fair words to his Indian allies and, after a few months, sailed back to his base. The reason he gave was the utter lack of supplies and of any means with which to make good the damage his ships would inevitably receive in action. It is indeed inexplicable and incredible that, with the lessons of Labourdonnais and d'Aché before them, the French should have failed in peace-time to build up the least reserve of marine stores for such small squadrons as they were likely to have operating in the Indian Ocean. Yet a commander of greater patriotism and with the least grasp of the situation would have made do somehow. By the time Suffren reached the coast Eyre Coote had overthrown the Mysore army, and, though the Indian princes still looked to France for aid, all had to be begun anew.

In fact the little India squadron was in poor disciplinary order, and Suffren's arrival showed this in an unpleasant row. The captain of the ANNIBAL having been killed at Praya, Suffren had given the command to her first lieutenant, Morard de Galles. This officer had fought the ship stoutly and had handled her well on her long sail under jury rig to the Cape. D'Orves' captains protested: some of their officers were more senior. The ailing d'Orves gave way, and moved Morard de Galles to the command of a frigate. The squadron sailed at the end of the year and ran into and took the fifty-gun *Hannibal*, which it incorporated under the name of PETIT-ANNIBAL. D'Orves was now so ill that he handed over the command, and he died a few days later. Suffren made for the Coromandel coast with his convoy of troops. He did not intend to return to the Île de France – 'that Cythera of the French navy' – for he found it too full of pretty women and of opportunities to indulge in trade. Lack of supplies and of replacements for gear notwithstanding, he would remain on the Indian coast till he was victorious or destroyed. Long splices and short commons would be the order of the day on board, and officers who wanted to return to base would have to come to heel.

In mid February he appeared off Madras. He had twelve ships to the nine of the English admiral, Hughes; but these were too well ensconced under the batteries of Madras to be attacked. Suffren considered his first necessity to be that of proving France's faith to her Indian allies; in other words, to land his troops. After that, if he was to stay on the coast, he must have some sort of base: not even his determination could dispense with that; and the best available was the good roadstead of Trincomalee in Ceylon which the English had recently taken from the Dutch. This was realized by the English admiral, who was as much a fighter as his opponent, and was determined to prevent him. Suffren sent his convoy southward along the coast and covered it from rather too far to seaward. Hughes came out from Madras, went after the convoy, and had cut off a few ships when Suffren came down to the rescue. The fleets thus came to grips off a little place called Sadras. Suffren had ordered his ships to form line as quickly as possible, regardless of the regular order. So they came into action as follows:

| Ship | Guns | Captain |
|---|---|---|
| HÉROS | 74 | bailli de Suffren (CE) |
| ORIENT | 74 | de Lapallière |
| SPHINX | 64 | vicomte Duchilleau de Laroche |
| VENGEUR | 64 | comte de Forbin |
| PETIT-ANNIBAL | 50 | chevalier Morard de Galles |
| ANNIBAL | 74 | de Tromelin |
| BIZARRE | 64 | chevalier de Lalandelle-Roscanvec |
| SÉVÈRE | 64 | chevalier de Villeneuve-Cillart |
| AJAX | 64 | Bouvet |
| FLAMAND | 50 | de Cuverville |
| ARTÉSIEN | 64 | Bidé de Maurville |
| BRILLANT | 64 | de Saint-Félix |

Hughes, with his flag in the *Superb* (74), had his fleet in more regular order, his own ship being fifth in the line. Suffren's plan of attack was tactically clever. He had the advantage of the wind, which was slight, and he stretched along the English line, firing on each ship as he passed, till he came before the beam of *Superb*, at which he backed his mainsail and engaged both the English flagship and, sufficiently to keep it occupied, the next ahead. This meant that the three leading English ships had no opponent whatever, so that the French could concentrate their twelve ships on the remaining six. Unfortunately, when each ship following the CE had engaged its opposite number, the first of them left over, without an adversary and free to overlap the enemy, was the ANNIBAL commanded by the man who had displaced Suffren's appointment of Morard de Galles. He not only failed to bring his own ship to leeward of the English, despite the flagship's signal to engage 'at pistol range', but checked the AJAX when it was about to leave the line to double on the English rear. Wedded to the idea of maintaining the line of battle, he was glad to use it as an excuse wilfully to thwart his admiral, for whose vigorous leadership, being a slack officer, he had taken a jealous dislike.

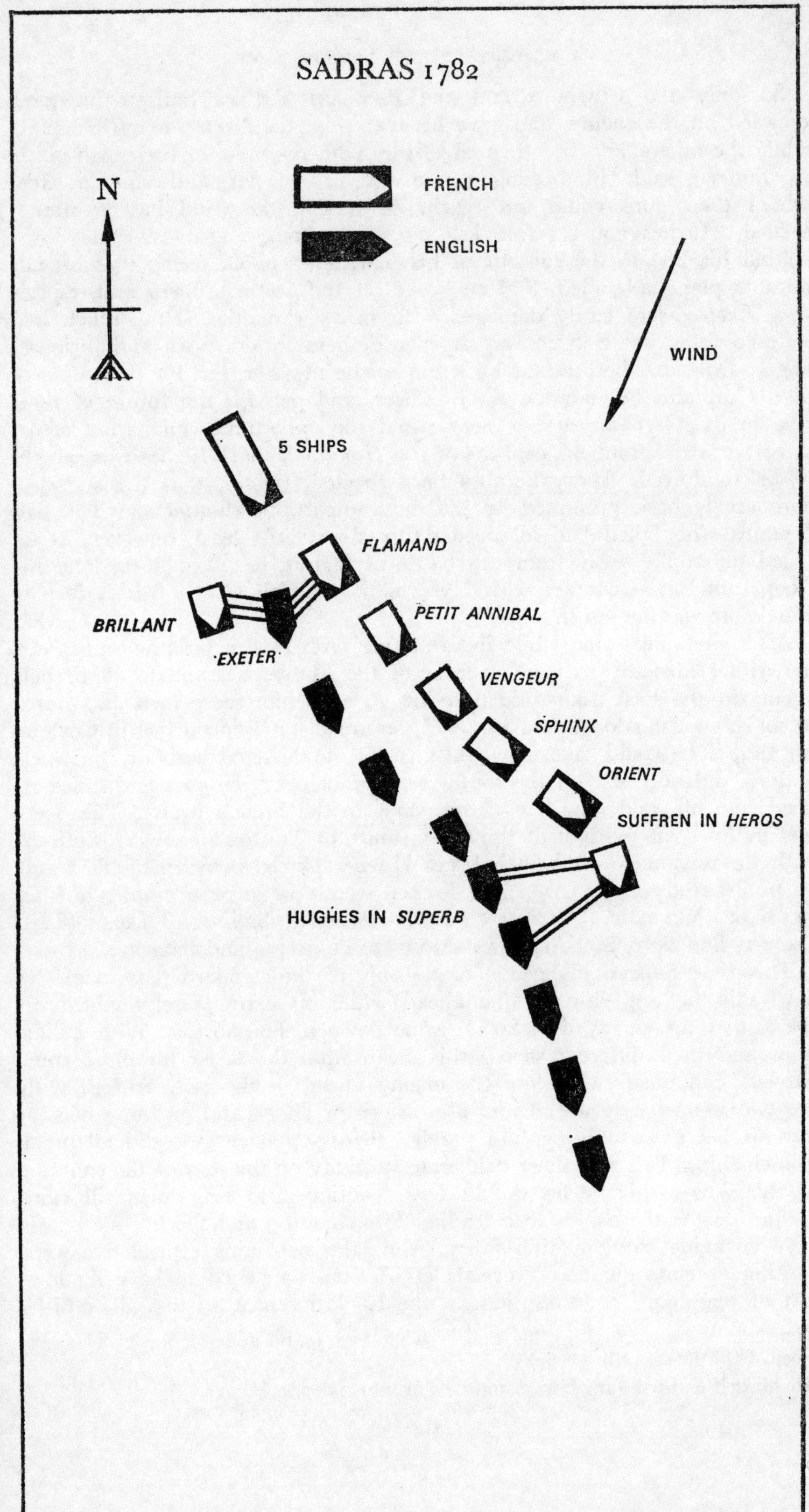
SADRAS 1782
N
FRENCH
ENGLISH
WIND
5 SHIPS
FLAMAND
BRILLANT
EXETER
PETIT ANNIBAL
VENGEUR
SPHINX
ORIENT
SUFFREN IN HEROS
HUGHES IN SUPERB

So only the FLAMAND and the BRILLANT did as Suffren intended, doubled on the enemy, and gave his rear ship, the *Exeter*, a terrible time[1] while the others 'knocked themselves silly with the noise of their own guns', as Suffren said, in blazing off a lot of powder and shot at foes whom their guns could not reach. At length, the wind having strengthened, Hughes got his free leading ships about, and they came down behind his line to the succour of his rear. Thereupon, seeing that his admirable plan had failed, Suffren broke off the action. *Superb* and, especially, *Exeter* were badly damaged with heavy casualties. The French had suffered little, but Suffren was at once desperately downcast and furiously angry. Downcast because, as he wrote to the minister, but for the stupidity of his captains he had the English fleet, and perhaps the future of their rule in the Carnatic, at his mercy; and the opportunity might not recur, since he knew, from his capture of the *Hannibal*, that Hughes was shortly to be reinforced. Angry because he suspected, rightly, that the stupidity had largely been prompted by the most unpatriotic disaffection. The five captains who failed had all been d'Orves' men. At least, however, as he sailed unhappily away from this battle of Sadras, he was able to land his troops and make contact with Hyder Ali. The fall of the British fort at Cuddalore was the result.

After spending some while in arranging with Hyder for provisions, and in writing earnestly to the Governor of the Mascarenes and to all French agents in the East asking them to try to send him some men and stores to make good his losses and wastage[2], he sailed for Trincomalee in Ceylon; for this, if he could take it, would give him a sheltered harbour, and such a thing did not exist on the south-east Indian coast. Hughes had sailed to head him off, and now had eleven ships to the French twelve. The fleets met in April off the islet of Providian, south of Trincomalee, when Suffren, with the advantage of the wind, got Hughes pinned between his fleet and an unpleasantly rocky coast. The French were now in proper order of line, to wit: VENGEUR, ARTÉSIEN, PETIT-ANNIBAL, SPHINX, HÉROS (flag). ORIENT, BRILLANT, SÉVÈRE, AJAX, ANNIBAL, FLAMAND, and BIZARRE.

This time Suffren planned a battle only of the standard pattern of the day – ship fighting ship in line ahead while his extra vessel doubled on the last of the enemy line. So he came down in line abreast. With sailing ships and their different speeds this means that the faster (or more combatively captained) will close the enemy ahead of the rest. Suffren with the two immediately ahead and also astern of HÉROS did so, but when he put his helm up to bring him parallel to his opponents, so did all other French ships. This was sheer deliberate stupidity on the part of the captains of the slower ships, who should have continued to run down till close to the foe. As it was, the two leading French ships, and the five sternmost were engaging barely within range, while the remaining central five were fighting at close quarters. Nevertheless Suffren happened to have made a sort of temporary concentration on the English centre, so that three-fifths

[1]Her commodore, asked by his master what was to be done, replied: 'Nothing, except fight the ship till she sinks'.

[2]He himself enlisted some lascars and bought some Negroes!

of the whole of the English casualties occurred aboard three only of their ships – *Monmouth, Superb,* and *Monarch.* HÉROS had brought up alongside the first of these, and her broadsides brought down two of the Englishman's masts. *Superb* pushed ahead to cover *Monmouth* and gave the French flagship a hard time till ORIENT came to his chief's aid and forced *Superb* out of line, whereat *Monarch* closed the gap, with BRILLANT coming ahead to join ORIENT and outweight her. The fleets were now getting dangerously close to shallows, and Hughes signalled to wear together and take *Monmouth* in tow. This done, in a seamanlike way, Suffren made a similar signal clumsily obeyed, and the brutal slogging match became more general, for the reversal of course had brought the five rear Frenchmen into closer action. A fire broke out in ORIENT, but BRILLANT boldly covered her. HÉROS lost her foretopmast and Suffren shifted his flag to AJAX.

The French fleet was, indeed, in some muddle, having only five ships still engaged, when the elements stopped the battle with a violent tropical thunderstorm and rain so heavy as to reduce visibility to nil. As it came up, Suffren signalled to cease fire, and the two fleets intermingled in great confusion, each captain concerned with keeping his ship from going aground – as some did, though only lightly and temporarily. The orderly handling of the British ships throughout the battle had contrasted strongly with the disorderly manoeuvring of the French, and, despite rather heavier casualties and the grave damage to the *Monmouth*, the enemy had had rather the better of things. So, next day, finding Hughes in a strong defensive position among the shallows, Suffren took his fleet down the coast to Baticaloa – which was still in Dutch hands. Here he made what repairs he could, insisting on remaining in Indian waters despite the almost mutinous attitude of many of his officers, who urged a return to the Île de France. They could do so the more justifiably since they knew that he had instructions to go there; but Suffren had given Hyder Ali his word that he would stay on the coast and, instruction or no instructions, he was not going to break it. Moreover, with his first-class strategic sense he knew that the only hope of re-establishing anything like the former French position in India lay in his remaining, to defeat the English fleet if he could, but at least to contest the command of the sea in the Bay of Bengal.

So Suffren went back to the Coromandel coast to meet Hyder Ali and plan with him an attack upon Negapatam, which the English had seized from the Dutch. He had been sent from France without any spares of gear, and with little reserve of ammunition and food; he had no depot from which to draw men, and no port save only an open roadstead; and after a long voyage and two battles his ships were in poor shape and their crews and supplies depleted. Yet with some supplies obtained from the Dutch and more from Hyder Ali, by stripping his prizes of spars and cordage, and by taking on board lascars and Hyder's sepoys, he got his ships re-victualled and munitioned and fit for sea, if slightly inefficiently manned, more quickly than his opponent who had a port at Trincomalee and reasonable reserves of equipment. The determination and resource that he showed in doing this are perhaps the most notable of his several claims to greatness. He was later to show them again in even more serious circumstances of destitution.

The meeting with Hyder Ali and the embarkation of troops delayed his sailing till early July and, arriving off Negapatam, he found Hughes in its roadstead waiting for him. The English admiral did not risk being attacked at anchor, but weighed and stood out to windward. AJAX had lost her main topmast in a sudden heavy squall and therefore could not be put into the line. So Hughes, being thereby equal in numbers to his enemy and having the wind, for once was the attacker. The order of the French line was: FLAMAND, ANNIBAL, BRILLANT, SÉVÈRE, HÉROS, SPHINX, PETIT-ANNIBAL, ARTÉSIEN, VENGEUR, BIZARRE and ORIENT. The English came down on a line of bearing, which they preserved perfectly, as the French were never able to do. (Indeed, throughout the long struggle between Suffren and Hughes the English fleet work was always admirable and far superior to that of the French.) Because the line was so perfectly preserved its rear ships never got truly into action, save at rather long range, for to get closer to the French rear, and they had the wind, would have meant breaking the sacred principle of line-of-battle. But the centre, and especially the van, of the two fleets became hotly engaged, and BRILLANT's mainmast was shot away, so that she drifted out of line to leeward. After some two hours of fierce gun-fighting by two-thirds of each fleet the wind suddenly shifted from south-west to nearly south-east, and blew the battle to bits by throwing both fleets into the greatest confusion.

The French ships paid off to port, with the exception of the damaged BRILLANT and the SÉVÈRE which were left between the two fleets, and each was attacked by the two English ships that happened to be nearest to her. Suffren, who had just signalled to re-form line, wore the HÉROS and came down himself to disengage BRILLANT, for he could not trust his captains to understand and act on signals quickly. The captain of SÉVÈRE, which was attacked by the *Sultan* and the *Burford*, lost his nerve and struck to the former, which ceased fire, but the French officers, furiously repudiating their captain, continued to fight the ship, and brought her clear. The fleets were now well apart, but Hughes felt that some of his ships had suffered too much for him to renew the action, and Suffren certainly could not do so; for he had had the worse of things, and his casualties were well over twice those of the English. The attack on Negapatam must be abandoned and he returned to Cuddalore. Here he received a formal demand from Hughes for the surrender of the SÉVÈRE, which he rejected with unwitting economy of truth on the grounds that her flag halliards had been shot away. When, on inquiry, he found what really had happened he sent her captain, Cillart, back to France. He also, stretching his strict legal powers (and no other French officer would have dared to do so), sent back Bidé de Maurville of ARTÉSIEN and the comte de Forbin of VENGEUR. The former had behaved badly in all the three battles that had been fought, as well as flagrantly disobeying an order on another occasion, and the latter was plainly incompetent. Cillart was

LE BAILLI SUFFREN DE SAINT-TROPEZ
(*by Pompeo Battoni*)

THE 'VILLE DE PARIS' De Grasse's flagship at Les Saintes
(off San Domingo, *by G Alaux*)

THE 'MONTEBELLO' A Napoleonic First-Rate
(*by F Roux*)

cashiered, and the others were briefly imprisoned before Bidé was dismissed the service and Forbin was 'invited' to retire[1].

The French ships were now in worse state than ever; for instance, no less than nineteen top-masts were needed. But the prodigious energy and resource of the formidable man who led them rose to the occasion. The entire mainmast was taken out of a frigate to replace that of BRILLANT, timber houses ashore were knocked down to provide scantling, long splices and poor native cordage made good the rigging, top-masts were improvised, and by August the fleet was ready to sail.

It made for Baticaloa where it could await and cover the arrival of a convoy that Suffren knew was coming from France to the southern point of Ceylon. One attempt to send a convoy went awry, for off Brest it ran into Barrington's fleet and had to bolt for port with the loss of two of its escort – a frigate, and the PÉGASE(74), whose slow sailing qualities belied its name and caused it to be overtaken by Jervis in the *Foudroyant*. The second convoy arrived safely, with the ILLUSTRE(74) and the SAINT-MICHEL (60), and it carried a number of troops, the vanguard of a force that was following from the Île de France under the marquis de Bussy, a veteran of previous wars in India whom the native princes held in great respect. Suffren made forthwith for Trincomalee, brushed past its batteries, put ashore the newly arrived troops with some landing parties from the ships, and was only too willing when the English garrison's commander proposed a capitulation with the honours of war. Hughes in fact arrived two days later; but Suffren had at last got his port. Could the English admiral get him out of it?

Hughes stood slowly away, trailing his coat. For he had now correctly gauged the nature of his opponent and the poor capacity of the French captains. His own captains would handle their ships like troops on a parade ground : the Frenchman would come out to fight and his fleet would straggle. Hughes hoped to get it to do so sufficiently to beat it in detail. He had judged correctly up to a point. Suffren had real superiority of numbers : by putting the newly arrived strong frigate CONSOLANTE into line he had fifteen ships to Hughes' twelve : he hoped to crush him for good, and he came after him in the following battle order :

| Ship | Guns | Captain |
|---|---|---|
| ARTÉSIEN | 64 | marquis de Saint-Félix |
| ORIENT | 74 | Lapallière |
| SAINT-MICHEL | 60 | d'Aymar |
| SÉVÈRE | 74 | Maurville de Langle |
| BRILLANT | 64 | chevalier de Kersauson |
| PETIT-ANNIBAL | 50 | chevalier Morard de Galles |
| SPHINX | 64 | vicomte Duchilleau de Laroche |
| HÉROS | 74 | bailli de Suffren (CE) |
| ILLUSTRE | 74 | comte de Bruyères |

[1]Bouvet of the AJAX – a veteran of the East India Company and of d'Orves' squadron – went ashore sick at about this time, and died a few weeks later.

| Ship | Guns | Captain |
|---|---|---|
| FLAMAND | 50 | Perrier de Salvert |
| AJAX | 64 | vicomte de Beaumont Lemaître |
| CONSOLANTE | 40 | Péan de la Villehunault |
| ANNIBAL | 74 | de Tromelin |
| VENGEUR | 64 | de Cuverville |
| BIZARRE | 64 | chevalier de Lalandelle Roscanvec |

Both in the battle that was ultimately joined twenty-five miles east of Trincomalee and in the approach to it, the impetuosity of the French commander, and the incapacity (or, in some cases, the unwillingness) of the French captains are seen at their worst. Hughes kept edging away with his fleet in perfect order: the pursuing French got into ever-increasing disorder, and signal after signal from the flagship only made this worse. At about 2.00 pm Hughes reduced sail and came into line. HÉROS and ILLUSTRE in the French centre backed topsails to enable their straggling colleagues to do the same. The result, owing to poor and un-prompt seamanship – was even greater muddle. The seven leading ships got too far ahead of the centre, and one or two of the rear ships overran it: of these AJAX had the good sense partly to remedy the blunder by coming into line ahead of the HÉROS. Suffren then signalled to run down on the enemy and engage at closest range, and emphasized it by firing a gun. Even on the flagship the discipline cannot have been too good, for at this she let fly her broadside, and, of course, other ships followed suit. In despair of getting his signals obeyed the bailli set the example by running down on the enemy himself, but only his immediate ahead and astern, AJAX and ILLUSTRE, followed suit.

He had intended to use his superior numbers by engaging each enemy in line while his sternmost vessels doubled on the English rear. VENGEUR and CONSOLANTE duly tried to do this, but were beaten back by the stout fighting of *Worcester*(64), the last English ship. VENGEUR had to fall out of the line on fire, and short of ammunition. Otherwise only HÉROS, ILLUSTRE, and AJAX were in close action and, with most of the English ships able to concentrate on them, they had a terrible time. Their ships' companies behaved with consummate bravery and suffered more than three-quarters of the losses of the whole French fleet; but they were fortunate that all the leading English ships did not come about and surround them. That they did not was due partly to the extreme lightness of the wind, and partly to the heavy, though over-distant, fire that ARTÉSIEN maintained on the English leader.

This battering of the three central ships lasted nearly three hours. At some stage HÉROS' flag was shot away amid English cheers and Suffren bellowed: 'Flags, flags, hoist them all round the ship'. Then suddenly the wind blew much stronger and from an entirely different quarter – from ahead of the fleets. This enabled the seven French ships that were ahead of their heroic three in the centre to come about to their rescue by passing between them and the enemy, who had also gone about. HÉROS and

ILLUSTRE had only their foremasts standing, and AJAX had her maintopmast down. They were in no state to resume the fight, but their stout resistance had done much damage to Hughes. Thanks to the great confusion among the French so that only two-fifths of their fleet had really come into action, he had had much the better of the battle, yet, with four of his ships a good deal knocked about, the French had still an effective superiority. So he stood away. Suffren had kept his newly acquired port; but the muddle of the battle had cost him dear and made him fail in his objective of a real victory. He returned, some ships under tow, to Trincomalee. There ORIENT was wrecked by bad seamanship in entering harbour. Her masts and gear were used in repairing HÉROS and ILLUSTRE and her crew filled gaps in the companies of other ships.

He wrote grimly to the Minister, throwing all the blame on the incompetence of his captains, apparently not realizing that his own impatient impetuosity had been a contributor to the failure. Once again, he wrote, he had had the English fleet at his mercy and they had been saved by the failure, to say nothing worse, of his captains: officers who had been some while at the Île de France ceased to be either seamen or fighters; he dreaded what would happen to the command should he himself be put out of action, and he begged that d'Albert de Rions should be sent in a frigate to support him and to take over if he should be killed; four of his captains had asked to throw in their commands, and he had been too displeased with them to refuse. Suffren must have used the rough side of his tongue on his officers; for while one of these, Tromelin, was a contumacious trouble-maker, later dismissed the service, and another, Lalandelle, was a slacker who then sent in his papers, the remaining two, Morard de Galles and Saint-Félix, were good officers who must have felt much provoked that they threw in their hands when, out in the Orient, officers were already short and there were no replacements for them. True, Morard de Galles pleaded a somewhat questionable ill-health; but Saint-Félix seems to have regretted the step he had taken, for he remained at Fort-Louis and returned to Suffren as second in command of a ship in some reinforcements. Suffren took his patched-up fleet to Cuddalore, where another piece of bad seamanship wrecked BIZARRE. The monsoon season was at hand. Hughes, deprived of Trincomalee, had sailed for Bombay; but Suffren doubted if Trincomalee could feed his fleet, and so went to winter at Achin in Sumatra.

He was therefore able to be back on the coast at the very beginning of the next year, 1783, well before Hughes, and did considerable execution to the trade. Bussy was at last approaching, having made a very slow passage from Europe, partly owing to an epidemic. He brought with him two seventy-fours, FENDANT and ARGONAUTE, and HARDI(66) followed a little later. This reinforcement was not adequate to balance that which England had sent under Rear-admiral Bickerton to join Hughes, who now had eighteen ships to Suffrens' fifteen. The French, throughout, had made the mistake of sending their reinforcements and convoys in little parcels, as they were ready, whereby they were liable to be cut up or driven back to port before they were clear of European waters. However, Suffren duly landed Bussy and the troops, for his energetic appearance on the scene had given him command of the Bay of Bengal, and then went to Trincomalee to give the fleet one of his energetic refits, or, more truly, repatchings. He had taken the frigate *Coventry,* which obligingly sailed into the midst of his squadron, but had lost one of his own few small vessels, the twenty-gun NAIADE, after a determined fight of five hours against the *Sceptre*(64). (The French corvette's captain was a junior officer named Villaret de Joyeuse.) On land things were going ill. Hyder Ali had died at the end of 1782, and his son and successor, Tipu, was being defeated by the British forces. Bussy, too, was a tired veteran, very different from the soldier he had been twenty years before. By June he was pushed back into Cuddalore, besieged by land and blocked by sea.

Suffren sailed to his succour. His fleet was a patchwork quilt of an affair – odd timbers had repaired the hulls, odd patches the sails, odd cordage and spars the rigging and masting; two ships had to keep the pumps going perpetually, few had a crew of more than two-thirds the right strength, and they would have to fight against odds. But the spirit of the commander was as high as ever. With clever manoeuvring he drew Hughes out of Cuddalore roadstead and got in there himself. Then, having shipped some soldiers and a number of sepoys to make up the gravely depleted numbers of his gun crews, he stood out to fight. A recent order sent out to him laid it down that in battle the flag officer commanding was to direct his fleet from a frigate (the French interpretation of the lesson taught by the cutting off of de Grasse's flagship at the Saints), so Suffren was in the CLÉOPÂTRE. The battle order of his ships was:

| Ship | Guns | Captain |
|---|---|---|
| SPHINX | 64 | vicomte Duchilleau de Laroche |
| BRILLANT | 64 | chevalier de Kersauson |
| FENDANT | 74 | chevalier de Peynier |
| FLAMAND | 50 | Perrier de Salvert |
| AJAX | 64 | Dupas de Lamoncelière |
| PETIT-ANNIBAL | 50 | Pas de Beaulieu |
| ARGONAUTE | 74 | chevalier de Clavières |
| HÉROS | 74 | d'Hémivy de Moissac |
| ILLUSTRE | 74 | comte de Bruyères |
| SAINT-MICHEL | 60 | vicomte de Beaumont-Lemaître |
| VENGEUR | 64 | de Cuverville |
| SÉVÈRE | 64 | Maurville de Langle |
| ANNIBAL | 74 | d'Aymar |
| HARDI | 66 | chevalier de Kerhué |
| ARTÉSIEN | 64 | chevalier de Vignes d'Arrac |

Hughes, with his flag still in *Superb*(74), his rear under Commodore King, who had also fought in all these battles, in *Hero*(74), and his van under Rear-admiral Sir Robert Bickerton in *Gibraltar*(80), hove to and waited for the French. Moissac, who had been the CE's flag-captain in all the battles, took HÉROS down against her old adversary, *Superb*. This left the first seven French ships to stretch and fight eight British and the sternmost seven to hold nine of them. Suffren kept the signal for closest action flying, and the British line, always in good order, continually gave way a little before the attack. The French line was steadily maintained, and after two and a half hours night ended the struggle. For a few days the fleets remained in distant sight of each other, each being wary of closing; then Suffren returned to Cuddalore. Purely as a naval action the battle had been quite indecisive, but the greatest credit must go the French for having fought an even fight against greater numbers with such worn and patched ships and with such depleted crews and their inexperienced sepoy helpers. Cuddalore and the French army were saved, and it was their English besiegers who were now in an awkward position. But like that of Wellington at Toulouse thirty-one years later, the battle with its slaughter had been unnecessary: Hughes shortly sent in a flag of truce to report that news had come to him overland that peace had been signed some five months before. In August a French frigate brought official confirmation, and with it Suffren's promotion to LG and the order to bring all but five ships and two frigates home.

Though he had won no actual victories and taken no ship of the line, he had fairly fought a stout opponent to a standstill, had landed his troops, and had saved the French and Dutch possessions in the east, and he had done this, despite occasional superiority of numbers, with all other odds against him. His crews had not been first class, for by the time he sailed France was running short of real seamen and gunners: his captains had been poor, for India was considered a side-show and the best officers were kept for the Atlantic: he had had no reserve of supplies and men, and he had had no port. His return home was a triumph. When he reached the Cape some of his English opponents were in the roadstead, and he was particularly pleased that they went out of their way to do him honour. In France, the king created an extra post of vice-admiral for him, which was to end with his lifetime. He did not go to sea again. In 1787 political troubles in Holland seemed likely to lead to war with England, and he went to command the main fleet in Brest; but the war-clouds blew over. It was his last service. In the autumn of 1788 he died of a severe attack of gout plus a possibly unwise medical bleeding. Later, one of his servants, anxious to attract attention, circulated a story that he had been killed in a duel: it received credence from some reputable historians, but it is sheer fiction.

There is rather more than a tendency among French naval historians to exalt Suffren above all their other admirals of sailing days. It seems to the present writer to be overdone. This is not to 'de-bunk' a fine officer. Patriotism, courage, verve, devotion to his Service, great professional skill: all these were his. So were determination and resource in daunting circumstances, tactical ability, and a sense of the proper use of sea power above that of his contemporaries. He was, too, the complete seaman. But he was not a lovable man. Determination when allied to harshness and egotism may command obedience and even a measure of respect: it does not command devotion. True that his severity was tempered with justice: if he dealt sternly with those who did ill, as we have seen, he also reported generously, as we have not seen, on those who did well or even adequately. Granted that some of his captains were bad and others merely adequate, yet he does not seem to have tried to weld them into a team by adding encouragement to his drive, by realizing that they knew only the old defensive rules in which they had been trained, and needed continual friendly explanation of the new ideas of attack and initiative that he intended to use. Such leadership might have enabled them to make good the excess of impetuosity that was his own undoubted fault. There must have been something wrong with a leadership that drove a sound man like Morard to resign his command when every able officer was needed. Certainly Suffren achieved wonders in the face of adversity, but he never commanded more than fifteen ships in line, and if he never suffered real defeat, he never won a real victory: there was no La Hogue, there was also no Beachy Head. That he was the greatest French admiral of his century is probably right: it probably is not right to put him on a pinnacle above Tourville or DuQuesne.

The terms of the peace that was signed at Versailles in January 1783 were that the independence of the new American states was recognized, that France gained Tobago, her former ports in Senegal, an augmentation of her fishing rights to the Newfoundland banks, and the removal of restrictions on the fortification of Dunkirk: Spain retained Minorca and Florida, both of which she had retaken, and gave back to France Louisiana, which had been ceded to her in 1763. Otherwise all conquests were restored.

Why had France, who had made such a great effort, signed a peace by which she gained so little? It was not because she despaired of successfully meeting England at sea after the defeat of the Saintes. That defeat had not cost many ships: combined with that of Spain her navy could still outnumber the English, and its spirit was still high. But the dominant position that she had held in the West Indies a year before would have to be won all over again, and Gibraltar, which seemed able to hold out interminably, would have to be taken if she were to make a peace that was triumphant. This might mean long campaigns; and her finances, now desperately overstrained, just could not stand further war expenses. A shift of the political scene in England had brought to office a ministry willing to cease hostilities if it was not required to surrender much besides the American colonies, which were recognized as lost: it was best for France to come quickly to terms with it. The result of the war was a disappointment. One of her two objects in making it, revenge on England, had been obtained by freeing the American colonies and partially destroying the Mediterranean base; the other object, recovery of her own losses by the Peace of Paris, had not been achieved, nor had she fulfilled her promise to her ally of recovering Gibraltar. Besides being a disappointment, though this was not yet realized, the war had been a disastrous mistake. Instead of peaceably reorganizing chaotic finances and adjusting social inequalities, France had incurred war expenditure that had brought her into bankruptcy, and the Bourbon despot who had betrayed his caste by aiding the establishment of a republic was to find his sin recoil upon his own head, not a little because of the example that the Americans had set to his own subjects.

Yet for the moment all seemed well where the navy was concerned. It was undoubtedly in a period of greatness as in the days of Colbert and Seignelay. It had fought its old enemy on equal terms and, apart from the battle of the Saintes, had, by and large, had rather the better of him. The peace had been made as between maritime equals. Castries, who was a forceful minister in the Colbert and Choiseul tradition from 1780 to 1787, was convinced that England would seek revenge and he set to work to prepare for that. When the war was ending it was decided to introduce flint-locks for firing the guns, and every new ship built, or older ship coming into dock for a complete overhaul, was to be fitted with copper sheathing. Ships of the line were put into three classes: first-rates of 110 to 120 guns, second-rates of 80 guns, and third-rates of 74 guns. All carried thirty-six-pounders in their first battery, and the first two rates had twenty-fours in their second battery, the third rates having eighteen-pounders[1]. The ships had a first-class designer in Sané. Castries pushed ahead with the restoration of Dunkirk to its former usefulness, and started an extensive rebuilding of Cherbourg, whose harbour works had been destroyed by the British in 1758; and he got the king – his majesty's one 'naval occasion' – to visit it. New ordinances in 1784–86 gave the service a new organization. Its ships and manning were divided into nine divisons, five based on Brest, two each on Toulon and Rochefort. This was possibly unwise, as was his shifting back to the administrative officers some powers that had been put in the hands of the seagoing branch.

In 1786 he created a corps of seamen-gunners, who signed on for eight years. The idea was admirable, and should have worked. As it was, these men, recruited partly from the ne'er-do-well riff-raff of the towns, largely contributed to the breakdown of discipline when the Revolution started. That he intended the old defensive tactics to give way to aggressive ones is clearly shown by the way he urged the king to create the fourth vice-admiralty for Suffren: the great honour and reward thereby given to that admiral made him an example for others to follow his methods. The American war had shown two weaknesses in the matter of personnel. One was that the system of *classes* had completely broken down. Men once called up were kept on without the release to which they were entitled, their pay, like succour from Austria, 'came late or never' because of the penury of the Treasury, and their service was such a wretched servitude that everyone sought to dodge it. His new system allowed registered men to opt out of seafaring, except in wartime, and graded the length of service differently for married and unmarried men.

[1]English ship ratings at this time somewhat belied their strength, for the carronade was at first treated as an additional gun to the normal rating, ie a ship rated at, say, a hundred guns would carry perhaps up to ten guns more, and so in proportion with lower rates. An admiralty official is said to have protested against this, in all seriousness, as unfairly misleading to the enemy.

The other weakness was old-standing and a greater trouble : the haughtiness of the noble officers, which showed itself not only with regard to the *bleus*, or non-nobles, but even among themselves. An instruction that put certain duties into the hands of the *bleus* was protested against and disobeyed because it was not seemly, forsooth, that these experienced and quite senior gentlemen, who were of the bourgeois class, should be in a position to give orders to young ensigns who were of the nobility. The nobles were unquestionably brave, and their training had made most of them professionally skilful, but they were quite insubordinate when it came to obeying any order of which they chose to disapprove. In the social climate of his time Castries could not alter this, but he did away with the *gardes*, devised a new system of cadet entry and grading for nobles and the sons of serving officers, and created a new rank of sub-lieutenant for the 'volunteers' (ie *roturiers* or bourgeois) who were admitted at an older age after sea-training in commerce.

It was amid this active reorganization that La Pérouse, who had a good war record, was sent with two frigates on an extensive voyage of exploration and scientific observation in 1785. He wrote, in 1788, from Botany Bay, and then disappeared. It was not till forty years later that it was discovered that his frigate had been wrecked on the then uncharted Santa Cruz Islands, east of New Guinea, and the survivors had been massacred by the natives.

Castries resigned in 1787 because of the continual attacks made on his department as being a costly luxury. He had done well, and was succeeded by the comte de La Luzerne, the learned military governor of the Leeward Islands. In 1789, the year which saw the start of the Revolution, France had seventy-one ships of the line, sixty-four frigates, and some eighty smaller craft. There was a large corps of officers, containing some dead timber but mostly well trained, and there was a reorganized arrangement for the provision of crews. In short there was a fine Service, despite one or two weaknesses, and one justly proud of its recent achievements. The Revolution was to reduce it to chaos.

CHAPTER XI

# Revolutionary and Napoleonic Wars: Part I – The Republic

The vast upheaval known to history as the French Revolution is our concern only insofar as it affected the navy – which, as we shall see, was greatly. Yet the effect on the navy cannot be understood without some knowledge of the phases through which the Revolution passed.

With his government bankrupt, the king summoned the States-General to meet in May 1789. They were a late medieval form of parliament, sitting in the three orders of Clergy, Nobles, and Third Estate (Commons), and they had not met for 175 years! Then each order had separately decided how it could help to meet the financial situation, and, this done, the Estates had dispersed. The king expected that they would now act in the same way; but the Third Estate, full of the idéologue teachings of the century and with the example of America before its eyes, was determined not to disperse until it had given France a constitution, and demanded that all three Estates should deliberate together for this purpose. It gained its point, and the joint Estates became a National (or Constituent) Assembly.

The Assembly's intention was peaceably to evolve an orderly constitutional monarchy out of a chaotic despotism. But the air throbbed with revolutionary ideas and there were many revolutionary agitators to propound them. So, as it has been said, this 'process of evolution was overtaken by a sudden picturesque event'[1]. The event was the storming of the Bastille by the Paris mob and some mutinous soldiers in July. That grim fortress prison at the east end of the capital typified the repression of the ancient régime, and its easy overthrow showed the mob its power.

In October the mob surged out to Versailles and brought the king to Paris, where he was really a prisoner, and where the Assembly, which followed him, was subject to all the pressures of the Parisian agitators. The fall of the Bastille had been followed by widespread attacks on property, with the result that 'communes' sprang up spontaneously all over France. These were local councils formed to protect property and preserve order through enrolling and paying an armed militia, the National Guard. They had no legal standing, but the king and his government lacked the power to suppress them, the more so as the Assembly, following the new American Constitution with its mania for election, had no wish to do so. Indeed, the Assembly was so anxious to make a complete break with the past that, while debating its constitution, it was also busy knocking

[1] Cf. Lord Elton, *The Revolutionary Idea in France* (Arnold).

away the props that supported the government, which was still the responsibility of the king and his ministers, by interfering with, and taking into its own hands, matters that were the government's concern. Order in the populace and discipline in the services therefore began rapidly to break down, the more so as control of the communes, especially in the big towns, began quite quickly to pass out of the hands of moderates into those of agitators who were backed by increasingly disorderly mobs. Such communes soon found that they could defy the orders not only of the royal government but even of the Assembly – for neither had any body on which it could rely to enforce them. Emigration of the noble classes, which had begun with the fall of the Bastille, swiftly gained momentum.

In the autumn of 1791 the Assembly produced its constitution for a limited monarchy and believed that the Revolution was over. But the elections for the first (or 'Legislative') assembly under this constitution showed that the feeling of the country had moved quite violently to the left. Abuse was hurled at the sovereigns of Europe, and they in their turn, worried at the position of the virtually imprisoned king and queen in Paris, made declarations that were not entirely wise; across the Rhine the *émigré* nobles were forming an army to invade their own country, exacerbation swelled, and in the spring of 1792 the Assembly forced the poor puppet king to declare war on Austria and Prussia.

The war thus started was to last, with short breaks in 1802–3 and 1814–15, for twenty-three years. It was to involve every European state; some were to be overrun and become French satellites, others were to make peace from time to time, and even to change sides; but there was one power that, entering the lists again after the first few months, was to remain in the fight to the end – the United Kingdom. The war therefore, whatever the great victories of the French armies, was always also a maritime one. Of course, with discipline in the French army at revolutionary discount, the early battles were a series of defeats and, with the fall of Verdun in September, Paris itself seemed to be threatened. This gave the leaders of violent revolution their chance. Under the cry of 'Treason' the Paris mob was roused, rushed the palace, massacred the Royal Guard and a number of other people, and the Assembly declared the king to be suspended.

This meant that the new idealist constitution was violated after a single year; but it had provided for its own amendment by a special sort of parliament – a National Convention. This was now elected, declared France to be a republic, and swiftly became the battle ground between two extreme revolutionary groups – the Girondins who represented the provinces and were de-centralizers, and the Jacobins who were centralizers representing Paris and a few of the large towns. The ill-fated king was executed in January of 1793, and war was declared on Spain, Holland, and England. The government was put in the hands of a Committee of Public Safety, and, to obtain control of this, the Jacobins, backed again by the Paris mob, proscribed the Girondins. France was now in a parlous position. Some provincial cities, notably Lyons and Toulon, were in revolt, as was the Royalist and Catholic district of La Vendée, while the armies of her enemies were on the frontier and should have overrun her but that they lacked any common plan or unity, and their generals lacked vigour.

Internally France was utterly divided. Only fervent unity could save the country; this was no time for argument or persuasion: those who disagreed or were even half-hearted must be frightened into union with the Jacobin Government. The result was the Reign of Terror whereby a special Revolutionary Tribunal, with paid and packed juries operating special laws which ran quite contrary to the normal rules of evidence, daily sent batch after batch of 'suspects' to the guillotine, many of them having committed no crime other than that of having been nobly born, a good number of others having even been leaders and makers of the original Revolution. 'Réprésentants en mission', Jacobin members of the Convention, were sent to the provinces, with complete powers of life and death, which some of them abused most terribly. One had to fall absolutely into line, or one was in grave danger of losing one's life. It was all not so very different from Nazidom with its concentration camps and gas chambers. Yet, because the Jacobins were tireless organizers, and because of other factors, such as the divisions and incompetence of France's enemies, it worked. France subdued her internal revolts, drove back her enemies, and even proceeded to conquest.

The relief from danger brought reaction. That strange moralist, Robespierre, who had come to dominate the scene, and who had already disposed of Danton (an originator of the Terror) and other leaders, now proposed to rid himself of such remaining who might stand in his way of imposing his own particular code of public conduct upon the country. The threatened politicians conspired quickly against him and in July 1794, after a famous scene, sent him and his associates to the guillotine to which he had sent so many others[1]. They found themselves hailed as having removed a tyrant and ended the Terror, and thought it wise to live up to that. So there came about what is known as the Thermidorian reaction[2]. Moderate members who had gone into hiding returned to the Convention,

[1]Among those guillotined had been d'Estaing, the admiral, who had been too loyal to France to emigrate, and whose only crime was that of being a noble who had held high command with courage if not with notable success. When sentenced by the tribunal he proudly exclaimed: 'Send my head to the English: they will pay well for it.'

[2]Thermidor – the hot month. The Revolutionary leaders were very 'classic-minded' and saw themselves as stern and austerely pure Roman republicans, nicknaming themselves Brutus and Gracchus and so on. When, first, the Church was disestablished and, later, Christianity was abolished in favour of a worship of Reason, a Revolutionary Calendar was introduced. Its brave new world started with 22nd September 1792 (the suspension of the king), the years from then onward being numbered I, II, III and so on in Roman numerals. The months were given names of classical derivation, eg Nivose (the snowy month), Brumaire (the foggy month) and so forth. (Incidentally, this must have made nonsense to those serving in, say, Martinique.) Since some of the actions fought by the revolutionary navy are known by dates from this revolutionary calendar, we will, when giving those dates, add that from the normal calendar. The use of the revolutionary calendar on *official* documents persisted until the end of 1805, but it had long since gone out of use in private correspondence.

the Jacobin Club was closed, a rising of the Paris commune was broken, and with it the power of its mob, the churches reopened, and if the Dauphin (Louis XVII) had not died of ill-treatment in prison, there would possibly have been a restoration of the monarchy.

As it was, the Convention got on with its proper business of devising another constitution and produced that of the Year III (Revolutionary Calendar) or 1795. This put the government under the control of a Directory of five, with a two-chamber legislature. Violated more than once in its four years of existence, this constitution was recognized by nearly everyone as a makeshift affair. The Revolution had failed and France was waiting for a man. In Brumaire (November) 1799 the successful general Napoleon Bonaparte, judging the moment right and having dodged his way from Egypt through English cruisers back to France, seized power. The Revolution had ended by putting France into the hands of a military despot. As First Consul (under yet another constitution) Napoleon was an absolute ruler, and the change that he made from Consulate to Empire in 1804 was one of name only. It enabled him to negotiate with other sovereigns as an equal and the imperial title and court pleased the French sense of grandeur.

It is against this background of, first, a powerless and collapsing monarchy, second a violent republic, bloodthirsty in its earlier years and always war-mongering, and, finally, a triumphant military despotism, that we have to consider the doings of the French navy.

Sailors were a rough lot and had a hard life. The *Inscription Maritime* – the 'classes' – as we have seen, had ceased to work well. There was cause for discontent, the naval ports rivalled the capital as centres of disturbance, the disorder prevalent on land spread violently among ships' companies, and discipline broke down badly. Perhaps the most notable instance was that concerning d'Albert de Rions (CE). He was held to be the most able flag-officer in the service, had been the admired friend of Suffren, was renowned for his bravery (we have seen him fighting his way through to offer a tow to the surrounded VILLE DE PARIS at the Saints), and was as charitable and upright in character as he was courageous and skilful professionally. The year 1789 found him in command at Toulon. When he went ashore to deal with a disturbance in the dockyard arising from the dismissal of two petty officers who had been preaching mutiny, he and his guard of officers were set on by the mob, insulted, wounded, and thrown into the common jail. The Assembly took cognizance of the matter and resolved, in effect, that there had been a misunderstanding and 'nobody was to blame'.

Of course d'Albert left his command, whereat his successor was seized on arrival and imprisoned for a few days before the mob-dominated commune allowed him to exercise his functions. The next year, with a dispute between England and Spain seeming likely to lead to a war which might involve France, d'Albert de Rions was sent to command the main fleet at Brest. He had to negotiate with the all-powerful municipality before it allowed him to take up the post to which the government had appointed him and then, when dealing with a contumacious coxswain according to a disciplinary code which the Assembly had issued against his advice, he found himself faced with mutiny aboard one ship. All the Assembly felt able to do was to send its crew to their homes (which was just what they wanted) and d'Albert de Rions, finding himself unsupported by his Admiralty for doing his duty, resigned his command and left the service. France had lost her best naval officer. (His successor at Brest lasted precisely one week before resigning.)

Similar disorders were taking place in all ports: officers were insulted at will, gallows were put up outside the doors of their houses, any disorderly seaman given some mild punishment by his captain lodged a complaint with the commune, and in some cases it was thought wise to recall an officer at the municipality's demand.

La Luzerne and his various short-lived successors at the ministry were powerless. The officers, even d'Albert himself, tried what is now called 'appeasement': they spoke to the insulting mutineers fairly, and they gave way to some demands. They might have been wiser to collect such troops and seamen as remained reliable and to have used force; but they found it hard to believe that the Service to which they were devoted was breaking up in this way, that the men would not come to see the folly of the stump-orators and return to duty, and that the Assembly and government would utterly fail to support authority. Therefore, as the Revolution moved to a violent left internally, so the breakdown of discipline in the ships grew worse. Some officers were hanged to street lamps by mob and mutineers at Toulon in 1792, and things reached perhaps their worst absurdity after the outbreak of war when the crew of a frigate in Leghorn deposed its captain in favour of a lieutenant of marines – entirely in the tradition and spirit of the buccaneer 'Brethren of the Coast'. The disaffection spread, too, to the colonial squadrons.

In these circumstances, with the greater aristocracy of the Court already emigrating from the end of 1789, it is not surprising that many naval officers, drawn mostly from the lesser aristocracy, emigrated too. Even by November 1791 the establishment at the chief naval port, Brest, was thirty captains and 160 lieutenants short, and the other naval bases were proportionately lacking in officers. Partly to meet the losses by emigration and resignation, and partly from egalitarian motives, the Assembly, in 1791, had issued a comprehensive new decree. The officer corps was abolished and a new body of officers appointed to consist of three admirals, nine vice-admirals, eighteen rear-admirals, 180 post captains, and an appropriate number of lieutenants and ensigns. (An ensign, who might be any age from eighteen to thirty, had more responsible duties than those given to a sub-lieutenant in the British service and must be considered as senior to him.) It will be noticed that the old titles of LG and CE disappear[1]. Merchant skippers could be given commissions in junior ranks to form a sort of naval reserve, in which there were to be continued promotions by seniority, and the officer training schools were abolished. Of course the new officers were appointed from the old corps, but room in the lower

[1]Flag-officers will therefore henceforth in these pages be abbreviated as VA and CA (*Contre-amiral* = rear-admiral). Certain captains were *chefs de division,* a permanent commodore-ship.

grades was left open for the entry of merchantmen and other non-nobles who could pass the necessary examinations. The distinction between nobles and *bleus* disappeared, indeed, in view of the emigration it was fortunate for France that she had the latter to fall back upon, and a merchant skipper or reservist officer of eight years' sea-time (including two in the navy) was eligible for posting as a captain.

In so far as it did away with class distinctions the decree was good, but it was bad in the way it destroyed the basis of granting commissions to fill the officer ranks with those who had only a merchantman's training. These could take a ship from one port to another safely enough, but there is far more than that in naval service with its fleet manoeuvres, its gunnery, its signals, and its discipline of personnel. Naval officers called the attention of the Assembly to this, but that body just would not understand that a naval officer needed special training. Nor did the decree work even to fill the gaps, for emigration and resignations were greater than recruitment. There was a modifying decree in 1792, by which, *inter alia*, the Legislative (or second) Assembly put civil administrators in charge of ports *and* of ship movements within them!

The Convention made a further modification in January 1793, admitting merchant captains with five years' sea-time; and further decrees in March, June, and July of that year really meant 'get any officers who are at all suitable'. Finally, in October 1793 came the extraordinary decree of 'purification'. This was largely due to one Jean-Bon St André, the member of the Committee of Public Safety with the navy as his special concern, to whom, as a Protestant preacher and the Jacobin son of a fuller, noble Catholic officers were anathema.

Lists of naval officers were to be posted in their ports and their home towns, and anyone who suspected an officer's *civisme* (ie devotion to the Revolution) was to denounce him. Then the council of the commune and the seamen of the arrondissement were to decide by vote whether the minister should be asked to dismiss the officer or not. It goes without saying that many were denounced (and often imprisoned) in order that the Pistols of the service – the strutting noisy types who could shout loudest for 'égalité, la nation, et la patrie' – should take their places. In this year, too, Jean-Bon St André persuaded the Convention to pass a decree abolishing the corps of seaman-gunners. It was, said he, to create an aristocracy of the sea that a man should serve only aboard ship: gunners from the land artillery, under their officers, could man and fight the batteries. The difference between serving a gun on firm land, on its wheels, and operating one, on its carriage, on the moving platform that was a ship was wildly ignored for the sake of so-called 'equality'. Moreover, many of the men drafted on board had no seamanship whatever.

In such circumstances what chance had France at sea in the war in which by 1793 she was already engaged? Her ships numbered roughly two-thirds of those of Britain; but as Britain would have to detach many to protect her colonies and her merchant shipping, the numbers would not be so disparate. Moreover, though in sum British ships of the line outnumbered the French by three to two, in total weight of broadside they were superior by only one-sixth. The French used heavier guns, were more stoutly built in the hull, carried larger crews, and were better designed, so that they were distinctly superior to English ships of an equivalent rate. Here their superiority ended. During the three and a half years from the start of the Revolution to the outbreak of the naval war the arsenals had once again been allowed to go to ruin. Stocks had to be obtained in any way possible, wherefore cordage was faulty, any old timber and iron had to be used for repairs, and sails were ill-tailored of poor cloth, so that not only were accidents aloft frequent but less spread of canvas was carried than the almost too powerful masts could bear. Rather oddly, this state of affairs persisted even under the Empire and, together with this inefficiency of crews, explains why French ships, though now copper sheathed, were often out-sailed by British vessels which were their inferiors in design.

More important was the difference in quality of the crews. Though she had to use the press-gang, Britain had a good supply of well-trained and well-disciplined seamen. The French crews were almost a disorderly rabble, with loud-mouthed revolutionary spouters as their incompetent petty officers, with main top men who went to their top for the first time in a storm or in action, and with ill-trained and unexercised gunners. Though this state of affairs improved under Napoleon's régime, French gunnery and seamanship throughout this long war were always markedly inferior to those of the British. Above all, the British had a large corps of well-trained officers, many of them with battle experience from the American war. The French officers were collected, as we have seen, from any possible source. The remaining nobles and the *bleus* of the royalist service were a minority; the rest were inexperienced, knew nothing of working in a fleet or of battle, and, for some years, till they were weeded out, were not infrequently unsuitable incompetents. When Villaret-Joyeuse took the French fleet to sea in 1794, he and one other flag-officer had been lieutenants only three years before: the third flag-officer had then been a sub-lieutenant, and of his twenty-six captains, nine had been merchant service officers, one a lower deck rating, and another a merchantman's boatswain. From this one can imagine the quality and experience of the more junior officers. France had the better ships: in all else the odds were against her.

At the outbreak of war in 1792 only one of France's opponents had even a small naval force. This was the kingdom of Sardinia, ruled by the House of Savoy whose principal territories stretched along France's Italian frontier. Orders were sent to Toulon to fit out a squadron of nine of the line. But already, only three years from the start of the Revolution, the destitution of the dockyard was such that no more than five could be put to sea. Others followed later, or came round from Brest. This fleet made a raid on the little port of Oneglia, which harboured corsairs, and detached some ships to bully an apology from the King of Naples for an alleged insult. (The LANGUEDOC was dismasted there owing to bad handling in poor weather.) Then at the end of the year it sailed for Cagliari in Sardinia to make an invasion of that island. The ships duly silenced the batteries and landed the troops. These were mostly fervent revolutionary volunteers from Marseilles and, after marching a short way, they panicked for no reason, and rushed frantically to the shore demanding to be re-embarked. The fleet had already lost one ship of the line through bad pilotage at Ajaccio, and it was now to lose another through having to hang on in Cagliari gulf in bad weather in order to take off the frightened soldiery. It crept back to Toulon in a disunited way at the start of 1793.

France followed this poor beginning by declaring war on all the maritime powers in February of that year. Spain, with the Family Compact operating, would surely arm after the execution of the king, and French troops had taken Antwerp and forced the opening of the Scheldt. This was a violation of Dutch territory and the breach of a treaty dear to England, who would never be content while France occupied Antwerp opposite her weak east coast and so held a pistol to her head. By taking the first step France hoped that her privateers would seize a number of English traders before they could get safe to port, where their crews would be available for naval vessels.

In the Atlantic the year saw little naval activity, but much naval disorder. Morard de Galles (VA) was in command at Brest. He had been a captain under Suffren and was a competent officer, loyal to his Service whatever might be the Government of the country. He took the fleet to sea in early spring. It ran into bad weather and there were collisions and dismastings which quickly showed the utter incapacity of many officers, and of crews who largely ignored orders and stayed below. One captain was killed by an accident when doing a job that should have been done by an able seaman, and the admiral himself was unable to wear his flagship when its headsails blew away because he could get no more than thirty men on deck.

The fleet returned, damaged and half water-logged, in penny numbers. When it had been rather sketchily refitted, Morard de Galles took it out again to guard the approach of a convoy, but, sighting Howe's Channel fleet, could only run for it to Quiberon, though he was superior in numbers. Howe did not attempt to pursue, no doubt considering, as did all the allies, that France was so divided and disordered that she must soon collapse, wherefore it was not worth losing lives in a battle. This was to reckon without the ferocious drive of the Jacobins who made up the Committee of Public Safety. They ordered Morard de Galles to lie off Quiberon to prevent any assistance from reaching the Royalist rebels, to cover all coastal trade, and to put out to bring in, or attack, convoys. This was an impossible combination of tasks and the admiral asked, first to be allowed to take the fleet back to Brest, and next to resign his command: he was told to remain in command and obey orders. When bidden in September to intercept a weak Dutch convoy, he found that his men would not obey orders to make sail. The ships' companies, cooped up on board, wretchedly clad and wretchedly fed, insisted on a return to Brest; and a *député en mission* agreed that there was nothing else for it.

At Brest there arrived several deputies headed by Jean-Bon St André. This member of the Committee of Public Safety dealt with the trouble on strange lines. He relieved the admiral of his command, solely and frankly because he was nobly born and so could not expect good citizens to obey him. (Later he was put under arrest for some time.) He sent some other officers and some ratings to Paris for trial by the Revolutionary Tribunal, where they were guillotined on lying charges. This appears to have been done on the demands of the local revolutionary societies who had had complaints from seamen that these men had done their duty. That was followed by a wholesale dismissal of almost all officers who remained in the fleet from the old royal navy, other than the *bleus*, whom Jean-Bon was wise enough to make a point of protecting and retaining. (Their places were filled with 'Pistols' nominated by the local so-called patriotic clubs.) There was not a flag-officer left; but Captain Villaret-Joyeuse was raised to flag-rank and given command. Having taken these arbitrary and unjust measures, Jean-Bon issued, and saw enforced, some stern disciplinary decrees. These brought the fleet to order, but not to efficiency.

If the Atlantic saw nothing but naval disorder, the Mediterranean scene was much more eventful. After its return from Sardinia to Toulon the fleet there was commanded by Trogoff de Kerlessy (CA), a former noble. Toulon, as we have seen, was the home of great revolutionary disorder and was dominated by a local committee of Jacobins who terrorized the majority. When the crew of a frigate mutinied that body arranged an entirely irregular court martial to try the captain (!) and sentenced him to death for no other reason than that he was an officer of the old royal navy. The whole fleet was in a mutinous state, considered that it did its duty by shouting 'Vive la Nation', and refused to weigh and put to sea.

But the proscription and execution of the Girondin members of the Convention (see p 202) altered everything. Like a number of others of the provincial towns whom the Girondins had represented, Toulon rose in revolt and expelled its Jacobins. When the Toulonnais saw that the risings in the provinces were failing, that the Jacobins of Paris remained in control and their armies were approaching to restore their authority and exact retribution, they declared for the little imprisoned king, Louis XVII, and called for aid to the allies. These had, in the Mediterranean, an English fleet under Lord Hood and a Spanish squadron under Don Juan de Langara. Hood, who had refused a request to succour Marseilles (which had also risen against the Convention), accepted the invitation to Toulon, but on conditions, such as the withdrawal of the French fleet to the inner harbour, which left him in control; and he entered the roadstead in August. The Spaniards arrived a little later. Trogoff seems to have let things drift. He had spent the previous months of the year in correspondence with the Minister of Marine, a former privateer named Dalbarade. To the admiral's requests for help in retaining discipline, and for supplies to get the fleet into order, Dalbarade had sent vague or even ridiculous replies; and Trogoff probably felt too dispirited and, perhaps, as a former noble, too anti-Jacobin, to do anything but go with the current. His second, St-Julien (CA), tried to whip up naval resistance to the hated English, but got little support and fled.

The events in Toulon had taken the allies by surprise. Though France had been at war for over a year they had failed to realize the patriotic fervour of the bulk of her people, thought that she must soon collapse, and had not put out their full strength to deal with her. Therefore they had not adequate troops ready to take advantage of the situation. Soldiers were sent: British, Neapolitan, Sardinian, and Spanish; but they were not enough to hold a sufficiently large perimeter, and there was no unity of command. Hood had carried out an inquiry among the French ships as a result of which he packed off some four thousand officers and seamen who, though they had not fought against their entry, were obviously opposed to the allies. These were put into four of the French ships of the line, which had been disarmed, and they were sent under Captain Bouvet to French Atlantic ports. (Some of those who landed at Rochefort were guillotined at the orders of a packed tribunal set up by two deputies *en mission*: a poor reward for refusing to join their country's enemies.) The removal of this potential fifth column could not make good the lack of troops at Toulon. The Jacobin armies closed in: the skill of their artillery officer, Bonaparte, drove the allies from their positions, and Toulon had to be evacuated in December. There was a dreadful scene of people fleeing the guillotine by trying to get aboard the ships, and all this amid falling shot and the flames and explosions of burning vessels.

Hood had been unable to carry off the French fleet during his occupation of the town, for the Toulonnais were his allies. Evacuating, he tried to destroy it and its dockyard. This was only partially done, for the Spaniards deliberately neglected their share of the destruction, Langara frankly admitting later that the sinking of the French ships would have weighted the naval balance of power altogether too heavily in favour of the English. So one ship of 120 guns, four of 80, and nine of 74 were saved for France, with some frigates. But nine 74s, including the HÉROS, Suffren's old flagship, and some lesser ships were burned, while one 120-gunner, two 74s and some smaller vessels were carried away by the English, and another, plus some frigates, by the Spaniards. Toulon was left to a dreadful fate and, what with flight, loss in battle, and executions, was more than half depopulated. Hood was reproached for not having made a more complete destruction; but the loss of ships to France was more than that of a battle. A naval disaster without a naval combat.

With the population largely dispersed, and the arsenal destitute, it was some time before Toulon could again serve adequately as a naval base.

During the year the British had begun to mop up French colonies by the easy occupation of Tobago, St Pierre and Miquelon, and Pondicherry. Among the single ship actions was an unusual one when two frigates of equal strength lying within neutral waters off the United States sailed out of them to give battle to see which was the better man. The English frigate had rather the worse of a slaughter that may have shown gallantry but was a pointless playing with men's lives. In San Domingo there had been strange happenings. The usual convoy of the French West Indian trade had been collected there, but two commissioners sent out from France refused to allow it to sail until the warships that were to be its escort had been replaced by the arrival of others. This led after a while to a kind of civil war between the governor, plus the merchants, and the commissioners, who armed the Negroes and criminals from the prison against their own compatriots. Eventually, when the season was too late for the Atlantic crossing, the ships all left for American ports. Here, in the next year (1794), they were gathered into a huge convoy together with a large number of grain ships which the home government, threatened with famine, had ordered its representative in the States to assemble. One of the new CAs, Vanstabel, was sent out with two of the line and some frigates to bring it home and prodigious efforts were made to get the Brest ships to sea to save it from being intercepted by the British Channel fleet under Lord Howe.

Nielly (CA) first sailed from Brest, with five of the line, as an advanced protection, and was followed on May 16th by the main fleet, twenty-five and some frigates, under Villaret-Joyeuse (CA) with his flag in the MONTAGNE (120)[1]. His second-in-command was Bouvet (CA) in the TERRIBLE (120). He was joined by the PATRIOTE (74), which had wandered from Nielly's squadron, just before he sighted Howe. This was on May 28th, roughly 400 miles west of Brest. Each fleet was twenty-six strong. Villaret was somewhat handicapped by the presence in the flag ship of Jean-Bon St André, who could override him. But the handicap was not great, for they seem to have agreed well enough, and it disappeared when it came to combat when, in Villaret's words, the *coquin* soon found an excuse to go below.

Villaret's instructions were to avoid a battle unless it were necessary to save the convoy, and he had the weather-gage. So he manoeuvred both to keep this, whereby Howe to leeward could not come to grips, and to draw Howe away from the path of the convoy. He had been heading northward before a strong southerly breeze, but on sighting the English he signalled to form line from columns, on the port tack, ie heading westward. The awkward slowness with which his ships carried out this ordinary evolution cannot have encouraged him. Howe, to the north of him on a parallel course, signalled a general chase and sent five of his fastest sailers to harass the French rear. These were engaged by the powerful RÉVOLUTIONNAIRE (once the BRETAGNE) of 110 guns, which dropped back to the hindmost position for the purpose. This formidable vessel successfully held off the lighter ships engaging her, and the *Audacious*,

[1]Those reading French naval History of this period in detail will find a large number of ships with names not previously encountered. These were not newly built ships, but old ones with their names changed to suit the shifts of the political winds. First every name connected with the royal government gave place to revolutionary titles, such as RÉVOLUTIONNAIRE or TYRANNICIDE or classical republican heroes, or even, somewhat unsuitably for ships, revolutionary dates, eg TRENTE-ET-UN MAI. After the Thermidorian reaction these were modified, to be modified again under the Consulate and Empire. The ETATS DE BOURGOGNE in the Royal Navy first became the CÔTE D'OR, then, with the Jacobins in power, it became the MONTAGNE – the popular political nickname for the Jacobin party in the Convention – next, on their fall, it was renamed the PEUPLE, and it finally settled down as the OCÉAN. Of course, names like DAUPHIN-ROYAL and ROYAL-LOUIS disappeared, to become for a while the SANS-CULOTTES and the RÉPUBLICAIN, but even the AUGUSTE had to become the JACOBIN, and Suffren, though a naval hero, had to be partly dishonoured by having the ship named after him rechristened REDOUTABLE – and so on. For some reason, perhaps because he was less recent than Suffren, the TOURVILLE retained his name.

which continued the fight till after dark, was so battered that she had to put back to England. The bravely fought Révolutionnaire, however, had to fall astern with heavy casualties and was so damaged that all her masts went overboard: she was fortunately met and helped back to France by the Audacieux from Nielly's squadron[1].

During the afternoon, to keep his ships together for they were beginning to straggle, Villaret had tacked, so that the fleets were running south-eastward; but next morning, May 29th, Howe tacked again, going westward and aiming both at an attack on the French rear and at getting to windward. Villaret promptly tacked also, and, as the superior seamanship of the British had brought them close to the enemy, the fleets engaged. The critical moment of these manoeuvres had now been reached. Howe signalled his fleet to go about and pass through the French line and, when his leading ship failed to execute the order and so threw his van into some confusion, he led his centre down and, with two others, broke through the French and got to windward of them. This done he bore down on the two rearmost Frenchmen, Indomptable and Tyrannicide, which, being badly sailed, had lagged astern and were now in danger of being cut off and surrounded. To save them Villaret signalled to wear in succession, and his leader, Montagnard, failing to do this owing to some damage, he set the example by wearing his flagship, the rest of the fleet following. He was now running down past the bulk of the English fleet which was in some disorder, but he did not close with them as his object was to rescue his own rear. This he did, but by passing the English on an opposite tack he had left them free to sail across the stern of his ships and join their admiral. Villaret had lost the wind, and Howe could now force an action.

With this the day's battle ended. Though saved, the two rear ships were much injured and the Indomptable(86) left for home under the escort of the Montblanc(74): Tyrannicide was kept in line, towed by her next ahead. Villaret's fleet was reduced to twenty-three (Howe having now twenty-five) and stood away somewhat north of west, taking Howe, in pursuit, clear of the likely route of the convoy. Fog now descended on the scene for a day and a half during which time, though Villaret lost the Montagnard which put back for Brest without orders, he had the great good fortune to be joined by Nielly with three ships, and by a fourth which had come from France, so that he once again had twenty-six of the line to the British twenty-five. Nielly shifted his flag to the Républicain and, battle now being unavoidable since Howe had the wind and was more than skilful enough to keep in touch, fog or no fog, Villaret arranged his fleet in the following order:

| Ship | Guns | Captain |
|---|---|---|
| Convention | 74 | Allary |
| Gasparin | 74 | Tardy |
| América | 74 | L'heritier |

[1]It seems that she struck but that it was unnoticed in the dark: *Audacious* was not too damaged to have boarded her had she known. Her officers were imprisoned in Brest for 'leaving the fleet'; but were soon released.

| Ship | Guns | Captain |
|---|---|---|
| TÉMÉRAIRE | 74 | Morel |
| TERRIBLE | 120 | Bouvet (CA) |
| IMPÉTUEUX | 74 | Douville |
| MUCIUS | 74 | Larréguy |
| ÉOLE | 74 | Keranguen |
| TOURVILLE | 74 | Langlois |
| TRAJAN | 74 | Dumoutier |
| TYRANNICIDE | 74 | J Dordelin |
| JUSTE | 74 | Blavet |
| MONTAGNE | 120 | Villaret-Joyeuse (CA) |
| JACOBIN | 90 | Gassin |
| ACHILLE | 74 | Lavillesgris |
| VENGEUR DU PEUPLE | 74 | J Renaudin |
| NORTHUMBERLAND | 74 | Etienne |
| PATRIOTE | 74 | Lucadou |
| ENTREPRENANT | 74 | Lefranck |
| NEPTUNE | 74 | Tiphaine |
| JEMMAPES | 74 | Desmartis |
| TRENTE-ET-UN MAI | 74 | H Ganteaume |
| RÉPUBLICAIN | 110 | Nielly (CA) |
| SANS-PAREIL | 90 | Courand |
| SCIPION | 74 | Huguet |
| PELLETIER | 74 | Berrade |

The fleets were over 400 miles out in the Atlantic and the action about to be fought, known to the French as that of the Thirteenth Prairial (revolutionary calendar) and to the English as the Glorious First of June, took place further from the land than any other major battle of sailing-ship days. The wind was south by west and both fleets were on the port tack, ie heading west. Villaret had made all sail during the previous day and the night in a last hope of getting ahead of the English to avoid battle, according to his orders; but having failed he reduced to short sail. Shortly after eight Howe signalled to bear down on the enemy, each ship to break through the line and engage its opposite number from leeward. The approach was slow, as usual under sail, but the French van opened fire at about nine. Some of Howe's ships, notably the *Caesar* in the lead, made no attempt to carry out his order and engaged from windward, others did so because they could not make a gap, like the *Brunswick*, which ran against the VENGEUR broadside on and fought her in a terrible corps-à-corps duel. In fact only seven broke through, and those not all before doing some fighting from windward, while an eighth got to leeward by going across the stern of the rearmost Frenchman; but those that did break through reduced the French line to great confusion.

Howe's flagship, *Queen Charlotte,* made for the stern of the MONTAGNE, which backed sails to block the way. At the same time the JACOBIN, her next astern, made more sail, to close the gap, and so overran the MONTAGNE and came up on her lee. The *Queen Charlotte*, squeezing through, and raking the French flagship with dreadful effect as she did so, was smartly brought up between the two Frenchmen, giving each her broadside. The JACOBIN replied, but instead of remaining to hammer the English admiral between the two ships, passed on and, firing a broadside or so at English ships as she went by, stood off to leeward and out of the fighting. MONTAGNE, also passing on, was engaged in turn with *Royal Sovereign*, *Valiant*, and *Marlborough*. The *Queen Charlotte* next fought with ACHILLE and then with JUSTE. Some French ships acted like the JACOBIN, fired a broadside or so at passing ships and then stood away to leeward: these included the two leaders, CONVENTION and GASPARIN, as well as TOURVILLE, TRAJAN, and TYRANNICIDE from the centre and some, such as PATRIOTE and PELLETIER, from the rear. Others remained, fighting stoutly if not always efficiently, though some from van and centre, early dismasted, wholly or partially, were overtaken by ships astern of them. The battle became, in fact, a general mêlée, and a heavy pall of smoke soon reduced visibility to a short distance.

At about half past ten this cleared and Villaret could take stock of the position. He found MONTAGNE almost alone, though a few ships were reasonably near and more or less in line on him. These included Bouvet's TERRIBLE with only her foremast standing. Eleven of his ships, mostly at least in part dismasted, were, in his own words, 'pell-mell among the English'. The remaining ten or eleven were in various stations to leeward, some more or less in line there. He signalled these to close the flag and, his signal not being obeyed, ran down towards them signalling them to form line on, and follow, him. This they did, and he led them towards the ships that were surrounded by the enemy. He did not intend to beat up to renew the action – some of his ships were too damaged and others too unreliable for that – but to give any of those cut off an opportunity to join him, if they could get clear. His course took him towards the *Queen* which, shot about in the earlier fighting, had engaged the PATRIOTE and then had had a furious duel with the SCIPION (which lay near her[1]) and was now too dismasted to move. To save her Howe, getting his flagship about with difficulty, for she had lost her main-topmast, signalled to form line, and steered to support her.

This manoeuvre was successful, but it could not prevent Villaret from recovering and towing off the SCIPION, and the withdrawal into line of many of the English ships allowed RÉPUBLICAIN, JEMMAPES, and MUCIUS to make some sort of sail on what stumps of mast were left to them and escape from the surrounding English to the shelter Villaret was providing. The last named of these had struck, but had not been boarded and so felt entitled to get away. (This would probably have been considered to be dishonourable in the old royal navy.) In the fighting that was still continuing while this happened, the most notable incident was the sinking of the VENGEUR DU PEUPLE. She had had a terrible hammering from the *Brunswick* which

[1]The discipline of the French navy was so poor that SCIPION's men had left their guns and hailed that they surrendered; but her officers kept her flag flying.

had hooked on to her and had a great advantage in that her gunners had flexible rammers and spongers on stiff rope with which to clear and load their cannon while the French had the old inflexible wooden implements and so, with the ships jammed together, had great difficulty in working their guns. Other ships had joined in the battering and only the TRENTE-ET-UN MAI had given VENGEUR even a little temporary support. At about four-thirty she struck and was plainly sinking. The victors sent their boats which took off many of the crew (who left their pumps and their wounded, as did her captain), and she went down before the boats could return, though many more of the crew, clinging on to wreckage, were picked up.

Villaret stood by till the evening when, the ships that were cut off having surrendered, he made for Brest. Howe has been criticized for having chosen to fight the French fleet instead of seeking the convoy. This discussion is not our concern, though it may be said in passing that it is hardly credible that the loss of one convoy, however large and however badly needed, would in itself have brought France to her knees. A more justifiable criticism is that he failed to pursue the retreating French. He was an old man of sixty-eight : for more than four days, culminating in a desperate battle, he had been under great strain, with practically no rest : he was utterly exhausted and does not seem to have been well advised by his flag-captain. Villaret was certainly fortunate in not being followed. Five of his ships were in tow and must surely have been taken; and he might also have been sandwiched between a pursuing Howe and a squadron of nine of the line under Montagu. These had been sent to seek the convoy and had missed it, but sighted the returning French fleet to windward fifty miles from Brest. The last thing that Villaret wanted was another battle, with so many of his ships damaged and full of wounded; but he put on a bold front, made for the English, and Montagu, facing odds of two to one, stood away. The French fleet and Vanstabel's convoy of 117 ships got into Brest almost together.

Villaret could view the result with a certain amount of equanimity. His instructions had been to fight only to save the convoy, and he had been brought to battle solely because of the slow incompetence of some of his captains in manoeuvring. Many years afterwards, when he had capitulated as Governor of Martinique, he was willing to discuss the campaign with his captors. He had been told, he said, to get the convoy home safely or come to Paris and lose his head : what did the loss of a few old ships matter : he had saved his convoy and he had saved his head. Barère, the orator of the Committee of Public Safety, made a notorious lying speech in the Convention in which, much as the Nazis dramatized the Stalingrad surrender, he dramatized the sinking of the VENGEUR DU PEUPLE, and thereby distracted attention from the disaster.

For disaster it had been. Seven ships were lost : VENGEUR sunk, and SANS-PAREIL, JUSTE, AMERICA, IMPÉTUEUX, NORTHUMBERLAND, and ACHILLE taken. Jean-Bon, as the politician in charge, had to justify himself, and did so by throwing the blame on the captains. Two were sent to prison and six others were cashiered. This was perhaps harsh; but they were no great loss, most of them being those who owed their appointments to strutting before the Popular Societies and denouncing their superiors in order to obtain their places. In addition to showing the superiority of British fleet work the battle had shown the predominant superiority of British gunnery: the English killed and wounded were 1 148, the French losses were over 5 000, which, even when prisoners are deducted, leaves a loss of roughly three to one in casualties: the *Queen Charlotte* had forty-two killed and wounded, the loss aboard the MONTAGNE was 300. The English had had damage to a few ships, but the damage to the French was far greater. This superiority in gunnery, though lessened later, was to endure throughout the long war.

Other events of 1794 pale beside this campaign and can be briefly mentioned. Most of such small ship actions as there were went against the French, usually because of poor seamanship and gunnery; and when Corsica rose in revolt the British, going there in support, mopped up the frigates and corvettes that were in the island's harbours. But there was a gallant action off the Île de France. This was blockaded by two English sixty-gunners and the only French forces were two frigates. These stood out and fought their two more powerful opponents and, though battered, did them sufficient injury to oblige them temporarily to raise the blockade. In the West Indies, too, a couple of frigates under Captain Leisségues took an active part in the expulsion of the English from Guadeloupe, which they had invaded.

There were only a few ships left in Toulon after the allied occupation and destruction of the previous year, but Martin (CA) took these to sea in June with his flag in the SANS-CULOTTES(120). Sighted by Hood's fleet, he ran for it into Golfe Juan, near Cannes, and put himself into a strong defensive position. It was five months before, a gale having blown the English blockaders away, he was able to dodge back the short distance to Toulon. Nielly (CA) put to sea from Brest with five of the line in October with his flag in the TIGRE(74) to intercept British convoys. He failed in this, but did run across two ships of the line returning from convoy escort and overtook one of them, the *Alexander*(74). This English ship beat off one of its assailants and damaged two others before striking. It was once again a case of poor gunnery, for the French casualties were three times the number that they had inflicted on their over-matched opponent. Even this capture was offset by the wreck of the powerful RÉPUBLICAIN in Brest roadstead owing to faulty gear.

Finally, this disastrous year ended in another disaster, at once almost scandalous and almost grimly farcical. The damaged fleet returning from the battle of the 13th Prairial (1st June) found the dockyard bare of means to repair the ships. Apart from Nielly's 'sortie', therefore, they had been unable to put to sea. Yet there was much talk, especially in the Convention, of the why-isn't-that-fleet-doing-something nature. So, as it had been decided to send some ships under Renaudin (the captain of the VENGEUR promoted to CA for his defence of that ship) round to Toulon to strengthen the squadron there, the whole fleet was ordered to see it out of port and to cruise for a fortnight. This was at the very end of December.

Nothing could better illustrate the sheer autocratic ignorance of the French politicians; for one did not send a sailing ship into North Atlantic winter weather unless there was need, and, even then, unless the ship was well found. Apart from Renaudin's division the ships were anything but well found, and, again apart from Renaudin's vessels, they were provisioned only for little more than the fortnight, so poverty-stricken was the dockyard. Three of the line sank, and it was only with difficulty that the bulk of their ships' companies was saved: another ran ashore just as it was sinking (fifty of its men were drowned inboard) and two more struggled into the nearest harbour in a battered state. Blown out to sea, the rest of the fleet had to reprovision itself from the ships destined for Toulon, which therefore could not make for that port; and it was not till 2nd February in the next year (1795) that the whole lot could get back to Brest, two of them barely still afloat. They had captured some merchantmen and a corvette. The Thermidorian Reaction (see p 203) had cleared out the old Committee of Public Safety and Villaret-Joyeuse could freely write his mind to the new one on what he thought of the officers left to him after the great purge of the 'uncivic' former naval officers carried out a year before. Some post-captains, he wrote, were less than mediocre, while the captains of frigates were purely out for plunder and had no proper naval sense whatever. Add to this the destitution of the dockyards and one can see that it is a far cry from the Service of Louis XVI's time. The egalitarian revolution had done its damage quickly: it was to take long years to repair it.

A nadir was now to be reached. In March 1795, Martin (CA) was ordered to take the Toulon fleet to sea to clear the British out of the way (so confident was the *réprésentant en mission* who was giving Martin his orders) before sending an expedition to expel the English from Corsica. Martin, who had received his promotion to lieutenant only in 1792 and then, in a year, had been further promoted to captain and contre-amiral, knew that he had not yet the experience or capability to command a full-sized fleet; and frankly said so. Nor was his fleet a good one to command, for he doubted if some ships would stand up to foul weather, and of his ships' companies of 12 000 men, 7 500 (including over 2 000 soldiers) had never been to sea before, such was the shortage of sailors. He had fifteen of the line, with his flag in the SANS-CULOTTES(120) which had been formerly the DAUPHIN-ROYAL, and was subsequently to be the ORIENT. He captured the 74-gun *Berwick*, which had been dismasted and was struggling along under jury rig, and then met the English Mediterranean fleet of thirteen of the line, plus one Neapolitan, which, under the command of Hotham, had sailed from Leghorn on hearing that the French were at sea.

Martin had the weather-gage, and tried to cut off four English ships that were ahead of the rest, but the wind was too light for three days for the fleets to close. Bad seamanship and fitting out now affected the French for, first MERCURE lost her main topmast and stood away for Golfe Juan, and second ÇA-IRA(80) and VICTOIRE collided, whereby the former lost her topmasts and fell off to leeward between the two fleets. Next day

Martin formed line of battle and steered to protect this lame duck, which was already being attacked. Her assailants dropped away and CENSEUR(74) took ÇA-IRA in tow. Martin had shifted his flag to a frigate, in accordance with an order then in force, and during the following night his powerful flagship lost touch with the fleet! Next day a light wind reached the English who attacked the lagging CENSEUR and ÇA-IRA. These beat off two ships, but two others came up. As the French got this wind their line fell into confusion, but Martin got four ships to bear down towards their two fellows in trouble. They did no more, however, than pass the English on an opposite tack, each fleet cannonading the other. CENSEUR and ÇA-IRA, both much battered, struck. Martin took the remains of the fleet back to Toulon, and Hotham, who was the reverse of an energetic commander, did not pursue – fortunately for the incompetent and disorganized French. *Illustrious*(80) had been damaged aloft in her fight with the two French ships and, breaking her tow to Leghorn, ran ashore and had to be destroyed.

So ended this 'battle' of Cape Noli: a frustrating sort of action because of the lack of wind. As, in return for losing two ships, the French had previously taken the *Berwick*, while the *Illustrious* went ashore, they could claim that losses were equal; but in fact they knew that their fleet-work had been deplorable and their gunnery bad. They had done some small damage to the masting of a few English ships, but this had been nothing to the battering received by ÇA-IRA and CENSEUR, and their casualties were much the heavier.

Back in Toulon, when he busied himself to get his fleet to sea again, Martin found himself faced with a full-scale revolt in the town. Ever since the expulsion of the allies the Jacobins had been in power there, and they did not view kindly the fall of their party in Paris. So these patriots seized the arms in the arsenal, called out the dockyard workers and, claiming that the traitors in Paris had ordered the fleet to sea to put it into the hands of the English, urged the sailors to desert. These, underfed, ill-clad, and irregularly paid, were only too ready to do so. It was to stop the desertions that Martin took the fleet to sea again in June. He had his old flagship (now named ORIENT) and nineteen of the line, for Renaudin's ships had arrived from Brest. Some of his captains were soon to become well-known names in the French service – Cosmao Kerjulien, Allemand, Richery, Infernet, Ganteaume, and Dumanoir – but the ships were in poor trim and sadly undermanned. (Two had to be sent back, one because of a collision, the other because she was unfit to be at sea.) Pursued by Hotham, who had twenty-one of the line and two Neapolitans, Martin had no choice but to run for home. His last five ships were engaged by the fastest sailers among the British and one of them, ALCIDE(74), was set on fire and blew up. She had struck, and British boats took off some of the crew. The sad truth was that Martin had not dared to risk a general action to save her, because of the state of his fleet. The rest got away, as Hotham once again forbore to press the pursuit.

This was all shameful enough; but worse was to happen to the Brest squadron. After the tragic absurdity of the winter cruise, only a dozen ships were in a condition to go to sea, and some of those had unsound masts and gear – for the dockyard storehouses were empty. Vence (CA) was sent with three of the line to protect the considerable coastal traffic, for in those days of horse traction and bad roads, goods were transported by water as much as possible. In June he and his convoy were pursued by five of the line under Cornwallis, whose flag was in the *Royal Sovereign*(100). Vence was not of the kind that forms line and sacrifices himself for his charges. Exchanging a few shots with the enemy he ran for Belle-Île and the safety of Palais roads, leaving the convoy to have some of its ships taken. He then sent overland for rescue and escort, and Villaret-Joyeuse was ordered to sea with nine of the line, his flag being in the PEUPLE(120), and a number of frigates. Having picked up Vence, Villaret was taking the whole twelve of the line back to Brest when he sighted Cornwallis's five and pursued them. As his flagship was a poor sailer Villaret shifted his flag to a frigate, and the position after two days became that the British were well together in line ahead, but the French were strung out, their faster sailers overtaking the enemy, poorer sailers dropping astern almost out of sight. It behove the faster ships to attack the British, who were keeping close line at the speed of their slowest ship, and delay them till the others could catch up. Villaret signalled the speedy ZÉLÉ and five close astern of her to do so.

ZÉLÉ opened fire on the last English ship, the *Mars*(74), then, on the plea that he was short-handed and unhappy about his masts (which had been good enough to put him in the lead), her captain backed sails and was content to keep up with the pursuit, out of range. The TIGRE, DROITS DE L'HOMME, FORMIDABLE, and a frigate engaged the rearmost ships, especially the *Mars* which, damaged in rigging and some canvas, began to drop astern. At this, Cornwallis brought his powerful flagship round and beat off the attackers, *Mars* making good her damage by first-class seamanship. The pursuit then continued for a while, but TIGRE and DROITS DE L'HOMME had to lay off, injured, his slow ships were out sight astern, the enemy's Channel fleet was known to be at sea, and Villaret felt that his squadron was getting altogether too strung out, so he called off the chase. In this action starting off Belle-Île Cornwallis had certainly conducted a masterly retreat; but the plain truth was that at least one French captain had funked and the gunnery of the other ships was not good enough for its purpose. The damage and casualties to the English were distinctly less than those that they had occasioned the French – the *Mars*, long under fire, had only thirteen wounded.

Four days later this French fleet was sighted by Sir John Borlase Warren, who, with three of the line, was escorting an ill-fated Royalist expedition which, landing at Quiberon, hoped to occasion a general rising in France. Sir John at once sent off to notify Lord Bridport who was covering the operation with the Channel fleet, fourteen of the line, with his flag in the three-decker *Royal George*. When Bridport came in sight Villaret decided that he must run for it to the shelter of the Île de Groix (off Lorient) and he and his two flag-officers shifted to frigates in accordance with the prevailing ruling when an action was imminent. His intention was to anchor in close line behind the Île de Groix, whereby he would be in a good defensive position and, not being under sail, his men could concentrate on the guns; but his immediate worry was the slow sailing of the ALEXANDRE. (This had led to her capture from the English in the previous year.) He sent frigates to tow her and REDOUTABLE, another slow-coach, and signalled the fleet to form line of battle on and ahead of ALEXANDRE. The order was imperfectly obeyed, the French captains being more concerned with flight. He therefore signalled to form line abreast on ALEXANDRE. This would have kept the fleet to her speed and enabled them, by an eight-point (ie right-angled) turn to come into line of battle on either tack. This order was ignored, despite hailings and separate signals to individual ships.

By the second day of pursuit the British, who were in a general chase, came up with ALEXANDRE. Villaret then signalled to form line of bearing on her, which would have put the fleet into port quarter line on its leading ship which, in this flight, it goes without saying was the ZÉLÉ. Only FORMIDABLE and TIGRE obeyed, whereby they and ALEXANDRE became hotly engaged, while some English ships reached up to fire on others, like the DROITS DE L'HOMME, which had not the speed of the rest. As a last desperate effort Villaret brought up the frigate in which he was sailing across the path of the English, hoping that his captains would rally round their admiral. Not they: flight and safety were their order of the day. ALEXANDRE, FORMIDABLE, and TIGRE, the last when a mere mile from Île de Groix, were all taken, after a stout fight against odds. The rest were behind the island by then, but Villaret, justifiably doubting if they would put up any sort of fight, took them on into Lorient, though Bridport, who was no very dashing commander, had in fact called off the pursuit. Eight British ships had been engaged and had suffered 144 casualties: the French casualties numbered over 800.

These ships in Lorient did nothing to interfere with the Royalist landing at Quiberon, and this concerns the navy only in that, in the fighting and the dreadful executions that followed its failure, many Royalist naval officers were killed who would have been the greatest use to France if, like many other émigrés, they had been willing to return when Napoleon came to power. The crews of the Lorient ships were, in fact, sent home for lack of food, only enough men being retained to sail three ships round to Brest after much-needed repair. As Bridport maintained no blockade, this was done, and the crews returned overland to bring back a second batch, and then a third. The French navy seemed to have reached bottom. The battle off Groix had been a most shameful example of cowardice and complete lack of discipline. Even granted that Villaret's orders may have been to a small extent face-saving, nearly all his captains had disobeyed all signals. Some, like the bar-parlour politicians that they really were, tried to save their bacon by quickly launching complaints against their admiral. This availed them little. The captain of ZÉLÉ was dismissed the Service and imprisoned: three others were cashiered.

Clearly things were badly wrong. In his reports, from the return from the foolish winter cruise onwards, Villaret had not minced his words. Gunners were lacking: 'our shots', he wrote, 'land in the water'. Above all, the officers were often incompetent and unsuitable. Nineteen out of twenty had merely sought their appointments, and thereafter intrigued for promotion, for the sake of the pay. They had been selected by the Popular Societies who had picked them because of their 'verbiage of patriotism': they were insubordinate and ignorant, and good merchant captains would not come forward to serve under these men whom they would not have trusted to keep a watch. Even where their patriotism was genuine 'patriotism alone cannot handle a ship': this was the case with some lower deck ratings of the old navy who had been commissioned though they could barely read or write and knew only routine.

So Villaret. Draconian measures were plainly necessary and one of the last acts of the dying National Convention was to take them. A new corps of gunners was to be created and trained. To get rid of the bad officers the existing officer corps was abolished once again. A new one was created with the establishment laid down of the number of officers of each grade. They were ultimately to come from *aspirants* in two classes – roughly corresponding to the former cadets and midshipmen of the British service. For the moment, the new corps of officers had to be formed largely by re-appointing the less incompetent of the existing corps, but vacancies were left in each grade. It was going to take some time, however, to sort out the officers and train the gunners. The National Convention dissolved at the end of October and, by the Constitution of the Year III which it had passed, the Directory took on the government (see p 204). It appointed as Minister of Marine Truguet, an admiral of the old navy who had been in command at Toulon till a little before that town had called in the English[1].

In the autumn there had been a little activity in Toulon. Richery, shortly to be raised to flag rank, having fitted out six ships with difficulty, sailed on a somewhat extensive mission – to see transports to San Domingo and trap English frigates on the coast of that island, to destroy the English settlements in Newfoundland, and, returning, to cruise off the Azores. In fact, once through to the Atlantic, he ran into a rather weakly protected convoy, took one of the escort – the former French *Censeur* – and nearly all the merchantmen. He led his prizes into Cadiz and spent the rest of the year selling them, and getting some of his ships damaged in harbour by a sudden storm. A further piece of Toulon activity was the sailing in October of Captain Ganteaume in the RÉPUBLICAIN(74), with some lesser vessels, for the Levant. Here he did little beyond leaving a dismasted frigate in the Dardanelles for repair, and returned early in the next year. There had been a number of small ship fights in which France had lost quite heavily. One provided an unusual feature. The corvette LIBERTÉ was engaged by a British frigate off San Domingo and reduced to a sinking state. The British captain hailed her to surrender. The French captain replied that he would not strike, but would not repel any boats that came to save his crew! The boats were sent, the crew was taken into them just in time, and LIBERTÉ went down with her tricolour flying. It seems a pity that

[1]A number of officers of the old navy who had been 'purged' by Jean-Bon St André or by the Popular Societies of the ports, and had not emigrated or been sent to the guillotine, did return to the service under the Directory.

the enormous development of weapons since that day does not permit modern naval warfare to be conducted with the same humanity (though attempts have been made) and the same chivalry.

The year 1795, then, had been disastrous and shameful for France at sea. The reverse had been the case on land. Early in the year her armies had overrun Holland and had taken the Dutch fleet by attacking across the ice, in which it was frozen, with cavalry and artillery. That fleet was now to pass into French direction, for Holland was turned into a satellite under the name of the Batavian Republic. Prussia, weary of the war and pursuing an unscupulously selfish policy, had made peace. Spain had followed suit, and was even shortly to change sides. Of the great coalition against her, only Austria, Sardinia, and Britain remained in the field.

The next year, 1796, saw many engagements between frigates and smaller craft. These all went in favour of the English, who were often stronger than their opponents. They were not always so, and there is one case in a fight between equal ships of the French running from their guns. Sercey (CA) commanding a division of six frigates based on the Île de France fought off two ships of the line, but did not attempt to do more, perhaps because, as usual for French commanders in the Indian Ocean, he had been sent out short of stores and without spare sails or spars. In the Antilles the English occupied Grenada and St Vincent, while Richery at last left Cadiz to make his destructive raid on Newfoundland and then to return to Brest. It was not till the end of the year that France attempted any major naval operation.

This was an expedition to occupy Ireland. The idea came from Hoche, the brilliant young general who had destroyed the Royalist invasion at Quiberon and had just stamped out the Vendéan rising. Bonaparte's armies were rolling up the Austrians in Italy: the other remaining enemy, Britain, must also be attacked by land; and Wolfe Tone and other Irish Republican leaders in Paris were urging that, if only an army were landed in Ireland, the whole country would rise. So, in early October, Villaret-Joyeuse was ordered to fit out the Brest fleet for the purpose. That vice-admiral did not like the idea. Truguet had put in a year's earnest work, especially with regard to personnel, and some of the old navy had returned to the fold. But Truguet in Paris could not fully realize in how poor a state were the ships, how lacking was good quality gear to put them right, how desperate was the shortage of seamen. The Revolution's fervent armies knew success: its navy knew only failure. Its discipline had been restored, but it was dispirited and nothing like the service that had fought the American War. As he made no secret of his adverse opinion Villaret was rightly removed from the command, which was given to the other vice-admiral of Suffren stock – Morard de Galles, though his trials during the Revolution had left him only a shadow of his former self.

This officer and his chief-of-staff, Bruix (CA), made the most careful preparations possible, and by mid December the expedition was ready to sail. It consisted of seventeen of the line, thirteen frigates, some smaller vessels, and twenty transports. It carried over 20 000 troops, and these mostly encumbered the warships, the transports being used for horses, stores, and military impedimenta. Morard de Galles decided that he, with Hoche and Bruix, should sail in a frigate, FRATERNITÉ, and ordered two of his three other flag-officers, Bouvet, second-in-command, and Nielly, to do the same. Though the dreadful mistake of the winter cruise, that of sending the ships to sea with food only for the time that they were reckoned to be away, was not repeated, they were not really adequately victualled for all eventualities.

Brest is a magnificent harbour but it is set in a dangerous coast. From the roadstead there are two ways to sea that avoid navigational dangers: one is the broad Iroise Channel leading westward to the open Atlantic, the other is the somewhat narrow and tricky Raz de Sein leading west of south into the Bay of Biscay. Morard de Galles had given orders to leave by the Raz in order to avoid the English blockading fleet lying north of Ushant; for the object was not to fight a battle, but to land the troops. From the start all went wrong. Weighing at dusk on December 16th, Morard de Galles, finding the easterly wind veering south so as to be foul in the Raz, signalled to go by the Iroise Channel and set the example by leading into it. His signals were not observed, and those by gunfire were not understood, partly because, first, the SÉDUISANT(74) had gone on the rocks and was firing distress guns, and, second, because Sir Edward Pellew in the *Indefatigable* frigate, having sent his companion ship to warn the British fleet, boldly entered the Channel in the dark and, firing cannon and blazing off rockets and blue-lights, produced what were apparently incomprehensible signals. Six of the line and six frigates, missing all signals, went by the Raz, the others got out in great confusion.

Next day Bouvet collected eight of the line and seven frigates. The French were fortunate in that, having taken the risk of a winter sailing, they had the easterly wind that they wanted and it was of no excessive force. Bouvet and his group ran westward before it and then bore north for Mizen Head, which was the rendezvous. Near this they were joined by Nielly and Richery (CAs) with seven of the line and some frigates, so that the bulk of the expedition had, by great good fortune, come together. But its commander-in-chief, Hoche, and its admiral with his chief of staff were missing. They had been scouring the ocean in their frigate looking for the fleet. They had picked up the NESTOR(74) but had lost her again, and then had been seen and chased by an English ship of the line. In altering course to escape her, the sailors had so mishandled the canvas as to lose some of it (this in the admiral's ship!) and only by jettisoning guns and running many leagues into the Atlantic was FRATERNITÉ able to get away.

Bouvet led the rest to Bantry Bay, which was to be the landing place, and as the soldiers were aboard the ships and only two of the line were missing, there was no military reason why the troops should not be put ashore. But there was a big nautical one. Bantry Bay is very deep for its width, and the ships had to sail a good way up it to any landing place. The easterly winds that had been so helpful were head winds for going up the bay and they now blew furiously, with driving snow and sleet. The sailors, half clad, and all unversed, with poor gear, had to make short tack after short tack even to hold such little ground as they gained. Only good seamen in good ships could have made headway. The French were not good seamen and the ships were not in good trim. The Shan Van Voght might say that the French were in the bay; but they could not get ashore. They made a little headway and decided to land the next day, but that night a gale blew from the east, and the ships that had remained under sail were driven well out to sea. A further gale on those in the bay followed, anchors dragged, the frigates IMPATIENTE and SURVEILLANTE were wrecked, and Bouvet cut his cables and stood away, signalling others to follow. Six of the line and two frigates remained, but there was no flag-officer, and General Grouchy, the senior officer of troops, knew that a landing with the remaining soldiers was pointless. So these ships stood back for France.

The reason that they did not hang on to see if the others returned was that food was now running short. This was a strong reason with those blown out to sea for not beating back into Bantry Bay: had they done so and then found that they must delay even a little longer before collecting others and landing the men, they would not have had provisions left to take them home. As it was some of them were on reduced rations for several days while getting back. Bouvet was not an over-energetic officer, and as second-in-command of a great expedition gave up rather readily; but the food question was the main reason. The NESTOR did not get to Bantry till New Year's Day, to find it empty: the last ships had left on December 27th. FRATERNITÉ with the C-in-C and the admiral, beating up for Bantry at last, met the RÉVOLUTION while that liner, herself badly damaged, was taking the men off the sinking SCÉVOLA, and, with food short, made for France, being driven by British pursuers to steer for Rochefort instead of Brest. The frigate TARTU and some transports were taken.

The DROITS DE L'HOMME (74), blown out to sea from Bantry, loyally made for a second given rendezvous. Finding no one there, she sailed for France later than the rest. She was sighted and pursued by Pellew in the *Indefatigable* with another frigate, the *Amazon*; and these two hung on to her on either side like borzois with a wolf. She was so ill-found that her fore and main topmasts came down, though the English frigates were able to carry all sail, and she could not open her lower ports to use her thirty-six-pounders in the sea that was running. With soldiers aboard, she was able to keep up a vigorous fire; but it caused few casualties, while her own losses were heavy and she was much damaged. The weather was too overcast for observations to be taken and the three vessels found themselves desperately embayed, one dawn, in Audierne Bay. With consummate seamanship Pellew got *Indefatigable* clear and out to sea. *Amazon* failed to get clear but, being a well-disciplined ship, got all but six of her men ashore when she grounded. DROITS DE L'HOMME hit a sandbank broadside on, heeled over, and her injured masts jerked out of her. Ill discipline and the fierce surf sank what boats and rafts were got away. Those people remaining aboard had a dreadful foodless and surf-beaten ordeal before the weather abated enough for boats to come off from shore; it lasted four days for some and five days for the rest. Her losses by drowning or

## NINETEENTH-CENTURY ADMIRALS

ADMIRAL DUPERRÉ

ADMIRAL HAMELIN

ADMIRAL BRUAT

ADMIRAL COURBET (*by Pirodon*)

'LA GLOIRE' The First 'Ironclad'

'RICHELIEU'; ONE OF FRANCE'S LAST TWO CAPITAL SHIPS
Main armament 8–15 in. Photographed at Dakar before being refitted in America with AA guns and equipment

exhaustion are variously given between two and four hundred men[1]. Otherwise those ships that got back to France had done so by January 14th. This failure of an expedition had cost France two of the line and four frigates as well as some lesser craft and transports. Winter weather, mismanagement, and poor seamanship in poor ships, and not her own exertions, had saved Britain from invasion.

The year we have now reached, 1797, saw but little French naval activity, yet there were some maritime events of major importance. The first of them was the evacuation of the Mediterranean by the British. The victories of Napoleon Bonaparte in his great Italian campaign had driven the Austrian armies away from the coast so that English ships could no longer assist them: the overrunning of Italy by French troops had deprived the British of ports there; Spain was now an active ally of France and the inefficiency of her large and powerful fleet had not yet been realized. Joined to the French ships in Toulon it vastly outnumbered the British Mediterranean squadron. So Jervis, who was in command, was ordered to evacuate Corsica, where the British occupation had become unpopular, and Elba, and base himself on Gibraltar and the Tagus – for Portugal had a longstanding British alliance. This evacuation was completed by the end of 1796 except for Elba, and Nelson brought away the troops thence early in 1797. This left Brueys (CA), commander of the Toulon ships, free to get to sea and, when the time came later in the year, he sailed round into the Adriatic to assist Napoleon's shameless destruction of the neutral republic of Venice by seizing its fleet (which was of little value owing to poor scantling and bad design) and its Ionian Islands. This was in midsummer: Austria had ceased to fight in April. Britain was now France's only enemy.

Thanks to the enthusiasm of Hoche, who realized that Britain could be beaten only by victory on her soil, the idea of invading Ireland had not been dropped. With the Spaniards willing to use their fleet more or less at French bidding, and with the Dutch fleet in French hands, there was a vague scheme for a union of these two forces with the Brest fleet – when that was refitted. This combination never became more than an idea because when the Spanish fleet sailed from Cartagena for Cadiz as a first step towards Brest, it was intercepted and heavily defeated by Jervis and Nelson off Cape St Vincent in February. As to the fleet in Brest, there was a great deal to be done and little or nothing with which to do it. For the heavier gear had perforce to be brought by sea, as had much of the lighter gear bought in the Baltic, and English cruisers were so active, even in inshore waters, that coastal traffic along the French coast was both slow and precarious. Therefore, though two of Britain's main fleets, those of the Channel and the Nore, were in a state of mutiny for a good deal of the cruising season, the Brest ships could do nothing. (In the third main fleet, blockading Cadiz, the stern hand of Jervis saw to it that discipline was preserved.) The mutinies had little that was revolutionary

[1]And women, for she had aboard some prisoners from English merchant ships. These prisoners strove so well in the common danger that they were sent home under flag of truce.

about them; the ships' companies would have returned to duty if there had been a certainty of fighting the French, and an attempt to invade without proper preparation and with an unseaworthy fleet would have been disastrous. Indeed, despite the mutinies, British frigates and brigs remained active, and most of the small ship actions of the year were fought in coastal waters that were nominally in French control: one French frigate was taken in the very Iroise Channel itself! Hoche died in September, but the idea of attacking Ireland was not quite abandoned. It was proposed that the Dutch army, twelve thousand strong, should carry out the invasion. But its general refused to move until the Dutch fleet had cleared the way. The defeat of that fleet by Duncan in a grim battle off Kamperduin put an end to the idea of a landing in Ireland.

The main French activity was a reversion to privateering in the old style of the 'Guerre de Course' of the closing years of Louis XIV. For in July, Truguet, at odds with the Directory, resigned, and his place as minister of marine was taken by Pléville-de-Pelley (CA) – who was to surrender it a year later to Bruix, a former noble officer who had been Morard de Galle's active chief of staff. Pléville was anxious to make good back and present payments due to personnel. Unable to get the money voted, he hired the smaller French warships to *armateurs*. Profits at the expense of English merchants certainly did accrue; but this was hardly the way to keep or train naval crews. The only oversea squadron of note was that of nine frigates under Sercey (CA) based on the Île de France. But he was bluffed out of attacking a rich convoy, had to send three of his frigates home to France for repairs, and was so hampered by the opposition of the island's council that he could achieve nothing.

Nevertheless, at the end of the year the idea of invasion was taken up again, this time to be direct across the Channel. General Napoleon Bonaparte, now much the most important figure in France, was given the command of an 'Army of England', had a survey made of all the little ports from Dunkirk via the Pas de Calais to Le Havre, and issued orders for the construction of a large number of troop-carrying gunboats. His preparations for the invasion of England six years later were undoubtedly genuine, but it is highly doubtful if they were so on this occasion. He had his eye upon the Orient, which never failed to attract him, and upon the occupation of Egypt as a first step to the expulsion of the English from India. To that, in the next year, he was to turn.

He had no difficulty in persuading the Directory to favour a great expedition against Egypt. That Government, precariously placed under a Constitution which it had violated, was only too glad to send well oversea a powerful and popular young general whom it suspected of playing already for his own hand. So, though the construction of small ships to take troops to England continued in the Channel ports, work at Toulon was given priority. It went ahead vigorously with the drive of Napoleon behind port administrators who were themselves both tireless and competent, and it was given a final impulse when the forceful admiral Bruix took over the Ministry of Marine well before the expedition sailed. This was a huge affair; not only was there a large army to be transported but there were to be civil officials and even savants. For Egypt, nominally part of the decaying Ottoman Empire, actually ruled by the emirs of its warrior caste, the Mamelukes, was to be conquered and permanently occupied. Some sections of the expedition were assembled in southern French ports other than Toulon, and in ports on the Italian coast, and when, having sailed in May, its various parts had duly met out at sea, it numbered well over 350 vessels of various kinds.

The escorting naval part of this immense spread of sails was under the command of Brueys d'Aigalliers (VA) with his flag in the ORIENT(120), an officer of the old royal navy, but of no strong personality. Flag officers under him were Villeneuve in the GUILLAUME TELL(80), Blanquet-Duchayla in the FRANKLIN(80), and Decrès commanding the frigates. There were thirteen of the line, six frigates, and about seventy lesser craft: there were also two more of the line and seven more frigates armed *en flûte*. The strength of the liners varied greatly. GUERRIER, PEUPLE SOUVERAIN, and CONQUÉRANT were old condemned ships that ought not to have been sent to sea, and the last named had to be fitted with lighter guns since the fire of her batteries at their proper weight would have shaken her to pieces. On the other hand some ships, such as FRANKLIN and SPARTIATE, were only just built and some others, like the flagship, had been so overhauled as to be as good as new, and were truly formidable units. Yet there were grave weaknesses. The Toulon officials had done amazingly well; but they could not achieve the impossible, and not only were supplies rather on the short side but, above all, the large number of seamen required just could not be found. The fighting fleet was some seventeen hundred men short of its proper complement of roughly ten thousand – an average of about one-sixth per ship, some, like TONNANT, being shorter-handed than others. This was almost an advantage on the outward voyage, for the ships were packed with soldiers, who could have fought the guns (in a way) in action; but it was gravely to affect events later, the more so as the seamen were often of poor-quality, ill-disciplined, and in a naval sense largely untrained.

Though the fitting of so huge an expedition could not be prevented from being noised abroad, the secret of its destination was, for those days, remarkably well kept. The British government, though advised by some agents on the Continent that Egypt was the aim, had its eye on both Ireland and the Kingdom of Naples, and refused to believe that the French would sail east of Italy. As we have seen, Britain had evacuated the Mediterranean, and its nearest fleet was that of Lord St Vincent (Jervis) which was blockading Cadiz. From this Nelson was detached with three of the line to watch Toulon, and reinforcements came to St Vincent with instructions that eleven ships were to be sent to join Nelson so that he could deal with the French. With the exception of the fifty-gun *Leander* these were all the lighter seventy-fours. Nelson did not see the French sail, for a gale had blown him away from Toulon (and had largely dismasted his flagship, *Vanguard*, which he refitted with remarkable speed). On his return off Toulon he found that his bird had flown and, joined by the eleven ships sent to him, set off in pursuit, seeking desperately for information of his enemy's whereabouts, course, and destination.

The French had in fact made for Malta, reaching it in early June. This island was still under the Knights of St John who refused, on grounds of neutrality, to allow the expedition to enter its harbour for water. Napoleon promptly landed his troops. Valletta was still a strong place and, had it been held, to take it would have delayed Napoleon inconveniently. But the Knights were not what they had been of yore: their fighting days were over, and they capitulated at once, most of them choosing to go into exile. Among those who did so was Suffren's youngest brother. A garrison having been installed, the expedition left for Egypt, steering a distinctly northerly course towards Crete. For this reason it was missed by Nelson who, having reached Malta and found what had happened, went hot-foot for Egypt as directly as the wind would permit. Reaching Alexandria on June 28th he found all quiet, and so set off again to scour the Mediterranean for his enemy. His sails could barely have gone below the horizon when those of the French appeared above it, to the northward on July 1st. The troops were landed, Alexandria taken, and the huge mass of merchantmen and *flûtes* entered its harbour. Brueys had brought his expedition safely to port. He was fortunate, for so huge a convoy at sea was immensely vulnerable. Nelson had, indeed, organized his fleet in three columns; two, totalling ten ships, were to deal with the escort, the other with the almost defenceless transports. But he was desperately short of frigates: had he not been so the French, who at one time were barely out of sight of him, would have been spotted, and a battle (which, had it led to the death or capture of Napoleon, would have altered the history of the world) would have been fought. Here is one of the great 'if onlys' of history: French historians think that, with soldiers still aboard the ships to help to man the guns, the result would have been favourable: in view of the strength of British gunnery (shortly to be demonstrated) this seems doubtful.

The expedition once in harbour, what was to be done with the fleet? For the British were bound to arrive sooner or later. Three choices were open to Brueys: to go into the Vieux Port of Alexandria, to go to Corfu (then in French hands), or to go to the shallow bay of Aboukir, a little east of Alexandria. He rejected the first on the grounds (a) that there was insufficient water to permit the entry of his larger ships (though a surveying officer reported that a little blasting would put that right) and (b) that, once in, he could be helplessly blockaded by a small force, which was true and a cogent objection. He disliked the idea of going to Corfu because his food supplies might not last out the voyage, and so he decided on the Aboukir anchorage. His several letters ashore to Napoleon at this time show him to have been anything but a man of firm decision and to have had continual fluctuations of mind as to whether, if his enemy arrived, he should fight him at anchor or under sail. Napoleon raised no objection to Aboukir for the moment, for, until he was well established in Egypt, it was desirable that the fleet should be near at hand; but when his victories had brought him to Cairo he sent positive orders for the fleet to go to

Corfu. The orders never arrived, for the courier was killed by Arabs.

Aboukir Bay is shallow as to water, and the anchorage is covered by a long sand-bank jutting out eastward from the northern point of the bay and culminating in an island, on which Brueys threw up a battery, though its fire could be no more than a deterrent, and not effective. The western shore of the bay runs north and south and Brueys drew up his ships with the leading one behind the island, on a line parallel to that shore in the following order :

| Ship | Guns | Captain |
|---|---|---|
| GUERRIER | 74 | T Trullet |
| CONQUÉRANT | 74 | Dalbarade |
| SPARTIATE | 74 | Émeriau |
| AQUILON | 74 | Thevenard |
| PEUPLE SOUVERAIN | 74 | Racord |
| FRANKLIN | 80 | Blanquet-Duchayla (CA) |
| ORIENT | 120 | Brueys d'Aigalliers (VA) |
| TONNANT | 80 | Dupetit-Thouars |
| HEUREUX | 74 | Etienne |
| MERCURE | 74 | Cambon |
| GUILLAUME TELL | 80 | Villeneuve (CA) |
| GÉNÉREUX | 74 | Lejoille |
| TIMOLÉON | 74 | L Trullet |

The four frigates with him (the rest were in Alexandria) lay inshore of this line of battle under Decrés (CA). There was a slight 'elbow' in the line, starting astern of the TONNANT, to bring the five rear ships a little nearer to the coast, so as to make it more difficult for the rear to be turned. It will be noticed that Brueys put his weak ships at the head of the line for, what with the deterrent of the battery on the island and the sand-banks jutting out from it, he did not expect the head of his line to be much assailable, so that any attack would fall on his powerful ships in the centre and rear. He seems to have pictured the course of events to be like that of the battle at St Kitts in reverse, with himself in the position of Hood, and Nelson in that of de Grasse, sailing past and cannonading and then coming round to repeat the process. He little knew his man.

At two in the afternoon of August 1st, the British were sighted coming in with the breeze astern of them. Watering-parties were recalled, and the frigates were ordered to send all hands that could be spared to strengthen the gun-crews of the ships that were the most short-handed – an emergency arrangement that hardly made for efficiency. There was time for a hurried council of flag officers at which Blanquet urged weighing to fight under sail and the others preferred fighting at anchor. There was much to be said for this. The fleet was already disposed to do so; with unskilled and short-handed ships' companies it would be easier to engage with little to do but fight the guns, and it was not dreamed that the English could possibly attack from the port or inshore side. Indeed, Brueys comforted himself that there would be no battle that day. By the time the English could get into the bay it would be evening; they would not want to fight a night battle in sand-banked waters that to them were uncharted. Nelson would surely lie off to consider the position, the more so as three of his ships were plainly about two hours' sailing astern. (These were *Alexander*, *Swiftsure*, and *Leander* – of which the first two had been detached to look into Alexandria.) Indeed, Brueys thought it possible that, though the wind was blowing from dead ahead down his line at the moment, he might, if he wished, slip away in the night – again as Hood had done.

Nelson's understanding with his captains, his 'band of brothers', was complete; and they had their ships in fine order. They knew that their admiral intended to attack the enemy's van, and the only signals that he had to make were to clear for action and to prepare to anchor by the stern. Though *Culloden* grounded on the bank off Aboukir Island and though three ships were well astern, Nelson came in unhesitating with the other ten. Foley, leading in *Goliath*, remembering his admiral's dictum that where there was room for a ship to swing to moorings there was room for another to squeeze past, sailed close across GUERRIER's bows before putting his helm up and the next four followed him, engaging where Brueys least expected it, on his port, or inshore, side. Nelson led his *Vanguard* and those following on to the starboard side so that, not long after 6.15, at about which time Brueys had signalled to open fire, the situation was that the five leading, and mostly weaker, French ships were under a concentration of eight British. All Brueys' calculations had gone wrong. He had immobilized himself and his strong rear vessels while his weak van was being destroyed. Even a prompt signal to all ships astern to follow him in weighing and beating up to the rescue would hardly have been in time to save the situation, and he was in no easy position to make it, for he was being engaged from starboard by the *Bellerophon*. The remaining British ship, *Majestic*, had fired first on TONNANT, who gave her a warm reception, and then engaged both HEUREUX and MERCURE. Though she suffered a good deal and her captain was killed, she must have been splendidly handled, for she did damage to all three Frenchmen. In the van, the old GUERRIER, having received the broadsides of the five ships that crossed her bows, was shot almost to bits, and her next astern, CONQUÉRANT, was soon in no better shape. PEUPLE SOUVERAIN had either cut her cables or had them shot away, and had drifted out of line, followed by her opponents, and so had left a dangerous gap ahead of FRANKLIN into which, when she came up, the little *Leander* tucked herself to make an enfilading fire on both FRANKLIN and ORIENT. The *Swiftsure* and the *Alexander*, arriving in the dark about eight, between them engaged FRANKLIN, ORIENT, and TONNANT. The French flagship had dealt faithfully with the *Bellerophon*, her powerful broadsides doing such damage that, after less than an hour's fighting, the British ship cut her cables and drifted down the bay, under fire of the rearmost French vessels, in almost a wrecked condition. But as the ships in the van were crushed, the British engaging them had dropped down before the wind to join the attack on FRANKLIN, ORIENT, and TONNANT, and it was seen that the flagship had caught fire. The flames spread with awful rapidity and it became clear that she was doomed.

Brueys was already dead. He had been twice hit, but refused to leave the deck, and a third wound killed him. Aboard TONNANT DuPetit-Thouars, almost dismembered, had had his dying body placed for support in a barrel that he might encourage his men; aboard FRANKLIN Blanquet was wounded and unconscious. The ships around ORIENT prepared to deal with scattered and flaming debris when she should blow up. The British held their ground, but the three French ships astern of her cut their cables. HEUREUX and MERCURE collided and drifted down the bay helplessly locked together for a while. TONNANT brought up dismasted. The remaining three merely veered their cables, but all semblance of a line disappeared. CONQUÉRANT had struck at nine o'clock, AQUILON and GUERRIER had followed suit by nine forty-five: at ten ORIENT blew up, and there was an awed silence of a quarter of an hour during which no gun fired. There seems to have been a *sauve qui peut* towards her end: one or two boats got away, Ganteaume, the chief of staff, being in one, and a number of men grabbed wreckage and got ashore or were picked up, but the loss of life was severe. The battle was renewed by FRANKLIN, but PEUPLE SOUVERAIN fired no more. SPARTIATE struck at eleven, and FRANKLIN at midnight. Both had scarcely a gun left on its carriage to fire, and had put up a brave resistance to odds. With this the battle really ceased. There was a certain amount of firing between the three sternmost Frenchmen and British ships that had come down the bay, by which TIMOLÉON was damaged, but it amounted to relatively little, for by then the British were exhausted, and the men were dropping to sleep at their guns.

Dawn broke on an appalling scene of desolation. Decrès went aboard TONNANT, HEUREUX, and MERCURE only to find their condition hopeless – the first totally dismasted, the others too damaged to sail. The last two struck as some English ships closed them. The dismasted TONNANT, her flag nailed to the stump of one mast, was the final surrender on the next day. Villeneuve signalled to GÉNÉREUX and TIMOLÉON to make sail, and though TIMOLÉON could not follow, having gone ashore, he led GÉNÉREUX and GUILLAUME TELL out of the bay, followed by Decrès with two frigates. There was no pursuit: most of the British ships were to some extent dismasted, three were down the bay taking the surrender of the last Frenchmen, and Nelson recalled the remaining ship, which alone was pursuing. TIMOLÉON got her crew ashore and was then set on fire; so was one of the two remaining frigates, the other having been sunk by broadsides from the *Orion*. Nine French of the line had been taken, two had been sunk on fire, two frigates had been destroyed. Not till Tsushima was there to be another defeat that was quite so complete.

The virtual inactivity of Villeneuve has been much discussed. Had he taken the initiative by bringing his rear ships into the fight at the start of the battle directly he saw its probable shape, and before the last English ships came into action, he might partly have restored the situation. Had he done so even as late as eight o'clock he might have salved a few of the ships[1]. His defence that weighing, getting under sail, and beating up to

[1]At about this time the captain of TIMOLÉON hoisted his topsails as a strong hint.

windward even a short distance into action would have taken so long as to be useless is only just about adequate. The truth is that Villeneuve, professionally competent and personally brave under fire, was not the man to take quick decisions and show initiative, but rather one to hesitate, 'letting "I dare not" wait upon "I would" ', till, in this case, it was too late. The jibe 'poltron de tête mais pas de coeur' would have been better applied to him than to Tourville. In any case the superiority of the trained British gunnery over that of the virtually untrained French was once again most marked. The French losses in killed were about seventeen hundred: those of the victors were just over two hundred. The damage to British ships was such as could be made good, rather more than less, where they were. Their French prizes were so shattered that three had to be burned (GUERRIER, MERCURE, and HEUREUX) and some of the others were taken home only with great difficulty.

Nelson landed most of his prisoners in Egypt, not out of kindness but because he could not feed them. The effect of the battle was great. It left Britain in command of the Mediterranean, put an end to any idea of an attack on India and, by shutting up a French army and some of the best French generals in Egypt, encouraged the powers of Europe to take the field in that Second Coalition which, but for the return of Napoleon, would have overthrown France. An aftermath of the battle was that the fifty-gun *Leander*, sent home with Nelson's despatches, came up against the GÉNÉREUX(74). This ship had separated from Villeneuve when that contre-amiral was leading the survivors of Aboukir to Malta, and she took *Leander* after that weaker ship had put up a strenuous resistance. Perhaps exacerbated by their losses, perhaps just acting on the principles of an egalitarian revolution, the French sailors behaved badly, plundering their captives and the wounded, their officers doing little or nothing to stop them.

While the battle of Aboukir (known to the English as that of the Nile) overshadows all other naval events of 1798, there were yet others of note. The most important of these was an attempt to support the rebellion that had broken out in Ireland. For all that France was watching for the opportunities that trouble in Ireland would give to her, she seems to have been singularly unready, despite the presence in the country of the Irish leader, Wolfe Tone; and the help that she sent was too little and came too late. In August three frigates reached Killala, carrying what was to be the advance guard of an expedition: five hundred men under General Humbert. They landed, found that the country had been all but 'pacified', and had to surrender. On their heels came a larger force in a ship of the line (the HOCHE, 74) and eight frigates, Captain Bompard being in command. They dodged out of Brest by way of the Raz de Sein, but were encountered to the south of Ireland by a superior force of three of the line and some frigates under Sir John Borlase Warren. They had to run for it and, in a squall, the usual accident happened, the HOCHE losing her main-top yard and mainsail, and her mizzen top-gallant mast. The British were coming up in a general chase and Bompard signalled to form line of battle. The signal was made too late for it to be properly executed, and in any case he was outweighted. The action took place fifteen miles north of the Scillies. Under heavy fire, the HOCHE was further crippled when her cro'-jack was brought down on the deck; and she had to strike. Three of the frigates were taken, and the rest scattered in flight. One of Warren's frigates overtook and captured one of them: the others got away for the moment. But two of them ran into other British frigates and had to yield, though one of these, LOIRE, fairly drove off two separate opponents by good handling and gunnery and exchanged shots with two others before, now damaged, she fell to a fifth. Finally, a third small expedition sailing from Rochefort for Ireland was seen and chased, and escaped back to France only by jettisoning guns. Granted that the previous attempt to send a large expedition had been a failure that militated against making a similar ambitious attempt, it was yet foolish to send such slight assistance and, at that, to send it in driblets. France made no further attempt on Ireland.

Among the single ship actions of the year two were of more than minor importance. The brand new HERCULE(74), commissioned in Lorient, was sent to Brest before, in the modern phrase, she was 'worked up'. She was intercepted off the Bec du Raz by the *Mars*(74), and was outmanoeuvred, outfought, and taken. As some small set-off to this, there was a gallant fight in which the corvette, BAYONNAISE, captured the much stronger *Ambuscade* frigate by boarding.

When Brueys had taken his fleet to the Aboukir anchorage he had left the bulk of his frigates and small craft safe in harbour at Alexandria. It was not hard for a little vessel to slip out and avoid the small blockading force. By this means Napoleon, especially after his further advance into the Ottoman Empire had been turned back by his repulse from Acre, sent urging the Directory to concentrate all remaining naval forces in the Mediterranean in order to bring back the troops marooned in Egypt. This the Directory was anxious to do, for things were going ill in Europe. The second coalition against France, formed after (and largely as the result of) Aboukir, was making headway, and Austro-Russian armies under that formidable eccentric, Suwarrow, were rolling back the French in Italy. There resulted, therefore, in 1799, one of the most oddly inconclusive of naval operations.

Orders were sent to fit out the Brest fleet, and Vice-admiral Bruix, the forceful minister of marine, himself went to stimulate the preparations and to take command. Capable and vigorous, he performed something like a miracle in getting the fleet quickly fitted out from an almost destitute dockyard – though the training of its ships' companies left much to be desired. A cutter, the REBECCA, was sent to sea to be captured, for she carried an officer with despatches to Irish rebels announcing the coming of the French, and Lord Bridport swallowed this bait, hook, line, and sinker, and sailed away with the Channel fleet to Ireland. Meanwhile Bruix came out by the Raz de Sein with twenty-five of the line, six frigates, and nine lesser craft, his flag being in the OCÉAN(120). His orders were to avoid battle, at least till he had picked up the Spaniards; and he hoped to be joined first by Spanish ships from Ferrol, and next to free the main fleet in Cadiz from its blockade. The ultimate objects of the voyage were to be the relief of Malta, whose garrison was blockaded by sea and besieged by angry Maltese by land, and then recovery of the army of Egypt.

He started with the usual accident, two ships failing to weigh so that one of them never caught up with him; but he suffered no interception. The Spaniards failed to come out from Ferrol, and the weather grew bad by the time that he approached Cadiz and sighted the English blockaders. Lord Keith, with fifteen of the line, handled his fleet ably off a lee shore to keep it between the French at sea and the Spaniards in the harbour. Preparing for battle the next day, Bruix lost touch with three of his ships during the night and, when they rejoined, what with his fleet in disorder, and the Spaniards pinned to port by the south-westerly gale, a heavy sea running and the danger of the lee shore, he decided not to use his considerable superiority in a combat, but to push on into the Mediterranean.

He was not putting his head into the lion's mouth, for the lion had opened his jaws so wide as almost to dislocate them. In other words, while Keith was off Cadiz, four ships were with Duckworth at Minorca, where they had supported General Stuart's capture of that island, one was at Tetuan, eight, under Nelson, were spread between Malta and Naples, and, finally, Sidney Smith had two blockading Alexandria. These scattered packages could have been mopped up in detail; but during his voyage Bruix had learned the difference between planning in Paris and handling an inefficient fleet at sea. The gale had damaged some ships; two others collided heavily: so he made for Toulon. Meanwhile Lord St Vincent, commanding from Gibraltar and alarmed by the situation, hurriedly called Keith away from Cadiz and sent him to join Duckworth, who, at Minorca, was particularly exposed. This freed Admiral Mazzaredo in Cadiz, and he followed Keith into the Mediterranean but, running into a gale, had to put into Cartagena with no less than nine of his seventeen ships in various degrees of dismasting. Bruix, meanwhile, had taken reinforcements and supplies from Toulon to French armies on the Ligurian coast. The British had been reinforced, so there was now no question of relieving Malta until his numerical superiority had been restored by a junction with the Spaniards. Bruix therefore sailed for Cartagena.

Keith missed him, but did capture a small division of frigates and brigs under Perrée (CA). These were originally some of the lesser ships of the Egyptian expedition which had gone into Alexandria. They had left that port to support Napoleon in his march on Acre and had landed supplies and ammunition for him at Jaffa. They then had only fifteen rounds per gun remaining and, pursued away from the Levant by stronger English forces, had sailed for home. When Keith cut them off they were a mere sixty miles from safety.

At Cartagena, Bruix found Mazzaredo anything but co-operative; for the alliance with France was wearing very thin on the Spanish side. The Spaniard refused to consider relieving Malta, much less going to Egypt, but eventually agreed to a joint campaign against trade in the Atlantic. The French did succeed in getting the Spanish king to order his admiral to accompany Bruix to Brest, and the joint fleet reached there safely, though a pursuing Keith was not far behind them. The sole result of Bruix's vigorously promoted campaign, therefore, was just to bring the main Spanish fleet into Brest, where it became a hostage for the continuing of a tenuous alliance. There is a nice ambiguity in the closing words of the journal kept by the French admiral's ADC: 'History will not allow such a campaign to lie in oblivion'. Was he thinking of its futility; or did he think that it was a great achievement to take a fleet to sea for some months and bring it safely back again? The latter type of self-congratulation occurs only too often among Frenchmen of the days of sail. For the truth was that the naval strength of the English was overwhelming: their cruisers were everywhere, and behind them could sail not less than a hundred of the line. That sailing, too, was superior. The English were not without accidents arising from faults of seamanship; but much less so than the allies. Two gales had badly damaged both French and Spanish fleets: Keith had sailed through both gales without losing a spar.

During the year French operations ceased in the Indian Ocean. At its start Sercey (CA) had three left of his division of frigates. Two (one of them handed over to a privateer) were taken. The third, PRENEUSE, having bravely fought off the fifty-gun *Jupiter*, found her return to Port Louis blocked. At the approach of English ships her captain ran her firmly ashore and abandoned her to be set on fire. Poor Sercey could only wait for peace to allow him to return home. When the Russians captured Corfu and Ancona the French ships in those places were taken. None was larger than a frigate, for the one ship of the line, GÉNÉREUX, got away and reached Toulon after an adventurous bombardment of Brindisi.

A fortnight after Bruix got back to Brest, Napoleon received from Sir Sidney Smith the malicious gift of a parcel of newspapers. These apprised him of French disasters in Europe, and he judged that the moment was ripe for him to return to France and seize power. Taking advantage of Smith's withdrawal for water, he sailed from Alexandria in the MUIRON frigate with another frigate and two cutters under the command of Ganteaume (CA) and, luckily missing any English cruiser, landed at Fréjus on October 9th. An anxious France gave a triumphal welcome to its ever-victorious general and, taking over a half-formed conspiracy in Paris, he threw out the Directory and its Assemblies on 18th Brumaire (November 9th), and gave France a new constitution with virtually all power in his own hands as First Consul. The Revolution was over: a military dictator ruled.

The next year he marched to restore France's position in Europe. He was fortunate in that Russia had by then broken faith with her allies: the half-mad Czar Paul had withdrawn her from the coalition. So Napoleon had only Austria to fight on land, and his victory at Marengo, followed by that of Moreau at Hohenlinden, enabled him to make a peace in 1801 that restored the triumphal position of 1797. But at sea it had been another matter. An army was cut off in Egypt, and in Malta the garrison was closely besieged and blockaded in Valletta. The army of Egypt was safe enough for the moment, in a country that could support it; but the Malta garrison was already on short commons and could be relieved only from the sea. The closely blockaded Brest fleet was powerless, but an attempt was made from Toulon. Perrée (CA) hoisted his flag in GÉNÉREUX(74), and sailed with three corvettes and a *flûte*, all packed with troops, munitions, and stores. The usual accident took place at setting sail, GÉNÉREUX losing a topmast, so that they had to return to port. They left again in February, when the blockaders had temporarily been blown out to sea. This Malta convoy was as hazardous a venture as any of the Malta convoys of the Second World War. Superior forces were waiting for it near the island, and it ran into them. The *flûte* was taken, the corvettes bolted (and got back to Toulon), and the GÉNÉREUX put up the usual courageous but rather inefficient and quite hopeless resistance before, with the admiral mortally wounded, she struck.

There now remained only one of the line that had been at Aboukir – the GUILLAUME TELL(80) – and she was blockaded in Valletta Harbour. Nevertheless it was decided to get her to sea: she could take away the sick and the non-combatants and so save rations to enable the garrison to hold out a little longer. Decrès (CA) hoisted his flag in her and made a bolt for it one dark night at the end of March. It had not been possible to conceal that she was getting ready to sail and the English were waiting for her. Despite skilful handling she was overtaken by two of the line and a frigate. Again there was a brave but hopeless resistance, which lasted some nine hours. Both the English ships of the line were much damaged, but GUILLAUME TELL was completely dismasted and could fight no longer. The only survivors of Aboukir were now the two frigates, DIANE and JUSTICE, both at Valletta. They too tried to get away before the island surrendered. DIANE was taken, but JUSTICE did outdistance her pursuers and gain Toulon. Malta capitulated in September.

In the Channel St Vincent had taken command. Howe and Bridport, to economize in ships and to spare the men, had followed a policy of distant blockade of Brest. They had kept the fleet in Torbay, and even at Spithead, relying on their frigates to tell them when the French were at sea, and on then coming up with them. This would not do for the stern old admiral from the Mediterranean. The coast around Brest was dangerous in a westerly wind, and a blockading fleet would have to beat out to sea; but such would be a head wind to the French who would not be able to get out of the inner roadstead. In other winds the blockade was henceforth to be close. So the French had the humiliation of seeing the British fleet anchored well to the south-east of Ushant, with the frigates even closer in, so that French ships could not be out even in Bertheaume or Camaret roads, much less in Douarnenez Bay. An attempt was made to get a few ships to Rochefort, from which it would be more easy to operate: it had to be abandoned. An attempt was also made to send a small expedition out to San Domingo: it found itself all but cut off and was glad to be able to get back to harbour. All along the coast, from Lorient north to the Texel, the English blockaded and ranged, cutting up the vital coastal traffic, and there were even instances (notably one at Dunkirk) of their coming in to attack and cut out ships in the very roadsteads. A division of three frigates did get away from Aix roads and crossed the south central Atlantic as far as Montevideo, doing some damage to English, Portuguese, and American traders[1], but, sighted by a ship of the line with some Indiamen, two of them were taken. Only late in the year did Latouche-Tréville (CA) come out of Brest with eight of the line for a brief cruise, and that had no object but to stop the ever-growing desertions.

As the army could not be brought away from Egypt some attempt must be made to enable it to maintain its hold on the country, for the ordinary wastage among troops under arms had reduced its numbers. So in January of the next year, 1801, Ganteaume (CA) got out of Brest, at the second attempt, with seven of the line and two frigates. His flag was in INDIVISIBLE(80), and his ships carried five thousand soldiers and much munitions. The gale that had blown the blockaders away scattered his ships, and though he collected them they had suffered the usual damage and there was the usual lack of any 'spares' to make this good. Therefore, on entering the Mediterranean and hearing, from one of some small British cruising craft that he captured, that Keith was ahead of him off Egypt in force, and Warren was near at hand to seek him, his nerve failed and he made for Toulon. He had some excuse for doing so, for, when ordered to sea again, he wrote to the minister that his seamen were in no state for a winter campaign, with their pay fifteen months in arrears, their bodies ill-fed and naked or clad in tatters, and their spirits utterly cast down by their wretchedness. His officers were in little better sort. However, he sailed again in March, but returned to port after some of his ships had been in collision. He was ordered out once more, this time to go by way of Elba to help the troops who were attacking that island. He did so, in April, and, having bombarded Porto Ferrajo, packed his large number of sick into three of his ships and a frigate to return to Toulon, replaced them with fit men from these vessels, and at last made for his objective. Finding that Lord Keith was off Alexandria, he prepared to land his troops at Derna. At this Keith's sails appeared above the horizon and Ganteaume could do nothing but flee. He had taken some small English warships on his way out and, returning, fell upon and took the *Swiftsure*(74). But this was a poor substitute for failing to reinforce the army of Egypt. That capitulated to an English force in August.

Napoleon did think of a last effort. Bruix was to sail from Rochefort with five ships carrying fifteen hundred troops. At Cadiz he was to pick up Dumanoir (CA) with six ships purchased from Spain, and Durand-Linois (CA), who was to come from Toulon with the three of the line that Ganteaume had sent back to that port from Elba (see above). The Rochefort ships never sailed, but Linois did sail and found that Sir James Saumarez was off Cadiz with a superior force. He therefore put into Algeçiras, on the west side of the wide bay whose eastern coast is the Gibraltar Peninsula. Here he moored his ships close, and parallel, to the shore

[1]Thanks to the activities of her privateers France was briefly at war with the United States. The Americans had built some powerful frigates. One of these took a weaker French ship that did not know that war was declared, another had an indecisive fight with a French frigate. Peace was then signed.

between two batteries in the following order: FORMIDABLE (flag, 80), INDOMPTABLE (80), DESAIX (74), and, behind them, the MUIRON frigate. Rear-admiral Saumarez came round, early in July, to deal with him with six eighty-gun liners, his flag being in the *Caesar*. Linois' position was strong, especially as he was so wise as not to rely on Spanish promises to man the batteries but sent men and munitions to them from MUIRON. (They arrived in the nick of time for the crisis of the battle, to find that Spanish 'manning' consisted only of a few incompetent militiamen with very little powder and shot!) Saumarez found it difficult to bring his craft into action, for there was little wind, and little room to use his superiority. One of his ships could never get close enough to be in action at all, and another, the *Hannibal*, only ineffectively. With the French ships at anchor their fire was more accurate than usual, the batteries were well worked, and the British began to suffer. Saumarez therefore made a boat attack on the southern battery, and ordered the *Hannibal* to get round the north of the line to deal with the other. The boat attack was repelled, but Linois, thinking that the *Hannibal* was aiming to repeat Nelson's Aboukir manoeuvre by getting between him and the shore, signalled his ships to cut their cables so that they drifted gently on to the mud, still with their broadsides to the enemy. *Hannibal* had in fact gone aground. Saumarez also signalled to cut cables and resume the fight closer inshore. But such wind as there was prevented the British from getting into good fighting positions and, when the battle had lasted in all for some five hours, Saumarez cut his cables again and put off for Gibraltar. *Hannibal*, unable to get afloat, struck her flag. The *Caesar*, which had to be towed clear, the *Pompey*, and the *Venerable* had all suffered heavily.

To have beaten off such superior forces, and to have taken one ship, was a distinct success for Linois, but he too had suffered damage, his flag captain and that of the INDOMPTABLE had been killed, and Saumarez, refitting with the customary quickness of the English, was still watching him from across the bay, ready to pounce if he left his strong position. So he sent to Cadiz for help and Don Juan Moreno (LG) came round with one French ship, the SAINT-ANTOINE (74), recently acquired by Dumanoir, and four powerful Spanish ships. To his annoyance Linois was ordered by the Spaniard to sail with him in a frigate, leaving his flagship, FORMIDABLE, to be commanded by a frigate captain. The captured *Hannibal* was unable to get clear of the bay, but the remaining eight (four French and four Spanish) sailed for Cadiz in the afternoon of July 12th, with Saumarez weighing in pursuit.

The grim comedy which followed undid Linois' victory. The allied ships got badly out of station in the dark, and the faster of the British pursuers got among them. The damaged FORMIDABLE and the SAINT-ANTOINE, which was a slow sailer with only a scratch crew, got badly astern and the latter was outfought and taken. Two of the Spanish ships, each taking the other for an enemy, opened a furious fire on each other, and both caught ablaze and sank, with heavy loss of life. At day the Spanish admiral made no attempt to turn his remaining ships to the rescue of the FORMIDABLE, which was pursued by three British of the line. Two of these came up with her. Her substitute commander, frigate-captain Troude, had grave trouble with his ship's company who, seriously short-handed because of her casualties in the Algeçiras battle, refused to fight, protesting that to do so would mean only a butchery. However, putting reliable men to guard the flag and announcing that she would sink with it still flying, he got them back to their duty. Indeed they fought the ship so well as to do heavy damage to her main opponent, the *Venerable*, which was reduced to such a state that the other attacker, *Caesar*, which had barely got into action, had to stand off to her rescue, lest she went ashore. So Troude was able to bring FORMIDABLE triumphantly into Cadiz. She was terribly battered, but it had been a brave achievement, and it was recognized as such.

Yet these attempts to help the army in Egypt were, in a sense, side-shows. Though peace with Austria was not yet signed, it was clearly coming, for she was plainly beaten. In reality Britain was already the only enemy. To deal with her Napoleon tried to find a fleet by promoting an Armed Neutrality of the Baltic powers. This league was to reject the British right of search, and as that right of search was vital to Britain, she would continue to apply it. That would mean, and did mean, war. Nelson's victory over the Danish fleet at Copenhagen in April, followed by the murder of the Czar Paul, plus the pacific attitude of his successor, brought the Armed Neutrality to naught.

In any case it was to have provided, at most, an important diversion. For already in March Napoleon had ordered the formation of a flotilla to carry troops to the direct invasion of England by the shortest route – across the straits of Dover. For this the boats built when he commanded the Army of England in 1797 were mostly available, and the shipyards of Holland, Belgium, and the French Atlantic coast were set busily to work to build others. Many fishing smacks and other small craft were collected, and the flotilla was to be completed by specially constructed flat-bottomed boats. There were to be thirty or so brig-rigged gun sloops each carrying three guns, twenty seamen, and about 130 soldiers; over 200 lugger-rigged gun boats, each having two guns (one light) with six seamen and a hundred troops; some huge rowing pinnaces; and some 'prames' – corvette rigged vessels of twelve guns[1]. The crossing was to be made towards the end of the year, the season of long dark nights, for, with France's only fleet closely blockaded with its Spanish allies in Brest, there could be no naval cover, and the giant expedition would have to rely on surprise. To command the flotilla there was appointed Latouche-Tréville (CA). It was a good choice. As an officer of the old royal navy he had shown vigour, professional skill, and devotion to his service. Nevertheless he had been expelled by the ridiculous purge of 1793, simply because he was of aristocratic birth. He had managed to avoid the guillotine and had returned to the Service when the popular violence of the Revolution was spent.

Never was a naval operation less naval than this, upon the French side. It was really a military affair. As his boats became ready Latouche-Tréville

[1]The figures given for the number of men to be carried by each type of landing craft vary greatly in different authorities. M Bottet, *Napoléon aux camps de Boulogne*, gives the much smaller figures of 44 and 28 soldiers.

mustered them in the Boulogne area. They were moored close to the shore in lines, and from each end of these lines others were moored at right angles till they reached the beach. They were mainly manned, and any fighting was mainly done, by soldiers who operated the guns and fired their muskets from floating wooden platforms instead of from Mother earth, who held a floating wall of fortification instead of a land redoubt.

The formation of the flotilla caused the liveliest alarm in England, however, and Nelson was called from the embraces of Lady Hamilton to deal with it. He made two attempts, in August. As the flotilla was moored in the shallows he could not get at it even with brigs; so his first attempt was a bombardment from bomb-vessels and gun boats. It had no effect. The second attempt was made at night by four divisions of ships' boats. Not all reached the lines of the flotilla: those that did so were driven off with loss. The great British admiral had unquestionably suffered a repulse.

Whether the invasion could have been made without naval cover is very much an open question. The flat-bottomed boats sailed well with a soldier's wind, but badly, almost unmanageably, unless the wind did happen to be dead astern, and they shipped a lot of water. As to the pinnaces, if they once fell off from being kept head to sea, their oars could not get them round again. The flotilla would have been fortunate if the necessary south-east wind, not common in the Straits, had chosen to blow when it was ready to sail, and it would probably have had to rely on its oars. Troops that had done nine hours' or so unaccustomed rowing through the night would hardly have been good material for a fiercely contested landing. And all this was to reckon without the English brigs and a frigate or so that would have been in the Channel and would have created havoc in at least some sections of the invading force, even granted that the French would have had some advantage of surprise. Be all that as it may, it is at least possible that Napoleon did not intend to invade. He almost certainly had not intended to do so in 1797, just as he entirely certainly did intend to do so in 1805. But in 1801 France needed peace, and Napoleon knew it, and knew, too, that he also needed peace to complete the establishment of his régime and that he could not afford a failure. England was tired of the war, and he had no longer to cope with the formidable William Pitt (who had fallen over the question of emancipation for Irish Catholics), but with the rather feeble Addington. It is at least possible that the formation of the flotilla was primarily intended to put the screw on that weak Prime Minister and bring him to negotiations.

It worked out that way: hostilities ceased in October pending the signature of the treaty of Amiens in 1802. The British negotiators had been no match for the triumphant Napoleon and his Foreign Minister, Talleyrand. England surrendered all her considerable list of colonial conquests except Trinidad (which had belonged to Spain) and Ceylon (which had belonged to Holland), so that France lost nothing overseas though the sea had given her nothing but defeats. Malta was to be restored to the Knights of St John under international guarantee, and France was to evacuate the south of Italy and the Papal States.

Though France had lost no territory her overseas trade had to start again from the beginning. It had been swept from the ocean and had given place to privateering. Some of the French privateers had been highly successful – notably Robert Surcouf, who operated in the Indian Ocean. Yet except for those who made money by it (including the French Government, which took its share) privateering was a delusion. England lost heavily; yet her mercantile marine increased, and she took many prizes, including a good number of the privateers. Though their letters of marque were usually for six months only, so as not to denude the French navy of seamen, the privateers embarked too many good men and, as we have seen, naval ships had to put to sea most inadequately manned. By 1801 there were no less than seventy thousand French prisoners in England, and nearly all of them were sailors. Nelson had landed his prisoners in Egypt after Aboukir, which had been the great naval capture of men, so while a number of those in England had come from single ship actions, many had been privateersmen. They would have been better value in naval ships, for it is in the use of the sea power of main forces, and not in mere commerce raiding, that wars are lost or won. France had now a breathing space in which to consider future naval policy and in which the vigorous impulse of Napoleon could rebuild her sadly diminished fleets.

Napoleon had no sooner seized power and established his Conseil d'État, which was his advisory Cabinet, than he ordered an enquiry into the organization of the naval bases. The decentralizing destruction of the old royal organization by the Constituent Assembly in 1791 had led to complete disorder and to almost complete disappearance of supplies. Napoleon reorganized local government by putting an all-powerful official, called a *préfet*, at the head of each *département*. The report on naval losses advised putting an all-powerful *préfet maritime* in charge of each base. He would correspond directly with the minister and would have under him the admiral commanding, the commissary responsible for pay, stores, and provisions, and the port director of naval construction. In other words, the old royal intendant was to be replaced by the maritime prefect, the same officer under another name. Napoleon adopted this report in July 1800, and of the six prefects appointed (for Brest, Toulon, Rochefort, Lorient, Le Havre, and Antwerp) three were CAs of the early revolutionary navy (Martin, Nielly, and Vence). The next month came a reorganization of the officer corps. There was once again a cutting out of dead timber; a number of officers were retired on half-pay. The establishment was laid down as so many officers of each rank and, each year, the minister was to pick those required for active service. The remainder were on an 'inactive' list and, if still on that list after three years, were treated as retired. No adequate arrangements were made for the proper training of future officers and nothing was done for the men. Despite his popularity with the troops, and despite his poses when among them, Napoleon did little for the welfare of his soldiers, who were just so many conscripts: still less was he likely to trouble about sailors. A vigorous impulse was given to shipbuilding, however, and to restocking the arsenals.

Just before he seized power there had been an interesting alteration in the gunning of French ships, occasioned by the adoption of the carronade. These upper-deck guns had been carried as an experiment in a certain number of ships since 1793. Now, in 1799, they became part of the official armament, replacing some of the ordinary upper-deck cannon. The replacements were almost always more numerous than the cannon surrendered, so that, as with many English ships, nearly all French ships carried more guns than the number at which they were rated. In ships of the line the extra number was six or eight guns, ie a ship rated at a hundred guns would carry 108 including her carronades. As his minister of marine Napoleon had at first a construction officer, named Forfait; but in October 1801 he gave place to Decrès, the CA who had commanded the frigates at Aboukir, and had got away with two of them, later commanding the GUILLAUME TELL in her brave attempt to break out through the Malta blockade. Decrès remained in office throughout the Napoleonic régime.

Directly hostilities with England ceased, and before peace was actually signed, Napoleon turned his attention to the colonial position. Most of the colonies had fallen into English hands, and were going to be regained by negotiation; but San Domingo was still nominally French. Nominally is very much the word, however, for in fact it had passed into the rule of an able Negro, named Toussaint l'Ouverture, and would have to be reconquered from him and his Negro forces. Vice-admiral Villaret-Joyeuse, who had returned to the command of the Brest fleet, was ordered to sail with ten of the line, some of the Spaniards in Brest, and some frigates, carrying a large army. He left in December and was followed by Latouche-Tréville (CA) with other ships, also carrying troops. This was a naval operation only in so far as the ships covered the landing of the soldiers at Cap Français and other places where they were put ashore. Otherwise it was a military affair, which turned out to be a disaster owing to a scourging epidemic of yellow fever that almost wiped out the army, and left its survivors holding just one or two strong posts. Before this position was reached Villaret and most of the ships had been recalled.

This was the last command of any note exercised at sea by Villaret-Joyeuse. Thereafter his services were used as governor of places of maritime importance, such as Martinique and Venice. He died in the latter in 1812. Historians have not, perhaps, done him justice – possibly because he suffered only defeats. As captain of a small ship under Suffren he had shown himself be a brave fighter against odds, nor was there any lack of courage in his handling of the MONTAGNE in the battle of Prairial. Alike in that action, in the manoeuvres that preceded it, and in the orders that he gave to the captains whose disobedience disgraced him in the affair off the Île de Groix, he showed himself to be a skilful tactician and a clever seaman. His objection to the Irish expedition, and his ability to keep his head and his command while telling the Revolutionary Governments the truth about their navy, mark him as astute and level-headed. The truth is that the fleets that he had to command were ill-found, ill-manned, and worse officered; and he did well to achieve as much as he did with them. Most of those who reached flag-rank in the Revolutionary and Napoleonic wars had been junior officers of the old royal navy, and Villaret was, perhaps, the best example of their merits. Napoleon, no bad judge of men, thought well of him. He was not a Tourville or a Suffren; but, apart possibly from Latouche-Tréville who died before he was fully tested, Bruix, and Cosmao Kerjulien who came to flag-rank too late to show if he could add an admiral's ability to his courageous vigour, Villaret seems to have been distinctly superior to the rest of the many and mediocre flag-officers who served France between 1793 and 1814.

CHAPTER XII

# Revolutionary and Napoleonic Wars: Part II – The Empire

The treaty of Amiens proved to be only a short breathing space. The British were glad enough to have peace, but somewhat disappointed at its terms. For their only gains were at the expense of France's Dutch and Spanish allies, and they considered that they had beaten the French. The two countries saw the position from irreconcilably different viewpoints. To military France the expulsion of the Duke of York's armies from the Continent, twice, meant victory: the reduction of their own fleets to impotence was comparatively unimportant. The British had expected nothing much of their army, but everything French that was at sea or overseas was or could be in their hands; and the French could not get at them. In their opinion it was they who had been the victors. At least they expected the peace to be observed in the spirit, that trade would be resumed, and that there would be no further changes in the map of Europe. But Napoleon introduced tariffs that were clearly directed against British goods, proceeded to remodel the map of Italy as it pleased him, and retained his forces in Holland. This being so, Britain refused to restore Malta to the Order of St John. Napoleon, already angered at the way that he was insulted in the British press, furiously demanded the execution of this part of the treaty. If the French retained Holland with, therefore, the exit from Antwerp, the British were not going to surrender the Mediterranean to them. Tension mounted, and Britain declared war in May 1803.

Of course, Napoleon had never thought, or intended, that the peace with the British would be permanent; but he did think that he would be able to keep them in play until some such date as 1806–8, by which time he would have built a large navy and would be ready to deal with them. His rage at being anticipated knew no bounds. He issued orders for the reassembling of the invasion flotilla and for extensive additions to be made to it, with Bruix (VA) in command[1], and he despatched Durand-Linois (CA), the victor of Algeçiras, to the Île de France with a properly found division of the MARENGO(74) and three frigates. But as France was taken by surprise, and Britain was ready, the year proved disastrous for the former in a minor way. For France was busy restoring her colonial position, for which she had a number of ships, mostly frigates and corvettes, at sea. The British swooped on these, and though the frigate POURSUIVANTE (Captain Willaumez) bravely fought off the eighty-gun *Hercules*, most of them were taken. The bulk of the captures were in the Caribbean, and

[1]He died in 1805 and the command passed to Lacrosse(CA).

included the DESAIX(74) which was surrounded by superior forces off San Domingo. The British also seized Tobago, St Lucia, the Banks islands, and the trading stations in India. The main interest, however, lay in the formation and fortunes of the flotilla.

The renewal of war with Britain had been greeted in France with a wave of naval enthusiasm rather similar to that evoked by Choiseul when he had taken office. Townships, districts, and bodies of people subscribed large sums. These were mostly used for extending the flotilla, at which all the shipyards from the Texel round to Rochefort were busily employed. There were 'prames' corvette-rigged, long for their beam, and mounting twelve 24-pounders[1]. There were gun-sloops, brig-rigged and mounting three 24-pounders. There were gun-boats, lugger-rigged with one 24-pounder and a lighter gun; and there were pinnaces, reduced in length from the clumsiness of those built in 1801, and lugger-rigged, but largely dependent on their oars. All these were flat-bottomed, so that they could be run up a beach, and the maximum draught, loaded with soldiers and their gear, did not exceed six feet. There was also a requisition of fishing smacks, and of coasters from fifty to a hundred tons. This caused no great hardship, for the British blockading and patrolling by smaller vessels was so close that little coastal sailing was possible in the North Sea or the Channel. The total number of vessels and transports in the flotilla varied somewhat from time to time, as beachings and other troubles caused accidents and withdrawals, but once it had reached full strength there were always the better part of two thousand craft available, and they would carry an army of not less than a hundred thousand men, complete with artillery, munitions, and with horses for most, though not all, of the cavalry.

With attention concentrated on this vast enterprise, there was little naval activity of the ordinary type. Such as there was took place largely in the Indian Ocean. In February 1804 Linois sailed from the Île de France to intercept the immensely valuable China convoy of the East India Company. His flag was in MARENGO(74), and he had also two frigates and a corvette. He lay in wait off Pulo Aor at the approach to the Malacca Straits, and he had good local information that the convoy would consist of seventeen ships with no escort. When it appeared it was sailing in most precise formation and the ships were not seventeen but twenty in number. Moreover, when the French were sighted, three large two-deckers stood out from the rest, flying man-of-war pennants, and formed line between Linois and the convoy. It seemed clear that there was, after all, an escort, and that it was more powerful than Linois could tackle. Being to windward, he tried to outreach the escort to get at the rear of the convoy; but the three ships made up towards him, seeking action, and so he had no choice but to stand away. In fact there was no escort. The three ships were East Indiamen of thirty to thirty-six guns, of whom the French ships could have made mincemeat. Dance, the Company's commodore, had brought off a first-class bluff which richly deserved the knighthood and more material rewards that it brought him; and poor Linois had more than lost the reputation that he had earned by his stout fight at Algeçiras. He failed also,

[1]Some carried up to twenty 36-pounders.

later in the year, to take the *Centurion*(60); but he and his ships did cause much loss to the trade.

Though there was no great French activity on the oceans, there was much planning, and that planning was Napoleonic. The flotilla would be complete by September 1804, and it would be in six divisions spread over seven ports stretching from Ostend down to Étaples. The problem was to get it and its soldiers across the Straits. The idea of a surprise invasion by using the long winter nights, which had been held, possibly not seriously, in 1801, was abandoned. For it was clear that the assembling of the whole flotilla near Boulogne from its various ports would take such a time that the vessels would inevitably be scattered by the winter gales. Nor, though it was fairly heavily armed, could it fight its own way across. While the boats were assembling and being moved from place to place, they had been attacked by small English warships. Where they had been creeping along the coast, in the shallows and with the cover and support of the numerous coastal batteries that Napoleon had had thrown up for that purpose, they had fared well enough. But on the one or two occasions that some had allowed themselves to get further out to sea they had suffered somewhat[1]. Further out still they would be under the attack of the larger frigates and some ships of the line. They would need an escort; and that escort would need to have seized command of the Straits. How was this to be achieved? When war was declared the British had over 120 of the line in commission and others in reserve. The French had thirty-six in commission in home waters with eleven in reserve and thirty still on the stocks. Twenty-one were under Ganteaume at Brest, five were at Rochefort, and Latouche-Tréville at Toulon would shortly have twelve. Yet the discrepancy in strength was not quite so great as it appeared; for the British had to maintain squadrons in the Antilles, off the St Lawrence, and in the Indian Ocean against possible French attack, had to protect convoys against privateers and to keep fleets spread widely on blockade from the Texel round to the Mediterranean. Some French ships would have to dodge the blockades and carry out diversions that would draw away the British while leaving themselves free to effect a concentration in the Channel.

The main force, which would have to be the striking force at the last moment, was the Brest fleet. But while Nelson, off Toulon, stood well away to sea in order to tempt the French to come out that he might give them battle, and while the Rochefort exits were hard to cover, Cornwallis blockaded Brest closely. Napoleon devised various plans, even toying for a while with the old idea that the fleet would land forces in Ireland and then come round the north of Scotland to perform its main task. This, and other plans, were pointless unless the Brest fleet could get out without a battle, for Napoleon did not want it to fight. Cornwallis' strength was usually at least roughly equal to that of Ganteaume, but when, on one occasion in 1805, the English numbers dropped to fifteen and Ganteaume signalled by 'telegraph' (a sort of inland semaphore) for permission to put out and fight, 'victory being certain', permission was refused. A naval

[1]In October 1804 the British loosed a number of 'machines' roughly akin to floating mines against the flotilla at Boulogne. They were a complete failure.

victory would serve no purpose, he was told, he must remember the mission (always the 'mission') for which his fleet was needed.

Nor was the Emperor (Napoleon had now assumed this title) helped by the death in summer 1804 of Latouche-Tréville. This vigorous and personable officer had trained and exercised the Toulon fleet keenly, as far as he could do in a roadstead, had even sent out a few ships from time to time to drive away Nelson's look-out vessels when these came in too close, and had generally got a good amount of *élan* into his squadron. To replace him Villeneuve (VA) was transferred from Rochefort in November. It is a little hard to understand this appointment. True, Villeneuve had put up a stout performance during the siege of Malta, whither, it will be remembered, he had gone after Aboukir: he was, too, a personal friend of the minister, Decrès, who knew him to be straightforward, personally brave, and professionally skilful: Napoleon, also, at this time, considered him 'fortunate'. Moreover, there was a shortage of tried flag-officers and, at this stage, it was not intended that the Toulon fleet should play a major part. But the appointment was unpopular in the service, which remembered his hesitations and failure to act at Aboukir; and, as his personality conveyed earnest worry rather than vigour, the *élan* of the Toulon fleet dropped sharply.

Towards the end of 1804 the situation altered. The Spanish Government, which by virtue of alliance with France was very much under Napoleonic influence, was pursuing a policy that was far from neutral where the British were concerned. A French ship of the line, AIGLE, was being sheltered in Cadiz, a few others, under Gourdon (CA), returning from the West Indies, had sought similar shelter from the British in Ferrol, and Spain was paying France a large subsidy for naval purposes. Britain, who had protested against all this, was not disposed to allow the bullion for further subsidy to arrive from South America, and intercepted the four frigates bringing it. These fought, and one was blown up. Spain now informed Napoleon that she would declare war; and though her fleet would not be ready till March 1805, this would permit fresh, and larger, combinations. Since he was a soldier, Napoleon believed in a mass of force. He was far from blind to the inefficiency of the Spaniards, and told his admirals that, when comparing the strength of an allied fleet with a British one, they should reckon two Spaniards as equal to only one enemy ship, but he never understood that, once fleets had closed, something more than a superior weight of metal was needed for victory. He was far too intelligent to disregard the uncertainty occasioned by winds and tides, as some earlier French landsman planners had done; and his orders almost always provided an alternative to allow for such things. Indeed, they were almost too permissive, too provided with alternatives, for admirals who, like Villeneuve, were over-weighted by their responsibilities. Yet he never (to use the apt words of a French historian) acquired 'an exact knowledge' of naval needs. The broken axle of a munition wagon would not be allowed to retard the rapid movement of a brigade, far less that of an army corps: why should the loss of a topmast gravely retard even a ship, much less a whole fleet? 'To a man with any strength of character', he wrote, 'the loss of some topmasts when a squadron leaves port is a very ordinary happening.' So the combinations that he devised for 1805, though grand in their conception, depended on things working out in a way, and to a timing that, even though it allowed some elasticity, could happen at sea only by an unusually favourable grouping of circumstances.

The main idea to divert the British from their close watch on the Channel was to make a serious attack upon the West Indies. For the Emperor rightly reckoned that the 'sugar interest' in the British parliament was so strong that powerful forces would be despatched to deal with it. To this end Bourgues-Missiessy (CA) was to slip out of Rochefort, and Villeneuve to sail from Toulon. Both were to carry troops, to take some of the islands held by the British, and to wait a certain time for the arrival of Ganteaume from Brest. If he did not appear they were to return to Rochefort, freeing the ships blockaded in Ferrol on the way. Missiessy sailed on January 11th with five of the line, his flag being in MAJESTUEUX (120). Arrived in the Antilles, his troops failed to take Dominica, which was his main objective, but he levied contributions on some of the islands. Villeneuve got away from Toulon on January 18th, with eleven of the line and seven frigates. His flag was in BUCENTAURE (80) and Dumanoir-le-Pelley (CA) was in FORMIDABLE (80). He sailed in a strong wind which turned into a gale. One of his big ships, INDOMPTABLE, parted company, as did some of the frigates; and most of his ships suffered damage which was more or less serious, and in the case of some of them serious indeed. So by the 21st he was back in Toulon. The trouble had been partly bad gear (one ship had to be withdrawn and two others were recognized to be not truly seaworthy) and partly bad seamanship. His men, Villeneuve wrote, had been well trained to work in a roadstead; they had not been trained to meet storms. In the same rising wind, Nelson, to save time on hearing that the French were out, had taken his fleet, by night, through a narrow and dangerous channel off Sardinia without the slightest mishap. Because Nelson, once he heard that the French had troops aboard, was somewhat obsessed with the idea that they would attack Sicily, or even Egypt, and steered to cover them, the fleets did not meet – to Villeneuve's frankly expressed relief.

Before sailing, the French admiral had written to Decrès in good heart about his squadron, though somewhat worried about its provisioning. But from his return to Toulon onwards his correspondence shows him to be very depressed[1]. He explained his coming back to port in a long formal report in which he complained grimly of the state of his ships and the incompetence of their crews, which, he said, were unfit for any enterprise and, had they met even an inferior force of British, would have been defeated. His confidence had clearly been not merely shaken but shattered by the unexpectedly poor showing of his fleet in the mere matter of heavy weather. He followed this report next day with a personal letter to his old

[1]Villeneuve's correspondence with the minister, from the time that he took command at Toulon up to Trafalgar, is given in fair detail as an appendix to Volume II of Jurien de la Gravière's *Guerres Maritimes*: see also Appendix IV of Desbrière, *Campagne de 1805*.

friend in which he wrote that his aim in the service had been to be a useful officer, not to win glory, that he did not want to become the laughing stock of Europe, and he virtually asked to be relieved of his command. He should have been removed forthwith; and it can only be supposed that it was Decrès' wish to see a friend, in whom he believed, justify himself that stopped this from being done.

With Villeneuve back in Toulon, orders were sent to Missiessy to reinforce San Domingo and then to return. (There was no difficulty in getting a small fast despatch vessel to sea from ports, such as Lorient, which, containing no ship of the line and none of the flotilla, were not blockaded.) A little later, the Emperor having devised a fresh combination, further orders were sent to him to remain; but he had already sailed, and made an uneventful return to Rochefort. The new plan was that Ganteaume was to sail with the Brest fleet for Ferrol to free Gourdon's blockaded ships there, and then to go with them to Martinique, where it was hoped Missiessy would have remained. Here he was to wait thirty days for Villeneuve, and it was presumed that a considerable British force would have been drawn off to follow him when Britain got the news of his arrival at Fort-de-France. As Villeneuve would have to wait for the completion of a new ship, the PLUTON, he would not sail till late in March, by which time the Spaniards should be ready. He was to go to Cadiz, collect the AIGLE and such Spaniards as could put to sea, and then make also for Martinique. If all went well so that Ganteaume had been able to pick up Missiessy, and Villeneuve also arrived within the stipulated thirty days, then the whole of the French fleets, and some Spaniards, would be joined under Ganteaume's command, whereon that admiral was to bring them back to the Channel and drive the remaining British away, so that the invasion flotilla could cross. Even if Villeneuve did not appear in time Ganteaume was to return and fight provided he had twenty-five of the line; if he was below this strength he was to go to Ferrol where the missing French, and the Spanish ships, would have been collected, and then to sail for the great battle to command the Channel.

This was a fine combination – on paper. Yet since Ganteaume had to get away from Brest without fighting, it depended on an easterly wind so strong as to blow the British blockade away while the Brest fleet got to sea. More than once that spring and summer Ganteaume's ships lay ready under the batteries in Bertheaume and Camaret roads, outside Brest; but the right wind never came and they had to return to the inner road. Villeneuve did not sail until March 30th; for when PLUTON was nominally ready for sea it was found that a mast was faulty (so much for French dockyards) and a new one had to be stepped. He had eleven of the line and six frigates, and he and Dumanoir flew their flags in BUCENTAURE and FORMIDABLE as before. He made first for Cartagena, hoping to be joined by the Spanish ships there; but their admiral had no orders to sail and refused to leave without them. Villeneuve rightly declined to wait and made for Gibraltar, passing through the Straits on April 9th. Sir John Orde's small blockading squadron off Cadiz stood away as he came in sight, and he was joined by the long-blockaded AIGLE and six Spaniards of the line under Don

Federigo Gravina (LG). Gravina, who was rumoured to be an illegitimate son of the late King Carlos III, was an excellent officer, very different from the Spanish admirals with whom the French had had to work in the past. He was professionally able, vigorous, notably loyal, and fully co-operative. He had installed something of his own spirit and capacity into the Spanish navy so that its officers, though still not as efficient as they should have been, were more worthy of their powerful ships than in the past: their crews, however, were still far below a proper standard of training.

On April 11th, Villeneuve left for the West Indies where he was instructed to wait for forty days for Ganteaume, and to fill in the time conquering some British possessions with the troops that he had on board. Meanwhile Nelson, when he heard that the French were out, and was failed for once by his frigates which lost touch with them, yielded to his old obsession and steered to cover the approaches to Sicily, Tunisia, and Egypt. Apprised that the French had passed Gibraltar, he then ran into headwinds and it was not till a month after Villeneuve had done so that he cleared the Mediterranean. He delayed briefly to see a troop convoy into safety and detached one of his precious ships, the slow-sailing *Royal Sovereign*, to strengthen its escort, and then made his best speed after the enemy. The French admiral made a slow passage. The Spaniards were poor sailers and four of them dropped astern out of sight, to join him later in Martinique. Some of his own ships were little better: the new PLUTON proved most unhandy, others needed the assistance of a tow at times, and it was not till mid May that he reached Martinique. So far fortune had favoured him, and his letters to Decrès take on a more cheerful tone, though he is still worried about the question of provisions, and still complaining of the poor quality of his ships, their gear, and their crews. (As he had had them under his command, at sea, for almost seven weeks, a vigorous admiral would have seen to it that they got more training than was in fact the case.) Nor, apart from seizing some merchantmen, was he active in making conquests of the alarmed British possessions; for all that he did was to capture the so-called 'HMS Diamond Rock'. This was an islet that stuck precipitously out of the sea, like a larger version of Ailsa Craig, six miles off Martinique. The British had seized and fortified it, and it was an unpleasant threat to French ships making for Fort-de-France. Cosmao Kerjulien in PLUTON efficiently directed its capture.

Now Villeneuve was joined by Magon (CA) from Rochefort with the eighty-gunners ALGÉÇIRAS and ACHILLE, and fresh orders[1]. He was to wait thirty days (more waste of time when time was precious) for Ganteaume and if he failed to arrive Villeneuve was to make for Ferrol, collect the French and Spanish ships there and sail north to drive away the Brest blockade and free the Brest fleet. This suddenly turned Villeneuve from being a diversionary into being the main executive in the great invasion plan, and threw on to his shoulders a load of responsibility that he was both ill-fitted and unwilling to bear.

[1]Villeneuve objected, with justice, that he could not both sail in sufficient force to capture British islands and yet be ready at Martinique to weigh at once when Ganteaume arrived – which were his orders.

Nelson reached Barbados on June 4th, having gained eight days on Villeneuve in his passage from Gibraltar. He landed the troops that he had brought and then sailed for Tobago on some false information that the French were attacking that island or Trinidad. On June 8th, Villeneuve seized a British convoy and learned of Nelson's arrival with, he was told, fourteen of the line: in fact Nelson had only ten ships. With the ships of the line already in the Antilles under Rear-admiral Cochrane this would raise the British strength to nineteen or twenty, a force equal (or superior) in strength to the allies. Already worried about his stores, on which the troops that he carried had drawn heavily, alarmed at the arrival of the famous English admiral, and judging rightly both that Ganteaume would not now appear and that the Emperor would be glad to have him back in European waters ahead of Nelson, Villeneuve abandoned all thoughts of attacking Barbados, hurriedly transferred his troops which he had embarked for that purpose to frigates which would take them back to Martinique, and made sail, on the 9th, for Europe. Nelson, having at last got sound information of the whereabouts of the French, deduced that Europe was their destination, and turned for it himself on June 13th, with eleven of the line, sending a fast brig, *Curieux*, Captain Bettesworth, with despatches for the Admiralty reporting his movements.

As the Mediterranean was his command, Nelson steered for Gibraltar, but Bettesworth, on a more northerly course, sighted the French, and he and his men alone had the vital information of where were the ships of Nelson and Villeneuve. Cramming on all sail for Plymouth, Bettesworth went post-haste thence to London which he reached at midnight on July 8th, and in the morning the Admiralty messengers were riding to the ports with orders for the fleets.

Vice-admiral Sir Robert Calder was to raise his blockade of Ferrol; the ships off Rochefort were also to raise their blockade and to sail to join him: Calder was then to lie in wait for Villeneuve off Cape Finisterre. This was done and the two fleets sighted each other on July 22nd. Gravina was leading the allies in the *Argonauta*(80), Villeneuve and Dumanoir were in the centre, and Magon (CA) brought up the rear in the ALGÉÇIRAS (80). The allies had the advantage of the westerly wind. After some manoeuvring, as a result of which both fleets went about on to the starboard (southerly) tack, cannonade was joined in mid-afternoon. The battle, known to the French as the fifteen-twenty, and to the British as Finisterre, was a nightmare one, not, this time, for lack of wind, but because fog came down. This, plus the smoke from the guns, was at times so dense that ships could not see their next ahead and had to fire, not even at the flashes, but at the noise of the enemy's cannon. The fight was a long, sometimes blindly conducted, gun duel. In the obscurity three Spanish ships fell out of line leeward. Cosmao Kerjulien in PLUTON saw their danger and steered to support them, but lost them in the fog. One got back into line, but the other two, dismasted, struck at about 8.00 pm. An hour later firing ceased.

When day broke Calder, now some miles clear of the allies, avoided further action. He seems to have been discouragingly surprised at finding the enemy twenty strong, when he had expected only seventeen, some of his ships had damage to masts (*Windsor Castle* had to be taken in tow), and he quite failed to realize that his duty was to lay into the French to the last ship in order to leave them too weak to play any part in the invasion planning. For this he was later reprimanded by court martial. Nor does Villeneuve come out of this indecisive battle more creditably. He was still superior in number to his enemy, though not greatly so in weight of metal, and his Spanish allies were naturally eager to recover their two ships that had been taken, but ATLAS had been knocked about, other ships were short of water, and his sick list, as usual, was heavy. So his pursuit was hesitant, and after two days the fleets lost sight of each other. Villeneuve then bore up for Ferrol.

With so much at stake, the action had been surprisingly inconclusive. Yet, by completing Villeneuve's discouragement, it was largely responsible for the failure of Napoleon's hopes for the invasion. Nothing more illustrates the difference between the two warring nations in their approach to matters maritime than the surprise shown by the Emperor at the interception of Villeneuve. How, he wondered, had the British known where Villeneuve would be: he had heard when Bettesworth had reached London, but he was sure that the British must have had their information before that, since they had obviously arranged a fresh distribution of their squadrons, and they must have taken not less than twenty-four hours to do so. In fact it had taken the Admiralty not much more than twice that number of minutes. Napoleon could not understand it, yet, where armies were concerned, he was used to making lightning decisions for fresh routes or alignment of large bodies of troops.

As winds proved adverse Villeneuve put into Vigo on July 28th. Here he landed not less than eight hundred sick and sent a despairing letter to Decrès. Everything had gone wrong: adverse winds: ships of poor quality with poor gear so that damages had been frequent: equally poor quality of officers and seamen: hence the failure to destroy Calder. The English would now have been strongly reinforced, and the French would be the laughing stock of Europe. Then, leaving behind the damaged ATLAS and two slow sailers of the Spaniards, he made for El Ferrol. An admiral so deep in the Slough of Despond should have been relieved of his command at once; but off Ferrol he received further orders. He was not to enter the port (which was difficult of egress) but to call out the French and Spanish ships there. This done, he was to sail to effect a junction with the five ships which had returned to Rochefort from the Antilles and were now freed by the raising of their blockade. (Their CA, Missiessy, having gone sick they were commanded by Captain Allemand – shortly to be given flag rank.) Then he was to make for the Channel, which would probably free the Brest fleet. If, which Heaven forbid, some unforeseen event should damage his ships, he might turn to Cadiz to rope in the Spanish ships there. 'Set out forthwith,' wrote the Emperor, 'sweep all before you, come boldly down on the enemy; if you give us control [of the Channel] for three days, nay, even for twenty-four hours, your task will be done: all is ready, Europe waits breathless on this great event: I put my trust in your bravery and skill.'

The exhortations, meant to encourage, had the opposite effect on Villeneuve. He knew all the work and money that had gone into preparing the flotilla and assembling the Grande Armée: he realized that the future of Europe hung upon his actions. Instead of stimulating, the responsibility depressed him even further. He distrusted the capacity of his fleet, and felt that the moment for even possible success had passed. Gravina, though loyally obeying orders, agreed with him, for the loss of time occasioned by foul winds and other mishaps would have allowed the British to collect their fleets. That being so, the temptation to Villeneuve to avail himself of the slight permission to go to Cadiz was fatal. Nevertheless, in Corunna, whither he had been obliged to go on being forbidden to enter Ferrol, he showed the greatest energy in getting his fleet ready for sea, though that port was lacking in resources. He got both his own ships and those in Ferrol weighed on August 10th, with some difficulty (Ferrol and Corunna are on opposite coasts of the same huge inlet), wrote next day to Decrès fearing that the state of some of the ships and the strength of the enemy would hardly leave him any choice but to go to Cadiz, sent the DIDON frigate to find Allemand, and then set a hesitant course for the Channel. He had twenty-nine of the line, of which eleven were Spaniards, plus some frigates and corvettes. On the 14th, he sighted some distant frigates, and could see that one was towing his own DIDON, which had had to strike to the superior gunnery of the British *Phoenix*, of roughly the same strength. So he was not going to make a junction with Allemand. Then a Danish ship which he stopped told him that the distant frigates were the outliers of a fleet of twenty-five of the line. This was, in fact, the well-worn trick of supplying the enemy with false information, but it was enough for the French admiral. Helms were put over and course set for Cadiz.

Allemand, with his five of the line and some frigates, duly sought Villeneuve, and twice narrowly missed him, once because of a serious omission by Decrès in the orders and information sent to him. Hearing eventually that Villeneuve was in Cadiz, which he had reached on the 20th, but was too heavily blockaded to be joined, Allemand embarked on an extraordinary cruise which ranged from the Penmarks in the north to the Canaries in the south. As he avoided the British who were seeking him, this earned his ships the name of the Invisible Squadron. He took the fifty-gun *Calcutta*, a brig and a corvette, and a number of merchantmen. Then, after 160 days, a long spell at sea for a French squadron, he got back into Rochefort, safe but with an appalling sick list, and having served little purpose.

The Brest fleet had been no more effective. Baron Ganteaume got his ships out to Camaret roadstead on August 22nd, but the British promptly came in so close that Cornwallis was slightly wounded by fire from the shore batteries, and, as Ganteaume had to avoid battle, he hurriedly got his ships back into the Goulet, the rearmost exchanging a broadside with the leading British.

While all this was happening, Napoleon had been waiting, as it were, watch in hand. Pitt had been recalled to office in 1804 and had taken up the old policy of forming a coalition against France. Prussia, pursuing an unscrupulously selfish course, would not join, but Austria was angered by Napoleon's ruthless remodelling of the map of Italy, and the Czar was angered by this, and even more by the seizure of the royal duc d'Enghien in neutral territory and his execution by a firing party. These two powers were now mobilizing. But they were notoriously slow, and the Russians, in particular, had a long way to march. In the invasion flotilla all was ready, it might truly have been said (as it was untruly said sixty-five years later), to the last button on the last gaiter. If the French fleet appeared there would still be just time to cross the Channel, overthrow Britain, and return before the slow Austro-Russians had broken through the reserve armies and the fortified towns of the eastern frontier. But Villeneuve failed to appear; it was clear that he must have gone to Cadiz. With no more fighting than a single indecisive battle the great invasion scheme had failed. The plan had been formidable: the instrument for its execution, the navy, had not. The Emperor gave vent to one great explosion of anger, then ordered the disbanding of the flotilla crews and the docking of its vessels, while his rapid troops were sent marching to the triumph of distant Austerlitz.

Though the invasion scheme had to be postponed, Napoleon had no intention of allowing his fleets to eat the bread of idleness. Later he was to order Decrès to send divisions from the Brest fleet to attack British colonies and trade, and to reinforce French possessions overseas. For the moment, Villeneuve was to get the combined fleet to sea. It was to make for the Kingdom of Naples to land there the troops that it had aboard so that they might further French operations in Italy: it was then to seize all enemy trade. As a use of sea-power this was absurd: the Emperor had refused to allow the Brest fleet to fight a battle on favourable terms, but he was now hazarding the combined fleet for a minor military objective. However, the instructions, though coming from Decrès, were couched in stinging terms, obviously in the Emperor's words and used at his bidding. It was shameful that the Emperor's ships should allow themselves to be blockaded by inferior or even equal forces: he wanted courage and daring in place of a depressing policy of defence; he wanted battles of extermination. So the enemy was to be attacked when inferior or even equal in numbers: there were to be audacity and the greatest possible activity. These orders arrived on September 27th.

Villeneuve had been having a difficult time in Cadiz. He was, in all, two thousand men short (seventeen hundred sick and three hundred deserters – but he had troops who could serve the guns): he had also great difficulty in supplying his fleet and in making good damages, for the arsenal at Cadiz had exhausted itself in fitting out the Spanish ships. As the allied fleet had approached the port, Collingwood, having only three of the line, had stood away; but he had been rapidly and strongly reinforced, and on September 28th Nelson had arrived in the *Victory* to take command of a blockading fleet equalling Villeneuve's own strength of thirty-three of the line. On October 1st, the French admiral summoned a council of fourteen flag-officers and senior captains to advise on the instructions that he had received. The Spaniards protested that three of their ships, being only just commissioned, were not worked up, whereat there was something of a scene, for the fire-eating Magon accused them of cowardice. Gravina tactfully asked for a vote; this was given against sailing, though they should be ready to do so in the event of a favourable opportunity[1].

The reproachful but 'spurring' orders to Villeneuve were sent on the 14th. The next day Napoleon, who had already described his admiral as a scoundrel and traitor, decided to replace him by Vice-admiral the comte de Rosily-Méros. The choice was odd, for Rosily had spent most of his service in administrative posts. Presumably a vice-admiral was needed to take command over Gravina, and none better was to hand. Decrès did not notify his old friend of his impending disgrace, but left Rosily to do this on arrival. Villeneuve heard unofficially of Rosily's coming, and wrote that he would be delighted to serve under him; but when Rosily, having reached Madrid, was preceded by strong rumours of Villeneuve's recall, that admiral wrote bitterly of undeserved disgrace and ordered the fleet to prepare to sail on October 18th. Almost all historians have considered that it was because he was stung by a reflection on his personal honour, that of cowardice, and was anxious to prove himself in battle, that he put forth on his last fateful voyage. It may be so, but it is more probable that this view does Villeneuve injustice. Watering his fleet was a problem for Nelson, and he had sent some ships under Rear-admiral Louis to Tetuan and Gibraltar for this purpose : he had also generously allowed the unlucky Calder to return for his court martial in the *Prince of Wales* : his ships of the line were thereby reduced to twenty-seven. Villeneuve's look-outs had reported this : he now had superior numbers and the Emperor's stringent orders were to break a blockade in such a case : the wind was favourable and by the time Rosily should arrive the opportunity might well be lost. Magon got some ships out on the 19th : the rest cleared on the 20th, and course was set for the Straits. In view of his orders Villeneuve's decision was justified.

The allied fleet was in the usual three divisions, the van under Alava (LG), the rear under Dumanoir, with also a squadron of observation (ie a reserve) under Gravina, which kept to windward of the rest. Nelson had followed his usual practice of keeping well to sea to tempt the enemy out of port, with a reliable signalling chain of look-out vessels. Though he had not thought that the enemy would sail so soon[2], he was quickly informed, made towards them, and his lights were seen by the allies late on October 20th. Villeneuve signalled by gun and lights to form line of battle, but, as it was dark, this was ill-done. He therefore repeated the signal at about six o'clock the next morning. There being little wind and a swell beginning to rise, the evolution was not complete when, shortly after eight, he further signalled his ships to wear together, which brought them on a northerly course. This may have been partly because the enemy seemed to be threatening his rear, as he had expected, and partly so as to have Cadiz on his lee as a refuge for damaged ships. For the latter purpose he would have been wise to make the signal earlier in combination with that for line of battle,

[1]Villeneuve's estimate of his own (French) ships was that twelve were good sailers, though six of these were not well manned, and the remaining six were unhandy.
[2]See a previously unpublished letter of Nelson in *The Times* of October 20 1939.

which might then have been properly formed. The reversal of course put Dumanoir's division in the van, and Gravina brought the reserve division into line to bring up the rear. But the line was never rightly formed. It was rather a shallow crescent than a line, concave to the enemy's approach; and even so some ships, unhandy or ill-handled, were to leeward of their proper station. These included three in the centre.

Nelson's captains knew his plan of attack. This was to be in two columns led by himself and Collingwood. (There would have been a third small reserve column had he had more ships.) One was to cut the enemy's line just astern of its centre and concentrate on destroying its rear, to which it would be superior in numbers. The other would attack the centre and crush it, while its sternmost ships would hold off the van, so far cut off from the battle, when it came about to the aid of its hard-pressed centre and rear. Collingwood was left complete discretion in the handling of his half of the fleet, and wide discretion was left to each captain, for, said the admiral, in the confusion of battle no captain could go far wrong who laid his ship alongside an enemy. Villeneuve, who was professionally shrewd, had partly divined the nature of his opponent's attack: it would be a concentration on the rear of his line. He had notified his captains of this in a memorandum, but devised no special means of meeting such attack beyond stressing a standard French signal to the effect that any ship not engaged with an enemy was to get into action by the quickest possible route, and adding that in battle one should not wait for signals from the admiral, who was himself likely to be hotly attacked and unable to see through the smoke what signals should be given. Once again it should be remembered that, in the days of sail, naval battles were a slow motion affair compared with those fought under steam: guns were truly effective only at fairly short range, and the working log of the *Leviathan* for October 21st shows that it was not until Villeneuve's fleet was clearly in sight that the ship began to clear for action, speed two knots! The light wind was coming from the west-north-west, and the swell, forerunner of the gale that was to come that night, increased during the battle.

The allied fleet was steering north in the following order, the ships that were so much to leeward as to leave serious gaps in the line being marked with an asterisk. (Other ships were a little to leeward, almost doubling their next ahead, but not leaving any real 'hole' in the line.)

| Ship | Guns | Captain |
|---|---|---|
| *Neptuno* (Span) | 80 | Don C Valdes |
| SCIPION | 80 | Berrenger |
| INTRÉPIDE | 80 | Infernet |
| FORMIDABLE | 90 | Dumanoir Le Pelley (CA) |
| DUGUAY-TROUIN | 74 | Touffet |
| *Rayo* (Span)* | 80 | Don E Macdonell |
| MONT-BLANC | 74 | Lavillesgris |
| *San Francisco D'Asis* (Span) | 74 | Don L Flores |

| Ship | Guns | Captain |
|---|---|---|
| *San Augustino* (Span)* | 74 | Don F Cajigal |
| HÉROS | 74 | Poulain |
| *Santissima Trinidad* (Span) | 130 | Don B Cisneros (CE) |
| BUCENTAURE | 90 | Villeneuve (VA) |
| NEPTUNE* | 74 | Maistral |
| *San Justo* (Span)* | 74 | Don M Gaston |
| *San Leandro* (Span)* | 64 | Don J Quevedo |
| REDOUTABLE | 74 | Lucas |
| INDOMPTABLE* | 80 | Hubert |
| *Santa Anna* (Span) | 122 | Don I M Alava (LG) |
| FOUGUEUX | 74 | Baudouin |
| *Monarca* (Span)* | 74 | Don T Argumosa |
| PLUTON | 80 | Cosmao Kerjulien |
| ALGÉÇIRAS | 80 | Magon (CA) |
| *Bahama* (Span) | 74 | Don D Galiano |
| AIGLE | 74 | Gourrège |
| SWIFTSURE | 80 | Lhospitallier Villemadrin |
| *Montañes*(Span)* | 74 | Don F Alcedo |
| ARGONAUTE | 74 | Épron |
| *Argonauta* (Span)* | 80 | Don A Pareja |
| BERWICK | 74 | Filhol Camas |
| *San Juan Nepomuceno* (Span) | 74 | Don C Churruca |
| *San Ildefonso* (Span) | 74 | Don J Vargas |
| ACHILLE | 80 | Deniéport |
| *Principe de Asturias* (Span) | 112 | Don F Gravina (LG)[1] |

When the fleets sighted one another the British were in no particular order and were making for the allied rear (then Dumanoir's division), but after Villeneuve had brought his fleet about the enemy formed into two columns. The port or more northerly one was led by Nelson in *Victory*(rated 100, actually 110) followed by *Téméraire*(98), *Neptune*(98), *Conqueror*(74), *Leviathan*(74), *Ajax*(74), *Orion*(74), *Agamemnon*(64) *Minotaur*(74), *Spartiate*(74), with *Britannia*(100, flag of

[1]The failure to form line properly makes it difficult to give the precise order in which the allied ships were ranged when battle was joined: nor is it made easier by the fact that both formal reports and personal recollections, jotted down near the time or written later, whether allied or British, are often so contradictory that it is impossible to reconcile them; eg the *Royal Sovereign*'s first raking broadside into *Santa Anna* is said to have brought down four hundred men, but though she fought for a long time after that, her total official casualty list does not reach that figure. As to the above order, in strict order of battle BERWICK and *Nepomuceno* should have been astern of *P de Asturias* and may have been so till the fighting in the rear became a mêlée: during this they went ahead. They certainly were ahead of Gravina and ACHILLE when they struck.

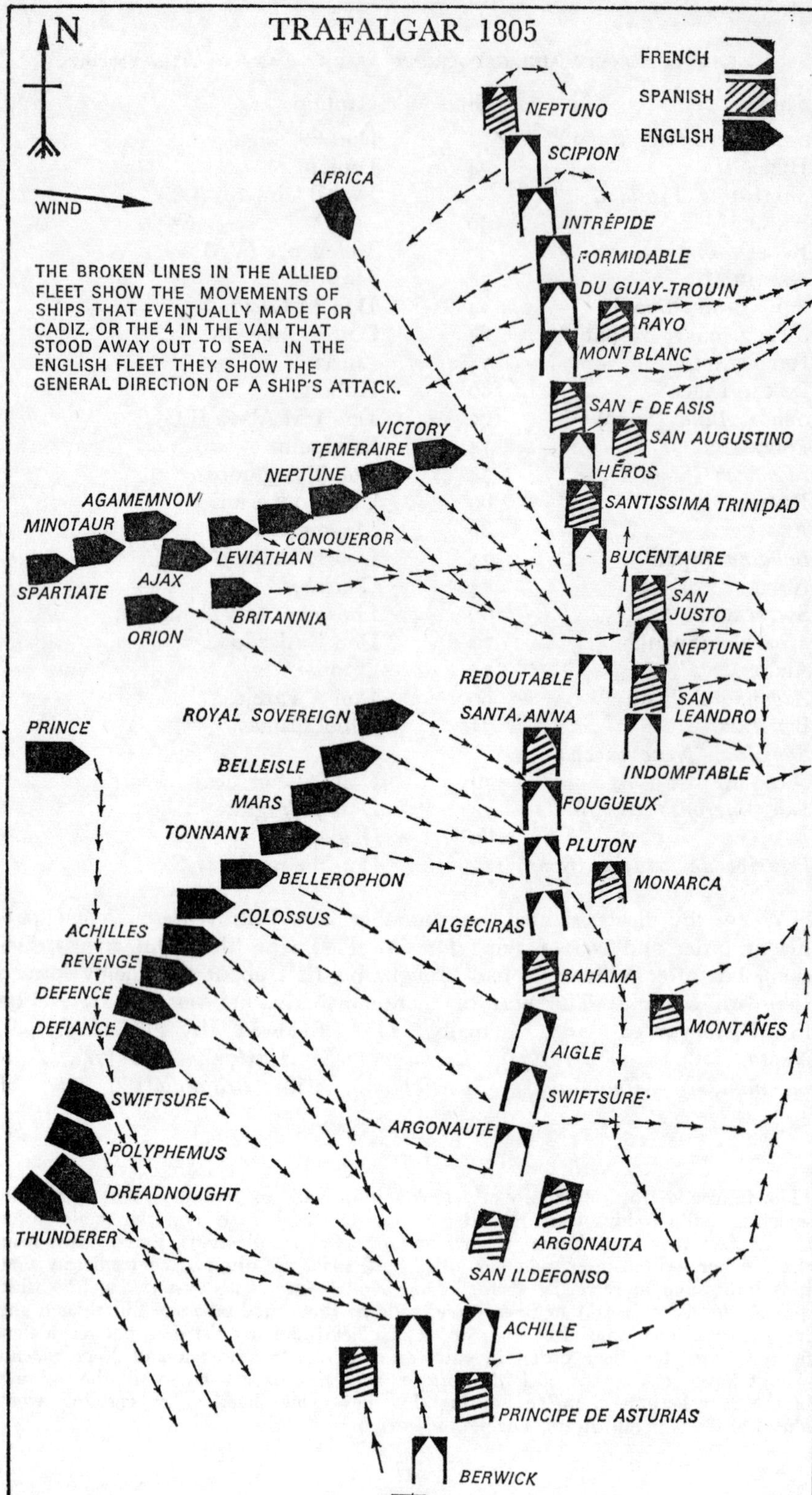

TRAFALGAR 1805
N
WIND
FRENCH
SPANISH
ENGLISH
THE BROKEN LINES IN THE ALLIED FLEET SHOW THE MOVEMENTS OF SHIPS THAT EVENTUALLY MADE FOR CADIZ, OR THE 4 IN THE VAN THAT STOOD AWAY OUT TO SEA. IN THE ENGLISH FLEET THEY SHOW THE GENERAL DIRECTION OF A SHIP'S ATTACK.
NEPTUNO
SCIPION
INTRÉPIDE
FORMIDABLE
DU GUAY-TROUIN
RAYO
MONT BLANC
SAN F DE ASIS
SAN AUGUSTINO
HÉROS
SANTISSIMA TRINIDAD
BUCENTAURE
SAN JUSTO
NEPTUNE
SAN LEANDRO
INDOMPTABLE
REDOUTABLE
SANTA ANNA
FOUGUEUX
PLUTON
MONARCA
ALGÉCIRAS
BAHAMA
MONTAÑES
AIGLE
SWIFTSURE
ARGONAUTE
ARGONAUTA
SAN ILDEFONSO
ACHILLE
PRINCIPE DE ASTURIAS
BERWICK
SAN JUAN NEPOMUCENO
AFRICA
VICTORY
TEMERAIRE
NEPTUNE
AGAMEMNON
MINOTAUR
CONQUEROR
LEVIATHAN
AJAX
SPARTIATE
BRITANNIA
ORION
ROYAL SOVEREIGN
PRINCE
BELLEISLE
MARS
TONNANT
BELLEROPHON
COLOSSUS
ACHILLES
REVENGE
DEFENCE
DEFIANCE
SWIFTSURE
POLYPHEMUS
DREADNOUGHT
THUNDERER

Rear-admiral Lord Northesk) and *Africa*(64). The starboard (southerly) column was led by Vice-admiral Collingwood in *Royal Sovereign*(100) followed by *Belleisle*(74), *Mars*(74), *Tonnant*(80), *Bellerophon*(74) *Colossus*(74), *Achilles*(74), *Polyphemus*(64), *Revenge*(74), *Swiftsure*(74), *Defence* (74), *Thunderer*(74), *Defiance*(90), *Prince* (98), *Dreadnought*(98) – twenty-seven ships, several being three-deckers, to the allies' thirty-three. The expression 'column' is more nominal than strictly correct for *Britannia, Orion,* and *Prince* were between the two lines and the last named was a notably slow sailer so that, though she was nearer to the enemy than some others of the division when fire opened, she was one of the last to get into action. As for *Africa,* that ship was never in column at all, for she had got separated from the fleet to its north, and came into the battle by the shortest course, whereby she ran the gauntlet of all the ships in the allied van, exchanging broadsides with them before joining in the attack on the *Santissima Trinidad*. Moreover, at about ten o'clock Collingwood signalled to his division to come into a line of bearing to starboard. This meant that his ships would fan out, that his fastest sailers could go ahead, and that each would try to find its own gap through the allies' rear, on which the British were concentrating fifteen ships to twelve. The nature of the battle, in fact, was that it became a mêlée in the rear, in which the British worked well together in supporting one another or engaging an enemy in pairs, while the allies fought their vessels for the most part as a series of single ship actions against one or more foes. Much the same is true of the centre, where the main weight of Nelson's attack was directed on to BUCENTAURE, REDOUTABLE, and the great *Santissima Trinidad*, a so-called four-decker and the biggest warship afloat.

As *Royal Sovereign* had recently been freshly copper-bottomed, she sailed well ahead of her division making for the stern of the *Santa Anna,* and came first into action. Nelson varied his course from time to time, to keep his enemy in doubt as to where he would strike, before he made for the French flagship. At this point, about noon, the plan of attack became clear, and if Dumanoir had been an admiral of any initiative, he would have commanded his division at least to prepare to come about, if not, indeed, actually to do so, in order to take part in a battle from which they would otherwise be excluded.

Villeneuve signalled to open fire when the enemy came into range and FOUGUEUX started the battle by letting fly at the *Royal Sovereign* at noon, an example followed by all ships near him. Gunnery was not good, for both *Royal Sovereign* and *Victory,* exposed for some time without being able to reply, sustained little injury. Collingwood broke the line astern of the *Santa Anna,* giving that ship a murderous raking broadside before hauling up to fight her on her lee. Here, the *Royal Sovereign*, under fire from FOUGUEUX, INDOMPTABLE, and *Monarca*, was largely dismasted before other British ships arrived. In the centre at twelve-thirty Nelson made for the gap astern of BUCENTAURE and gave her a raking broadside, but could not get through because Lucas had got more sail on to REDOUTABLE and closed the gap. The *Victory* therefore ran broadside on against that ship, and the weight of the collision carried both vessels, grapnelled together, to leeward, leaving a gap through which others could pass. Lucas, a flamboyant and boastful person, was yet a good officer and had trained his men especially in musketry and boarding. Though the cannonade below was blasting in his ship's side, his musketry mortally wounded the great British admiral and almost cleared the *Victory*'s upper deck. He had therefore called for his boarding parties and lowered his mainyard to make a bridge for them when they were swept away by a dreadful broadside from *Téméraire* which came up against his starboard side. Between the two three-deckers his position was hopeless, though it was not till half past three that, totally dismasted, he struck. Other British ships had passed through ahead of this trio to engage BUCENTAURE and *Trinidad*. NEPTUNE and the two Spanish ships that had got to leeward of their stations in the centre, instead of bearing up to the aid of their admiral had put their helms over and made for the rear of the line. This may have been a misinterpretation of Villeneuve's orders that all ships should get into action, for when they did this it was on the rear that the main attack seemed to be coming; but the captain of NEPTUNE was later removed from his command and left on the beach for five years (as was also the captain of ARGONAUTE). FOUGUEUX had made extra sail in vain to close the gap between herself and the *Santa Anna*, and so was raked by Collingwood's starboard broadside as he came through. She returned this as she came up alongside the *Royal Sovereign* (which was now broadside on to the Spaniard) but then ran bows on into *Belleisle*'s starboard beam and so was hard hit by an almost raking fire from that ship. This did not prevent FOUGUEUX from giving a heavy and destructive broadside to *Mars*, but by the time she got clear of the *Belleisle* she was seriously damaged aloft and so drifted against *Téméraire*. A fearful broadside from that three-decker completed her dismasting and, with heavy casualties, she could make no attempt to resist boarders, and struck at about two-thirty. She was the first French ship to do so, but three Spaniards had already surrendered. Nevertheless, despite their deplorably untrained crews, the Spanish ships as a whole, being well officered, were far better fought at Trafalgar than they had been for many a long year.

Long before FOUGUEUX's surrender the mêlée in the rear had become furious and general, some British passing astern of the allied line to engage from leeward, most breaking through gaps. PLUTON had fired on Collingwood's advancing ships and Cosmao had handled her skilfully to block the *Mars* and rake her; she was, indeed, preparing to board that ship when she had to engage *Tonnant*, and then, luffing up, she came on to the port quarter of the damaged *Belleisle* and fired on her. So far she had been notably well handled and had done much harm to two enemies but, *Polyphemus* coming on the scene, PLUTON paid off. Her casualties were heavy and she made slowly for the surrounded *Principe de Asturias*, whom her presence helped to disengage. ALGÉÇIRAS was much injured in her masting as *Tonnant* came through the line. She closed that ship as it turned on PLUTON to disengage the *Mars*, but *Tonnant* was the better handled: Magon's attempt to save the situation by boarding was blown to bits, he himself was killed and, totally dismasted, ALGÉÇIRAS struck at a quarter to four, having had much the worse of a long grim fight. AIGLE, SWIFTSURE, and BERWICK had already struck. The first had had a fierce struggle with *Bellerophon* in which her captain was killed and both had heavy casualties, had followed this with a long engagement with the *Revenge*, and then the arrival of the *Defiance* overwhelmed her, for she was all but a wreck. SWIFTSURE was not well handled. She had the worse of an engagement with her first antagonist, *Bellerophon*, was quite out-manoeuvred by her second opponent, *Colossus*, losing her mizzen and her main-top mast, and the *Orion*, coming up astern, settled matters by bringing down her remaining masts: she had five feet of water in her hold. BERWICK had lost her bowsprit by collision in attempting to bar the road to *Defiance* and then had been outfought by *Achilles*, being largely dismasted and her captain being killed. ACHILLE was engaged by *Revenge* and *Belleisle*, and lost her mainmast. Drifting away, she came under fire from several ships and then was in close action with her British namesake and with the late-arriving *Prince*. Her last mast was brought down, her captain was killed, and she caught fire. The flames spread and, though her flag was not struck, she ceased fighting, so that British ships that were near by sent their boats to save such men as they could. ARGONAUTE had been engaged, in the smoke, by *Colossus* and *Bellerophon*, and had suffered casualties from the superior British fire before sailing clear; but she had little material damage and really stood away from close action.

As six of the Spaniards, counting from *Santa Anna* aft, had also struck, the rear half of the combined fleet had been annihilated; and this was largely because most ships had had to undergo the attack of two, and usually more, ships in succession and often had had to fight two enemies at once. Much the same happened in the centre; the *Victory* had fired one raking broadside into BUCENTAURE, who then was dreadfully raked by *Neptune* as that vessel came through the gap that the *Victory* and *Téméraire* had made. This caused havoc to her 36-pounder gun crews and dismounted a number of her 24-pounders. *Leviathan* engaged her as well before she and *Neptune* passed on to attack the *Santissima Trinidad*, in which *Africa* later joined them. The *Conqueror* then came on his starboard to be the French admiral's main opponent, while *Britannia* later attacked his port side. So the two allied ships had five opponents to deal with, after having received the passing fire of others. HÉROS, the leading ship of the centre, was only partially engaged. She exchanged heavy broadsides with *Victory* and *Téméraire* as they came up, but thereafter, perhaps because her captain was killed, she just remained in her station ahead of *Trinidad* till that ship yielded.

Meanwhile, what of Dumanoir and the van? When the attack on the rear was made, and ships had broken through the centre to engage BUCENTAURE to starboard, Villeneuve made the signal (flag No 5) which he had emphasized in his memorandum to captains, ie ships not engaged to get into action quickly. Dumanoir repeated it, but made no move. His explanation that this was because of the difficulty occasioned by the light wind and heavy swell was accepted by the subsequent court of inquiry; yet the British ships, having broken the line, were thereafter able to manoeuvre reasonably readily. Later, when the battle had raged for well over an hour, Villeneuve hoisted a direct order to the van to wear and come into action. Till then the ships had done nothing but exchange broadsides with the *Africa* as that sixty-four passed them on her way to join in the fight in the centre. Dumanoir did then move, but it was a slow business, for some ships had to get their boats out to tow their heads round. Nor was the evolution orderly, for FORMIDABLE, SCIPION, MONT-BLANC, and DUGUAY-TROUIN tacked instead of wearing: the others wore, and the division did not keep together. INTRÉPIDE and *Neptuno* made for the heart of the firing in a last desperate attempt to rescue the admiral. This was 'magnificent but it was not war'. The Spanish ships other than *Neptuno* stood away to leeward. Nevertheless, the picture given by many accounts that Dumanoir brought down his four French of the line well to windward, exchanged a few long range shots, and then stood out to sea is unjust. In fact he made to cut off *Spartiate* and *Minotaur*, the last two ships of Nelson's column which, in the slow sailing conditions, were only just coming into action. He failed to do so. They outsailed him and, crossing his bows, raked FORMIDABLE and did damage to others of his line before passing on to fight *Neptuno*. He also engaged, at moderately close range, other British ships which worked out of the mêlée to form some sort of line against him, and his four vessels incurred a good deal of injury. When at last he came to the centre, his flag-captain told him that his battered masts would not stand a change of course. Only then, seeing that all was clearly lost, did he put out to sea; but he should have got into the main battle long before.

With all BUCENTAURE's masts down and many of her guns dismounted, Villeneuve, saying 'The ship has finished her task, but I have not', ordered a boat so that he could shift his flag. But the boat kept for such emergency had been smashed by the falling masts, and towards four-thirty, after more than three hours' fighting against odds, he struck, the dismasted and badly holed *Santissima Trinidad* surrendering a few minutes later. In the rear, the *Principe de Asturias*, which had been well fought against several opponents though Gravina had a wound that eventually proved mortal, managed to make enough sail to struggle clear. She obtained a tow from a French frigate and, seeing that BUCENTAURE had yielded, signalled for ships to join her and steered for Cadiz. She was followed by ten others: PLUTON, HÉROS, NEPTUNE, INDOMPTABLE, ARGONAUTE, and five Spaniards. It was then about four-thirty. To the north INTRÉPIDE and *Neptuno* were continuing their hopelessly gallant resistance to odds – they struck at about five o'clock – to the westward Dumanoir's four ships were making out to sea, nearby ACHILLE was burning fiercely – she blew up at half past five – and on the actual field of battle lay a large number of ships of which many were at least partially dismasted. In sum, seventeen of the allies (eight French and nine Spanish) had had to strike, and one (ACHILLE) was sunk.

Once again the superiority of British seamanship and, particularly, of British gunnery had been marked; but it had been a fierce fight, for some time an even one, and by no means a painless victory[1]. The scene on some of the British ships must have been grim enough; but that aboard many of the allies' must have been a ghastly shambles below deck, with the dead and wounded lying in heaps amid the swirling smoke of the guns, while that on the upper deck can have been little better, with bodies strewn among the debris of fallen masts and trailing shrouds. It had been a dreadful battle, and it was to have a dreadful aftermath, and a prolonged one.

'Or ever that evening ended, a great gale blew,' and the dismasted ships, which included almost all the prizes, were hard put to it. On several of the prizes the tow could no longer be maintained. When the tow was cast off BUCENTAURE and ALGÉÇIRAS their men, threatening to try to overwhelm the prize crews, though unarmed, came to an agreement with them. The French were to take back the ship, both they and the prize crew would work together for her safety and the British would then not be kept prisoners. ALGÉÇIRAS managed to struggle into Cadiz, but the battered BUCENTAURE went aground off point San Sebastian, the southern arm of the Caliz roadstead, and was wrecked. Her men were got aboard INDOMPTABLE which, having only one mast, had been unable to get into the roadstead when she came away from the battle; but the gale finally drove her ashore with heavy loss of life, both of her own people and of those saved from the wrecked flagship. The tow had also to be cast off AIGLE, BERWICK, FOUGUEUX, and the Spanish *Monarca*. AIGLE managed to beach in the entry to Cadiz harbour; the others drove ashore. REDOUTABLE, which was so shattered that she probably should have been evacuated the evening that she was taken, sank while still in tow the next morning; and the great *Santissima Trinidad* followed her on the 24th. Despairing of keeping them afloat in the foul weather, the British destroyed the INTRÉPIDE, *San Augustino*, and *Argonauta*.

On the 23rd Cosmao Kerjulien, the most senior unwounded officer, beat out of port with PLUTON, HÉROS, NEPTUNE, *San Francis d'Asis*, and *Rayo*, which were the only ships fit to go to sea, to try to retake prizes that were still near the coast. He recovered the *Neptuno* and the *Santa Anna*, whose tows were cast off at his approach. This sortie was a valiant attempt to retrieve something from a great disaster, but it proved profitless. For though the *Santa Anna* was got back into Cadiz, both *Neptuno* and *Rayo* were wrecked off Rota. Of course all these wreckings and sinkings occasioned fearful loss of life, especially among the helpless wounded.

Dumanoir's four ships had hurriedly made such repairs as were possible before meeting the oncoming gale and, having ridden it out, endeavoured to get back to France. His flagship, FORMIDABLE, was making so much water that her upper deck battery and some other guns were jettisoned to lighten her. The DUGUAY-TROUIN was also in a poor state. He was sighted off Finisterre on November 2nd by Sir Richard Strachan who, with *Caesar*(80), *Namur*(74), *Hero*(74), *Courageux*(74), and some frigates, was seeking Allemand – and thought that he had found him. With his flagship reduced to the value of a sixty-gunner, Dumanoir had no choice but to run for it. But by November 4th two frigates, with two of the line at their heels,

[1]Sixteen British ships were so damaged that they had to be sent home.

had come up with him off Cape Ortegal. He formed line in the order: DUGUAY-TROUIN, FORMIDABLE, MONT-BLANC, and SCIPION. The last two were soon under heavy fire, and to relieve them he tacked in succession. This gave only a momentary rescue, for it enabled the other British ships to close. SCIPION dropped astern with main top mast and mizzen down, and struck to the frigates. Next FORMIDABLE, and then the others, were overwhelmed, being attacked in turn from both port and starboard. Once again British gunnery made the French ships unmanageable, and their casualties heavy, while Strachan's ships suffered no real damage and his casualties were negligible.

This epilogue rings down the curtain on the great battle of Trafalgar. It was absurd for Napoleon to direct only a brief official report that 'some ships had been lost in a gale following an unwisely undertaken engagement'. The truth of a decisive defeat could not be hid. The campaign of 1805 had cost the French twelve of the line, and her Spanish allies ten (plus two taken by Calder off Finisterre); while the French ships that escaped to Cadiz were promptly blockaded. The British casualties at Trafalgar were approximately 250 killed and 1 200 wounded. Those of the allies cannot be reckoned accurately because of the appalling number drowned, to say nothing of contradictions in the various reports returned by French officers; but, with Spanish losses officially returned at just over 2 400, the total must have been well over 5 000[1].

The invasion of England was now out of the question, so that Napoleon was forced to adopt the policy of the Continental System which contributed to his downfall. He was not the man passively to accept such disaster: he soon had most of the shipyards of Europe in his control, the axes of the shipwrights were set ringing, and at his abdication nine years later the numerical strength of his fleets was greater than it had ever been. But they were not operative. The spirit of the service, which had begun to revive after the chaos of the Revolution, was broken; and Britain was left with a domination of the seas that was unchallenged for over a hundred years.

Dumanoir came before a court of enquiry in 1809. His defence of his actions at Trafalgar was accepted but, somewhat oddly, he was censured for his conduct of the hopeless battle against Strachan. This censure was quashed by the court martial for which he then asked.

The unhappy Villeneuve returned to France on parole in 1806. He landed at Morlaix, in the north-west corner of Brittany, and made his way to Rennes. Here, finding that he was generally execrated and blamed for a great failure, and having heard nothing from his old friend, Decrès, he took his own life. A story circulated, and has been accepted by some historians, that he was murdered by order of Napoleon. The Corsican was not above such petty vengeance but, triumphing after Austerlitz, and busy remapping Europe and dealing with Prussia, he was not likely to

[1]The French were unable to make an official return of losses, but REDOUTABLE had 578 killed and wounded of a ship's company of 672! Thomazi (*Napoleon et ses marins*) gives slightly different figures: he also states the total allied casualties to have been nearly 7 000, and puts the British killed at approximately 450.

have bothered about Villeneuve, and probably did not know of his recent return. In any case, the rumour is dispelled by the pathetic letter Villeneuve wrote to his wife, announcing his intention to put an end to his wretchedness. 'How fortunate that I have no child to be loaded with the weight of my name as its awful inheritance. I was not born for such a fate, I did not seek it, I have been dragged into it despite myself.'[1] That sad farewell is a not unfitting epitaph to the 1805 campaign and its disastrous conclusion.

The event of Trafalgar was so big, and the death of the famous British admiral in his moment of victory gives it such a sense of dramatic finality, that subsequent naval happenings have an atmosphere of anti-climax, and one is even likely to forget that the war against the British was continued by Napoleon until his abdication nine years later, and its naval activity was considerable.

Trafalgar overshadows everything else that happened at sea in 1805. But earlier in the year, off the Malabar coast, there had been a grim fight between the British frigate *San Fiorenzo* and the PSYCHÉ of the same strength which ended on the unusual note at sea of an agreed capitulation of the latter. Also in the Indian Ocean, Linois (CA) had sailed from Mauritius to make an attack on a convoy near the Cape of Good Hope. Its escort was the eighty-gun *Blenheim*, and, after a while, Linois broke off the attack. One of his frigates, ATLANTE, went ashore in the Dutch colony and was wrecked; but its crew remained to be the stoutest fighters in the force with which the Dutch governor tried, in vain, to repel the British attack on the colony in the next year. In the Antilles, too, the frigate TOPAZE had successfully fought and sunk the *Blanche* of similar strength.

It will be remembered (p 254) that, when the invasion of England was called off, the Emperor had ordered that divisions should be sent to attack British colonies and commerce. Two such squadrons were prepared from the Brest fleet. One was of five of the line and two frigates under Leisségues (CA), an officer who had done well in command of frigates and now flew his flag in the great IMPÉRIAL(120), and the other was of six of the line and two frigates under Willaumez (CA) with his flag in FOUDROYANT(90). Leisségues was to take reinforcements of troops to Santo Domingo, where the French were still holding out against the otherwise triumphant natives, to follow this with a brief attack on the British West Indies trade, and then to return. Willaumez was given an extensive commerce raiding programme. He was first to go to the Cape of Good Hope, then to lie in wait on the convoy route off St Helena, thence to sail to the Antilles for an extensive attack on the trade there, and finally to go north to destroy the Newfoundland fisheries.

The two divisions, seizing a favourable weather opportunity, left Brest together in December 1805 and then separated. They had scarcely done this when a British convoy passed between them. Each division pursued it in turn but achieved no more than the capture of two ships carrying some troops, while its escort was able to give the alarm that two French

[1]The letter is given near the end of the appendix to Vol II of Jurien de la Gravière, *Guerres maritimes*.

fleets were at sea. Squadrons under Borlase Warren and Strachan were sent out to deal with them. Meanwhile Vice-admiral Sir John Duckworth, blockading Cadiz, was told by a frigate that she had sighted French ships of the line. Duckworth promptly raised the blockade, which he could safely do since the destitution of the Spanish arsenal made a slow business of refitting the allied ships that had escaped from Trafalgar, and set out to pursue, as he thought, the elusive Allemand, who got back to Rochefort. So there were now five squadrons, two French and three British, playing hide and seek in the North Atlantic. Duckworth sighted Willaumez and pursued him. But for once the British were outsailed, and when his fleet got so extended that all but two of its vessels were out of sight of him astern, the British admiral called off the pursuit and, being now too short of water to beat back to his station off Cadiz, went on to the West Indies. Here he collected two of the line under Rear-admiral Cochrane and, hearing that some French had arrived in the northern islands, made for them, believing them to be Willaumez's squadron.

They were in fact that of Leissègues. That admiral, sent to cross the North Atlantic in winter, had naturally run into foul weather. His ships were scattered and, their seamen being untrained to the conditions, mostly suffered a good deal of damage. However, all got to Santo Domingo and, having landed their troops and stores, proceeded to repair their injuries. They had scarcely done so by 5th February 1806 when the DILIGENT corvette on look-out came in flying the signal for the enemy in sight. Leissègues signalled to cut cables and stood out to sea south-westward along the coast with the wind at east-north-east. Some of his ships, notably the BRAVE, had their batteries still encumbered with gear that they had used in refitting and had not yet stowed back in the hold. His somewhat irregular line was: ALEXANDRE(90), IMPÉRIAL(120), DIOMÈDE(74), JUPITER (74), and BRAVE(74). Duckworth was coming on in two columns, one much ahead of the other. One column was *Superb*(74, flag), *Northumberland*(74, flag of Rear-admiral Cochrane), *Spencer*(74), and *Agamemnon*(64); the other was *Canopus*(80, flag of Rear-admiral Louis), *Donegal*(74), and *Atlas*(74). Leissègues, crossing ahead of them, hoped to double a concentration on the leader of the nearer column and made the appropriate signal to the ALEXANDRE. But in making the necessary alteration of course that ship's rudder failed and she came up helplessly between the two columns. The line disintegrated and ALEXANDRE came under the heavy fire of several ships, which then passed on to deal with the others, leaving the *Spencer* to finish her off.

The battle had opened a little after 10 am and at 11.35 ALEXANDRE, largely dismasted, lowered her colours. *Donegal* and *Atlas* had gone for the BRAVE, which struck ten minutes later in a shattered state, and then passed on to deal with the JUPITER which yielded after distinctly slight resistance. The *Agamemnon* had gone, fruitlessly, after the frigates, and the other British ships had concentrated on the IMPÉRIAL. Hopelessly outnumbered, her main and mizzen down, and with heavy casualties, Leissègues ran his great three-decker on to the coastal rocks, and was followed there by DIOMÈDE, which had loyally supported her admiral. The difficult task was then undertaken of getting the ships' companies ashore. It had not been completed on DIOMÈDE when the British boats approached the next day, and her captain and some men were taken prisoner. The British burned IMPÉRIAL and DIOMÈDE, and BRAVE sank when they got her to Jamaica. Except that the French fleet was under sail the battle was not unlike a lesser Aboukir: there were the same proximity of the shore, the same quick attack and partial surprise, the same superiority of British gunnery (Leissègues had over 1 200 killed and wounded to Duckworth's 340), the same annihilation. In these circumstances it seems strange to find a great fuss being made over a point of punctilio. But Captain Keats refused to accept the sword of the captain of the DIOMÈDE on the ground that he had run his ship ashore after striking. As his colours had been shot away, not lowered, the French captain (Henry) made no end of a song and dance about this, until Duckworth wrote formally withdrawing the aspersion.

Though Willaumez fought no battle, his cruise had almost equally disastrous an ending. He had the FOUDROYANT (90, flag), VÉTÉRAN (captain, Jérome Bonaparte), CASSARD, IMPÉTUEUX, PATRIOTE, and ÈOLE, all seventy-fours. The prisoners from the two troopers that he had taken were put aboard a frigate which was to land them at Teneriffe and then to join him at the Cape. It duly arrived there, dropped anchor and took in sail, and then saw the Dutch flags lowered and British ones hoisted. The British had occupied the Cape and the frigate had walked right into the spider's parlour. Willaumez had heard of the capture and therefore cruised off St Helena. He had no luck, and so made for the Antilles, by way of Brazil where he took in water. Going with three ships to Martinique, he sent the others to capture merchantmen; but all were chased into Fort Royal by Cochrane's West Atlantic squadron. His arrival being no longer a surprise he made for Tortola, where the big home convoys were wont to collect. He missed them, and so went on to intercept the convoys for North America off the Bahama Channel. These, too, he missed, but he found the usual storm and suffered the usual results to French ships. Jérome Bonaparte parted company in the VÉTÉRAN. FOUDROYANT, PATRIOTE, IMPÉTUEUX, and ÉOLE were wholly or partly dismasted, the last two had to jettison some guns (the French seemed over-prone to this measure of safety), and all were scattered. FOUDROYANT went under jury rig to Havana where she refitted, having beaten off the 44-gun *Anson* on the way. IMPÉTUEUX was pursued into the Chesapeake by two British ships and ran aground. The British, disregarding American neutrality, boarded and burned her. PATRIOTE, ÉOLE, and the remaining frigate also went into the Chesapeake. Willaumez got the FOUDROYANT back to Brest early the next year (1807), where CASSARD, which had escaped all these misfortunes, had preceded her. VÉTÉRAN, which had made for home after her separation, was barred from Lorient by a British ship of the line and pursued into the little port of Concarneau, which no ship of her size had ever previously entered. It was three years before she could be got out. PATRIOTE limped back in 1808 (!) while ÉOLE and the frigate were so damaged that they were sold in America. The sole result of this long, ambitious, but deplorable commerce raiding cruise was the meagre capture of seventeen merchantmen.

A raid by three frigates on British whalers in the Arctic did them some damage; but one frigate was taken, and the others returned with a desolating sick list. Linois (CA), after failing against the convoy off the Cape of Good Hope, made for France with MARENGO and the BELLE POULE frigate, as he was short of food. He had the ill-fortune to meet Sir John Borlase Warren's squadron and was taken by the three-decker *London* after a fight so stiff that it took the latter three days to effect essential repairs. A division of five frigates with troops for Martinique ran into a squadron under Sir Samuel Hood who took four of them – three putting up a brave resistance. Another gloomy year in which the only slight rays of light were the capture of a lone, but stoutly resisting, Indiaman and the escape of the CANNONIÈRE frigate in action against the liner *Tremendous*, which was escorting a convoy.

The year 1807 saw nothing more than a few combats between small ships, and it is a sad commentary on the French navy that one of its most reliable historians can only write: 'Nevertheless there was great activity in the ports, and if the Brest division did not get to sea it came out of the *goulet* and manoeuvred in the roadstead'! The war at sea had really shifted from naval operations properly so-called to an attack on commerce. For if there were to be any coalition against the French Emperor it would be Britain who would have to finance it, and it was by her seaborne trade that Britain filled her money bags.

The death of Pitt early in 1806 had brought to power a ministry in which Fox was the leading figure. Fox had believed in the original French Revolution, considered that Napoleon had brought it back from its excesses into its true path, and was prepared to accept his professions at their face value. Tentatively negotiating for peace, he was soon disilusioned. The complete overthrow of French sea-power meant nothing to the Corsican: he was not prepared to accept even an equable peace, but only a triumphant one. He was now unquestionably master of the Continent. Austria had been pushed eastward, Prussia had been smashed and reduced to little more than a third-class power tied to the Emperor by a treaty that made her little more than a satellite, the rest of Germany, Italy, and Holland were satellites in fact, and he had charmed the impressionable (and defeated) young Czar Alexander into an alliance that committed him to a hostile attitude to the United Kingdom. There were two ways of attacking British trade. One was that of raiding, mostly by privateers. This would, and did, cause Britain great handicaps by making it necessary to move her ships in convoy, and by raising maritime insurance rates to great heights. The other was by excluding her trade from Europe and thereby stifling it. This was to be done by the Continental System, established by the Berlin and Milan decrees. The United Kingdom was declared to be 'blockaded', all Continental ports were closed to her goods, which were liable to confiscation if brought by neutral ships, and any ship which submitted to British licence and to British search at sea (which they could not readily refuse) was 'de-nationalized' and therefore liable to seizure. Napoleon failed to realize that her early industrial revolution had made Britain the world's workshop and that while French industry, encouraged by the exclusion of British competition, might cater for the needs of France, it could not supply her satellites and any neutrals who needed British goods, and these would fall back on angry smuggling to get them. This 'system' eventually played a large part in rousing Europe against him.

Britain replied with Orders in Council declaring Continental ports that were under French influence to be blockaded – a blockade that she was in a position largely to enforce – and directing all neutral traffic to come through Britain whence some of it might be allowed to proceed under licence. To back this Britain seized the Danish fleet lest Napoleon might use it, and persuaded the regent of Portugal, who was threatened with war if he did not exclude British trade, to emigrate to Brazil taking his fleet with him.

The naval war therefore pursues henceforth a curious course in which convoy and search are two of the main operations. Blockade by Britain was the third; for the carriage of bulk or heavy goods by water was almost essential, and British small ships were all-pervading, close to the Atlantic coast, to intercept it. So the Emperor had to erect coastal batteries at frequent intervals, employ hundreds of men in manning them, and little coasters and sailing barges could no longer cut across a wide bay but had to hug the coastline ready to pop under the shelter of a battery when a British brig or armed cutter appeared. True that the British, navigating in so daring a way close to a hostile coast, lost a number of small warships each year by grounding and shipwreck; but consorts could often take off their crews, and the loss was a small price to pay for the overall success of their blockade.

In 1808 Napoleon made what was to prove a grave blunder. The feeble king of Spain and his poisonous son and heir, Ferdinand, were at loggerheads; and they unwisely asked Napoleon to arbitrate in this serious family quarrel. He was triumphantly at peace everywhere, except with Britain: he had an army in Portugal to enforce the Continental System and had been given the right to garrison certain Spanish towns to ensure its lines of communication. A military occupation of Spain seemed, therefore, a simple matter: he could not resist the temptation to add a further satellite to his Empire. So he coerced the Spanish king to become a pensionary of France, abdicating, and disinheriting his son in favour of the Emperor's unwilling brother, Joseph, then king of Naples. Prince Ferdinand was clapped in prison, while French troops flooded Spain and installed the new king in Madrid. But Napoleon had not reckoned on the pride of the Spanish people, who were already disgruntled with their government's subservience to France. At first thunderstruck, when they had recovered breath they rose against the usurper: a provisional Junto took over the government in the name of Ferdinand and declared war on the French. The Peninsular War which, supported by British sea-power, was to prove such a long drain on Napoleon's resources, had begun.

The immediate effect of this national uprising was to put Rosily (VA) and his ships in Cadiz in an impossible position, with Spanish batteries around them, and the powerful British squadron outside the port now free to enter. When he realized that he had something to face that was much more grave than an unusually serious local riot, Rosily moored his ships across the entrance to the channel that led to the Carraca arsenal, beat off an attack by Spanish gunboats, and entered on negotiations, which he tried to spin out in the hope that French troops would get through to save him. They did not arrive and, his position being hopeless, he had no alternative but surrender. So the last of the French ships that had fought at Trafalgar were now lost, and with them the ATLAS which had remained at Vigo, where she had been left after Villeneuve's engagement with Calder.

With new ships building fast, especially in Antwerp, Napoleon had tried to devise combinations of his naval forces; but the blockade was too close. Cosmao Kerjulien (CA) got free from Toulon to escort some transports to Barcelona, and one other reasonably extensive operation was carried out with the usual mishaps. Allemand got away with his five 'invisibles' from Rochefort and came successfully through the Straits to join Baron Ganteaume (VA), who was now in command at Toulon. This gave that admiral ten of the line with which to sail to revictual Corfu. He ran into bad weather off Sicily, whereby four of his ships under Cosmao separated and went into Taranto, and his own flagship, the COMMERCE DE PARIS(110), lost all topmasts and yards. Taken in tow, she and the remaining ships reached their destination and discharged their cargoes. Then, leaving his flagship to repair and refit, Ganteaume sailed back to collect Cosmao and attack Sicily. Finding the British in force off that island, however, he made a somewhat painful return, through gales, to Toulon.

There was no other activity in 1808 beyond single ship actions between smaller vessels, which usually ended in favour of the British. The year did, however, see an important, if somewhat military looking, change in organization. To meet the manning difficulty the Emperor ordered the formation of a number of marine battalions. Each contained 456 men and fourteen officers under the command of a post-captain, and was intended to form the solid body of the ship's company for a 74-gunner. The remaining 160 crew, soldiers if necessary, would be drafted aboard when a ship was being fitted out for sea.

The French colonies were almost entirely at Britain's mercy, to be mopped up at will; but the Emperor naturally decided on a last effort to save them. Martinique and Guadeloupe were near the end of their resources and needed reinforcing and revictualling. So Willaumez (CA) was ordered to sail from Brest, pick up three of the line which were in Lorient under Captain Troude, go on to Basque roads where the Rochefort ships would be awaiting him, and thence, without anchoring, to go to Martinique. It is almost incomprehensible that an expedition so important and, in view of the preponderance of enemy forces, so full of risk should not have been more carefully planned. The blockading force off Lorient was small (three of the line) and that under Stopford off Rochefort was not large, but Lorient was a difficult port for ships of the line to leave, as they needed high tides and something like a soldier's wind to do so. True that Troude had alternative orders in case winds should be adverse, but at least one could have made sure of the high tides. As to the Rochefort ships, their commander, Captain Bergeret (their admiral had gone sick), was to be told the time of Willaumez' arrival by 'telegraph', so that he and Willaumez might nip Stopford between them; but there was apparently no arrangement for a 'telegraph' the other way, so that Bergeret could inform the minister that his squadron was inoperative.

Willaumez sailed in February 1809 with eight of the line, his flag being in the OCÉAN(120). A north wind enabled him to leave by the Raz de Sein, thus avoiding Lord Gambier's blockading squadron off Ushant. The blockaders of Lorient withdrew in face of his superior force and he waited a day for Troude; but Troude had not the right conditions of wind and tide to let him leave port. So Willaumez continued to Basque roads, Stopford standing away as he came, but sending a despatch vessel to warn Gambier. In Basque roads to his surprise Willaumez found not Bergeret's ships but a message from that officer that they could not move because of an epidemic. Willaumez had an order, if it proved impossible to collect either the Lorient or Rochefort ships, to proceed forthwith to Toulon. This he ignored and sailed further in, to anchor in Aix roads, losing the JEAN BART(80) in the process, for that vessel ran irretrievably aground on the rocky Palles shoals. Now Stopford resumed his blockade and, though joined by ships under Bergeret, Willaumez did not risk coming out to fight him – perhaps not unreasonably, for Stopford had been reinforced and the Rochefort ships were useless, because their crews had been tumbled into them just before sailing and were still being allotted their sea and action stations. He and Bergeret were then called to Paris to give an account of themselves, and Allemand (VA) was brought from Toulon to Rochefort to take command.

Meanwhile Troude in Lorient, who had three of the line, three frigates, and two *flûtes,* had managed to get his frigates to sea. These were ITALIENNE, CALYPSO, and CYBÈLE, all of forty-four guns, and under Captain Jurien of the first-named. They rightly made after Willaumez; but Stopford, warned of their sailing by a look-out, pinned them to the coast on their way, with three of the line. All Jurien could do was to make for Sables d'Olonne, a tiny port for coasting trade, with a tidal harbour and a dangerous little roadstead. Here he anchored his frigates close to the shore, broadside to the enemy, and prepared to offer, and did offer, a truly stout resistance. The *Defiance*(80) risked coming in close and anchoring, the other two British fought under sail. Shots carried away the cables of the frigates and they drifted ashore, but still broadside on, and still full of fight. As the tide ebbed Stopford was obliged to stand away to sea, *Defiance* getting clear only with such difficulty that at one point the French cheered, thinking that she had struck when her colours were shot away. The veteran CYBÈLE was hard aground and began to break up, but the other two, discharging everything into boats, managed to drag themselves on to the mud in the little harbour before the attack could be renewed. Though one old frigate had to be left to break up, and the other two were later sold to merchants, this courageous stand against odds reflects great credit on the French service, especially as the nearby coastal batteries, being short of shot and served by untrained militia, could give no support.

Troude eventually got his three of the line to sea from Lorient, and sailed to carry out his alternative orders – to take his two *flûtes* to revictual Martinique. Learning, on the way, that that island had fallen and that Guadeloupe was heavily blockaded, he went to the tiny Saintes islands, still in French hands just south of Guadeloupe, to which he sent a boat for orders. Rear-admiral Alexander Cochrane soon appeared with five of the line (one a three-decker) and some troops, which he landed on the Saintes. Troude was therefore going to be bombarded from land as well as assailed from sea. His boat having returned with orders from the governor of Guadeloupe to let his *flûtes* make a dash for that island, and to do what he thought best with his ships, Troude made a daring bolt for it by night. The HAULTPOULT(80) was a poor sailer, steered away from the others, was pursued and was taken; but Troude, with COURAGEUX and POLONAIS, both of eighty guns, got clear, and reached France safely.

Back in Aix Roads Allemand had taken command of the combined Brest and Rochefort squadron of eleven of the line and four frigates. Lord Gambier lay outside also with eleven of the line and a number of smaller vessels. No French admiral at that time would have sallied forth to fight even a slightly weaker force, much less a slightly superior one, and Allemand's attitude was entirely defensive. Nor, though he had fought his ship notably well in Howe's battle of the First of June, was the sombre Gambier at all the dashing kind of leader to launch an attack in a roadstead which had strong currents and was encumbered by dangerous shoals. Both sides thought of fire-ships and though such vessels, specially designed, had not formed part of a fleet for many decades, they were easy to fit out. Gambier disliked the idea; but in England that active young officer, Lord Cochrane, was busy urging their use, and meanwhile Gambier converted to that purpose some transports that he had with him. He also ordered careful night patrolling in advance of his fleet lest the French should be first with this form of attack.

Allemand, however, was concerned only with making dispositions to guard against it. The roads lie to the south and a little east of the fortified island of Aix that gives them their name, between it and the island of Oléron. The Charente river, up which is Rochefort, flows into the south-east corner. It debouches between a long tongue of land, on the north, which ends in rocks stretching out to the islet called Enette, and a long bank of rocks and shoals, on the south, which stretch out past Île Madame, and are known as Les Palles. As is common with tidal rivers, these rocky banks on each side merge into large stretches of mud, well covered at high water but barely submerged at low tide. He ranged his fleet in two lines, running south-eastward from within gun shot of the Aix batteries. The ships were two hundred yards apart from each other, and there was two hundred and fifty yards between the two lines. They were moored by a cable forward and another aft so as to be bows on to the enemy, and the lines were arranged so that the ships in the second line were not dead astern of those in the first line, but faced the gaps between them. The lines were:

FOUDROYANT(90, flag of Gourdon, CA), VILLE DE VARSOVIE(80), OCÉAN(120, flag), RÉGULUS(80), CASSARD(80), and CALCUTTA(54); then TONNERRE(80), PATRIOTE(80), JEMMAPES(80), AQUILON(80), TOURVILLE(80), and ELBE (frigate).

They were ordered to strip everything down to their lower masts, to lessen the danger of fire. A quarter of a mile ahead of them were the three frigates, INDIENNE, HORTENSE, and PALLAS, not stripped down, and ahead of them again was to be a boom covering the fleet's front. For the materials to construct this Allemand applied to the *Préfet Maritime* of Rochefort, who was Vice-admiral Martin, the commander of the Toulon fleet in Revolutionary days. Either Martin was casual in giving his orders, or his immediate subordinates were slack, for only a part of the necessary materials was sent, and rather late at that. Allemand had to make good by using a number of spare spars, cables, and anchors from his ships. The ships were also ordered to man their sloops and long-boats, putting a gun in some of them, and to send them to the boom on the night that an attack was expected so that they might divert, or even sink, the fire-ships. Looked at with hindsight, Allemand's dispositions were criticized; but really they were as good as any that could be made, save for two things. First, though it was wise to get the boats manned and into the water, it would have been better to keep them by their ships, to pull away a fire-ship or take out an anchor, than to send them to the boom on a night of strong wind and with a sea running, in a harbour whose currents made boat work difficult. (Few of them served any purpose and many did not reach their stations.) Second, he did not allow for the panic that was to seize his captains.

Lord Cochrane had arrived from England in the frigate *Impérieuse* with a number of fire-ships to follow, and with rather tactless instructions to Gambier to let him take charge of the attack. (These instructions gave even greater offence to the senior officers who felt that young Cochrane had obtained, over their heads, command of a task they were willing to do: they were not therefore very disposed to urge their admiral to attack in support of him, and Cochrane's signals to Gambier at the critical time were neither as tactful nor as informative as they might have been: this, as we shall see, was fortunate for the French.) On April 11th it was fairly clear to the French, who could see the British preparations, that the attack would be made that night, and the boats set out for the boom. At a gun-signal after dark the twenty-eight fire-ships came on, driven by a strong wind and a high following sea. They were preceded by four 'explosion vessels', filled with gunpowder, whose business was to blow the boom to bits to clear the way for the incendiaries. They duly did so, and their tremendous explosions completed the alarm of the French captains.

The three advanced frigates were the first to cut their cables, because they were being hit by their compatriots who were firing wildly at the oncoming fire-ships; but nearly all others quickly followed suit. Some did so not to avoid fire-ships but to avoid collision from those of their own ships that had already cut loose: they can hardly be blamed for not realizing that it would have been better to endure the collision and thereby save themselves and the collider from worse trouble. There were two exceptions to this. FOUDROYANT quietly waited events, and the incendiaries passed her by. CASSARD did cut a cable when threatened, but her captain had an anchor ready for such an eventuality, and dropped it to bring her up as soon as she had swung clear of danger. The rest were less circumspect, with the result that nine of the line and four frigates, most with only their lower masts and no yards, short of anchors and without the aid of their large boats, were desperately trying in the dark to enter a river into which it was difficult to get in daylight with anchors and all sail available. The ELBE frigate was the only one to succeed.

The detailed happenings of that wild night almost beggar description. Only two of the French were in grave danger of being burned. RÉGULUS, under Lucas, the brave captain of REDOUTABLE at Trafalgar, was set alight, cut cables to get clear of her assailant, collided with the flagship while putting out her fire, and brought up on the edge of the Palles, where she grounded. She got afloat next day only to ground again on the shoals on the opposite side of the roadstead, where she had a grisly time. The OCÉAN was seriously assailed and would have been destroyed had not the TONNERRE's long-boat bravely pulled a fire-ship clear: she cut cables to get rid of two others, collided with RÉGULUS then with PATRIOTE, and hoisted a jib and foresail; but this insufficient canvas could not keep her from running on to the Palles, and when she got off these next day it was to ground again on the opposite shore. PATRIOTE had cut her cables to avoid the drifting OCÉAN, but nevertheless hit her and TONNERRE before getting a jib hoisted, whereat she hit the flagship again and then went on to the Palles. TONNERRE, cutting one cable to try to avoid OCÉAN, and then the other to try to escape collision with PATRIOTE, was driven hard on to a rocky part of the Palles and bilged. And so it went on.

When day broke the two Frenchmen who had kept their heads were making sail to get into safety from attack, INDIENNE was stranded near Enette, and the remaining nine of the line with two frigates were aground on the Palles. Most of them had run ashore bows on and so were at the mercy of the enemy, even if they had not jettisoned at least a part of their artillery, as all but one did at some stage of the proceedings. Cochrane signalled for the British fleet to come on, and Gambier sailed in closer but then anchored again. He was not disposed to risk his ships of the line in a narrowing roadstead with tricky currents and scarcely known shoals. He sent in his gunboats and some small vessels, and later supported them with his three lightest ships of the line. These gathered round the most exposed of the grounded Frenchmen, VILLE DE VARSOVIE, AQUILON, and CALCUTTA, which were all on the northern end of the Palles. The first had cut her cables in the night and had gone ashore; she had got clear and made some sail, but only to ground again in a position where she could bring only a few guns to bear on her opponents. AQUILON had cut her cables during the panic in order to avoid the drifting RÉGULUS and had anchored: she had then unwisely weighed, and so had grounded in the dark. CALCUTTA had also cut her cables in the panic, and the wind had blown her on to the Palles. Her crew broke discipline under fire and made for the boats, her captain, losing hope, going with them, and the ship was abandoned and set ablaze. The captain of VARSOVIE, wishing to save his crew, started to send them away in such boats as he had available: the boats were not brought back. He was firmly ashore with few guns bearing, and signalled that his position was hopeless. He received approval to do as he thought best, and he then struck his colours. AQUILON also got some crew away, but then, powerless to resist, struck with three hundred men still aboard. The British set fire to both ships.

Meanwhile some of the stranded ships had got afloat, and were getting into the Charente for safety. Panic had not entirely disappeared from those that remained. The captain of TOURVILLE, seeing two fire-ships approaching, as he thought, abandoned ship. Having got ashore he realized that his alarm had been occasioned by the firing of the two captured ships by the British and returned. Some of his junior officers had already brought their boats back and were engaging such small British vessels as were approaching. Four men had never left the ship at all: one claimed deliberately to have stayed aboard and to have fired a gun on his own at a British boat, the other three were just asleep! TONNERRE was also attacked, and was abandoned and fired: she was in any case irreparably damaged by her grounding. By dusk on the 13th all the remaining ships were in safety except INDIENNE and RÉGULUS. The former signalled repeatedly for help, but in vain. She was steadily forced further on to the rocks and began to break up. On the 16th she was abandoned and set alight. RÉGULUS, though too far up the roadstead for attack by big ships, was under fire for days from small vessels, and Lucas had to blow holes in her stern to reply with such guns as he had not jettisoned. At last on the 29th, she got into harbour, and Gambier then stood away for England.

The British casualties were only thirty-two killed and wounded, nor were the greater French losses heavy, though there were many taken prisoner. But the fleet was almost as destroyed as it could have been by battle. Much of its artillery was lying on the shoals or sinking into their mud, and the ships had used up gear and were battered. It could be considered as out of action for months. Had Gambier been more willing to take risks it might have been out for good. In France some officers were court martialled. The captain of INDIENNE was put under open arrest (by one vote) for three months for having abandoned his frigate precipitately; the captain of TOURVILLE was sentenced to two years' imprisonment and dismissed the service for abandoning ship; and the captain of CALCUTTA was sentenced to death by a single vote for leaving his ship in the presence of the enemy, and was shot.

The affair of Aix roads had cost France four of the line and a frigate; a disaster without a battle worth mention, and a grave blot on the French navy.

Another attack on a roadstead in this year, however, went very differently. In July Britain launched the well-known and formidable so-called Walcheren expedition. Its object was the capture of Antwerp and the destruction of the shipping installations there. Antwerp had been the busiest of the ports building vessels for Napoleon's Empire, and it now held a squadron of ten new ships of the line under Bourgues-Missiessy (VA). The attack took Louis Bonaparte, Napoleon's feeble brother whom he had made king of Holland, somewhat by surprise. Missiessy was the life and soul of the defence. Against the great forces which the British deployed, he could do nothing to prevent their capture of Flushing and the approaches to the Scheldt, but his arrangements for the defence of Antwerp were admirable. Half of his crews were manning the forts, and were well instructed in their duties, the other half were on board the ships, which covered the river approaches behind booms much stronger and better defended than that which Allemand had been able to erect at Aix. Vice-admiral Sir Richard Strachan and general Lord Chatham were hesitant in their approach. They withdrew because the army was decimated by Walcheren fever; but they had plainly not liked the look of the Antwerp defences, which would certainly have given them a hot reception.

In October came another disaster without a battle, though a small one. It was due to the absurd French habit of sending out inadequate forces, in a way that invited destruction. A convoy had to go to Barcelona, and Baudin (CA) was ordered from Toulon to escort it with only three eighty-gunners and two frigates, though it was known that Collingwood was cruising with a powerful fleet. Of course it was sighted. Baudin sent the convoy into Rosas and tried at once to lead the enemy away and yet escape. Collingwood detached six of the line to deal with him. He was overtaken and pinned against the coast in bad weather because his inexperienced men were slow in taking in a reef. One liner and the frigates succeeded in getting away, the other two went aground and had to be abandoned and fired. True that it would have been costly to turn out a whole fleet to escort a convoy; but that would have been better than just wasting two of the line, the more so as Collingwood sent his ships' boats in to Rosas, destroying some of the convoy and carrying off others. The discouragement to French seamen at being thus used must have been great.

Single ship actions between frigates and small vessels were fairly numerous during the year, and generally went to the stronger force (which was usually the British) though sometimes only after a brave defence. The BELLONE(40), however, took a heavier armed Portuguese frigate. During the year Martinique, Cayenne, Sénégal, and Réunion all were taken by the British, the small warships in their harbours being mostly destroyed before the surrender.

Though the war was to last a further five years, and though the building of warships went on apace, the disaster of the fire-ships in Aix Roads was the last of anything like major movements of French naval forces. In a sense one can see why. The idea of invading Britain had been dropped in favour of the Continental System. French commerce had long since disappeared from the seas, except for a little coastal traffic, and this hugged the shallows along the shore where big ships could not go and brigs and gunboats were more suitable escorts. Nor were there colonies needing support for they had all been taken – Guadeloupe fell in February 1810 and the distant Île de France (Mauritius) was to go near the end of the year. Yet it is a little surprising that, with the lesson before him of a British fleet removing Sir John Moore's army from Corunna, and the further lesson of that fleet maintaining Wellington in his defensive lines at Torres Vedras outside Lisbon, Napoleon did not make a naval effort to operate off Spain's long coastline, if not in support of his own forces at least to handicap those of his enemy in the Peninsular War. He was content, however, to impose upon them the simpler handicap of exposing their ships and men to the vagaries of weather in blockading the naval ports of the Empire he had made in Europe.

Yet, before the French flag disappeared from the colonies, there was one last lively flicker of naval activity around Mauritius. The French frigates there were a down-at-heel lot, relying on captured brigs and Indiamen for the cordage and spars to replace worn-out rigging and cracked yards. But they were efficient in their raiding. Captain Duperré, their senior officer, in BELLONE(44), took some Indiamen after a fight, and turned one or two of them into warships. He was reinforced by the VÉNUS(44), Captain Hamelin, in 1809, and the ASTRÉE(44) early the next year. They too did some successful commerce raiding, and also took the *Africaine*(48), which rashly attacked against odds – though they had to release her as other British ships came in sight.

In short, the frigates from the Île de France became a serious menace to trade, and Captain Willoughby went after them with four forty-gun frigates. He saw Duperré entering Grand-Port with BELLONE(44), MINERVE(52), CEYLON(30), and VICTOR(18), and followed him in. Duperré and his next senior officer, P Bouvet, moored their ships close to the shore and prepared to fight it out against odds. Willoughby came in somewhat rashly, seeing that he did not know the harbour, but perhaps also somewhat contemptuous of his enemy, forgetting that these French crews were trained by their long spell at sea, and were not the usual inadequate men. As a result, the battle of Grand-Port gave the French one of their few naval victories. Two of Willoughby's frigates went aground and had to be evacuated and set on fire, having been unable to beat down the well-served French gunnery; a third had to strike, and the fourth, working its way in retreat out of the harbour, found itself cut off by VÉNUS and ASTRÉE, coming to the assistance of their colleagues, and had to surrender. Later, at sea, VÉNUS took the *Ceylon*(40) but was so damaged in the action that she could offer no great resistance when she and her prize were overtaken by another British frigate and two corvettes. Commodore Rowley now appeared in force and with troops, to deal with this nest of raiders, and the island capitulated in December. A group of three French frigates arriving too late, early in the next year, went to Tamatave in Madagascar when they found that Mauritius had fallen. Here they were engaged by three British frigates: two Frenchmen were taken, one by capitulation, but the third struggled back to France. (Her captain was rewarded by imprisonment and dismissal from the service, but was later reinstated.)

In that same year, 1811, a frigate action, almost sufficiently considerable to be called a battle, took place in the Adriatic. Captain Dubourdieu was sent with three French frigates and three others from Napoleon's kingdom of Italy to attack Lissa, an island at the south-eastern end of the Adriatic which the British were using as a base from which to control that sea. On closing the island he found himself opposed by four British frigates. These formed line on the starboard tack; but Dubourdieu, full of verve and confident in his greater number of ships and guns, and having the advantage of the wind, came in to attack with his ships all higgledy-piggledy. His own frigate, FAVORITE, being a good sailer, arrived well ahead of the rest, was well hammered, and Dubourdieu was killed. When the combatants drew near the land, both lines went about in succession, but FAVORITE was too damaged to make this simple evolution and ran ashore. Once again the British guns, inferior in numbers, were superior in handling, and the British casualties were less than half those of their opponents. Two of the Italian frigates struck and the remaining one, with the two remaining Frenchmen, had to flee. Once again, the year saw a number of small ship actions, which were notable as being mostly close to the shore of France, being fought by the brig and gunboat escorts of little convoys creeping round the coast.

By 1812 the shipbuilding programme of the Napoleonic Empire had reached a point where there were fifty-six of the line in commission, with others laid up in reserve. But they did no more than 'exercise' in their roadsteads, or just outside them. In this way, the Toulon fleet, under Count Émeriau (CA), more than once exchanged shots with the British blockaders. This was a pure waste of ammunition. Indeed the term 'in commission' must be interpreted as the ships being ready for sea, but having no crews to man them. When Émeriau wished to give a proper exercise to part of his fleet he could only do so by taking the men from the ships that remained at anchor to make up the crews of those 'exercising'; and those crews were often 'only novices and conscripts', as one frigate captain complained in reporting an unsuccessful action. The eighty-gun RIVOLI, built in Venice and sent into the Adriatic, was easily mastered, and taken dismasted, just for this reason, by the *Victorious* of the same strength. For Napoleon had withdrawn most of the marine battalions to support him on land.

Here the tide was beginning to turn against France. Napoleon had overthrown Austria, when that power went to war again, in 1809; but that was the last of his triumphs. The European peoples who had once welcomed the French as some sort of liberators were now tiring of their domination, and of the calling up of their young men for imperial wars. The Continental System weighed heavily on the satellite states and Napoleon had had to annex the satellite kingdom of Holland and all coastal states round to the Danish frontier to stop the extensive smuggling that was rendering the System ineffective. The merchants of Russia's Baltic provinces were complaining bitterly to the Czar that they were being ruined, and the Czar was also unhappy with Napoleon's policies in Poland and Turkey. In 1811 he refused to maintain the System, and that meant war. The grim struggle in Spain seemed to have no end; and now the Emperor was to march into Russia on that adventure which ended in the disaster of the retreat from Moscow.

When the scope of that disaster became known, Austria and Prussia not only broke their subservient alliances with France, but changed sides.

For a while the Emperor held his own in central Europe; but the generals of his outlying armies were defeated, the allied armies closed round him, and they finally overthrew him in the 'Battle of the Nations' at Leipzig. He fell back behind the Rhine, while the allies entered Holland and Italy, and Wellington crossed the Pyrenees into France itself.

During this time the only naval activity had been attacks upon commerce made, usually, by frigates sent to sea in pairs. They had had some success, but had also suffered losses. Now, as the allies pressed on France, the navy had to do what it could to defend it. The powerful Antwerp fleet under Missiessy, which had never been to sea, was responsible for almost all the defence of the great port, fifteen of the line being moored to protect the river approaches, and the seamen manning the forts. The Brest sailors were landed to serve in the army, the RÉGULUS was sent to help to keep the British troops from the Gironde. The raiding pairs of frigates fought their last actions. One pair had to strike to the *Venerable*(80) off Azores: the ÉTOILE and SULTANE fought off two British frigates, as did another, CLORINDE, but all three had later to strike, for the British ships were everywhere. SCIPION(80) had to escape from threatened Genoa and Baron Cosmao Kerjulien (CA) put to sea with three of the line to bring her to Toulon. Attacked by superior British forces, of whom he had a good start, he handled his retreat well, and only his rearmost ship, ROMULUS(80), was in grave danger. Her captain and, after he was wounded, his first lieutenant showed really clever seamanship in keeping her too close to the land for the British to get inshore of her, and she got into Toulon, albeit with casualties and damage.

All was now collapse. Duperré (CA) had to yield with three of the line when Venice fell: another surrendered in Genoa. RÉGULUS, hemmed in off Bordeaux, was fired by her captain to avoid striking. Napoleon, having fought a last vain campaign in France with the brilliance of despair, abdicated in April 1814. When he returned from Elba to France the next year, for that period known as the Hundred Days which ended with his second abdication after Waterloo, there was no naval event worth mention. The twenty and more grim years of almost uninterrupted war were over.

For the French navy those years had been one long humiliating tale of defeat and disaster, relieved only by the small defensive victories of Algeçiras and Grand-Port, an occasional success in a single ship action, and by the bravery of those who were at least willing to face death at the hands of an enemy who was by no means always in greater strength but was almost always more skilful. Even the commerce raiding, the *Guerre de Course*, in which France had previously shown such ability and success, was a failure, despite the triumphs of one or two practitioners like Surcouf. It was widely encouraged after Trafalgar, and some *armateurs* made fortunes out of it; but though privateer captures rose to a peak in 1810, thereafter they declined sharply, while privateer losses were so heavy that each year some 2 500 men, mostly prime seamen, joined the swelling masses of prisoners in England, and by 1814 few privateer ships were left. Commerce raiding ships of war also suffered heavily. Indeed, during the long struggle both sides lost many warships, of different grades, but the losses of Britain from her larger fleet were little more than a quarter of those of France and her allies; and it is notable that, whereas the British losses were mostly by shipwreck or other accident of seafaring, the French losses were almost all by capture, sinking in, or because of, battle, or destruction by fire to avoid surrender. Yet in the American War France had been able to fight Britain on at least equal terms.

The reason for the decline lies partly in the Administration. The Revolution knocked this to pieces, and for two of the Revolutionary years it was left without authority, the only authority exercised being that of *réprésentants en mission* who were largely, if not entirely, ignorant of naval matters. The seizure of power by Bonaparte restored authority, but for almost all his time as head of the state Decrès was minister of marine. He was conscientious and a capable seaman but, like others who had been overwhelmed at Aboukir, he did not really trust his fleets to meet the British. We find him more than once advising Napoleon to abandon the idea of invasion, and he favoured the attack on commerce. This attitude did not make him an inspiration to his admirals, nor a vigorous promoter of training in fleet manoeuvre. Another reason lay in the false economy which, even under Napoleon, supplied the ships with poor quality gear. In the chaos created by the Revolution the dockyard stores were emptied, suppliers went out of business, and ships had to try to manage with what could be haphazardly scraped together. Even during the Empire gear was second-rate. It is not only poor seamanship that accounts for beautifully designed vessels being outsailed by the British: it is also because of sails that split, cordage that parted, and yards that cracked. There were, too, many instances of stores, both marine and of foodstuffs, being in short supply. The British sailor (and the numerous foreigners – Scandinavians and others – who came under this heading) lived under a harsh discipline, but was cared for. He did have enough to eat, a supply of warm clothing, and a hammock to sleep in. The French sailor might find himself on short rations, and with no change of garments and no hammock, unless he brought these things with him. The sick list on British ships was usually small: that on French ships was often appalling.

Above all the difference between the two navies lay in personnel. Here France was in trouble from top to bottom. The outstanding figure of Nelson overshadows all his contemporaries; but there were several other British admirals who were of first-class quality: the average flag-officer was well on top of his task, and there were not many that were feeble. France produced few admirals that were of the least mark, and a number were moderate indeed both in personality and capacity. As to other officers, the Revolution broke up the fine corps that had been developed since Choiseul's naval revival and hardened by the experience of the American War. The failure of the early Assembly to maintain discipline led to the emigration of a number, but it left many who were keen on their Service, and loyal to it and their country. The absurd decrees of later assemblies enabled 'popular societies' of tub-thumpers to drive these out of the service for no better reason than spiteful class prejudice, combined with a wish to find lucrative posts for their wind-bag friends. Some of these officers returned when order had been restored, but others had gone abroad, and some had gone to the guillotine for no better reason than their gentility of birth. The idea of making promotion from the ranks available was excellent in theory, and in the Revolutionary armies it often worked; but it takes different qualities and a longer time to make a naval officer. Those that did not have to be weeded out of the Service were usually more than brave enough, but their lack of training and experience was often costly.

Equally foolishly the Revolution broke up the corps that provided the trained nucleus of crews. Volunteers and conscripts made conquering soldiers, therefore, it was argued, they should make conquering seamen. Of course they did not do so. France had not a large supply of prime seamen; but only too many of those that she had were allowed to go into privateering, and others moved inland, preferring conscription for the army to service in incompetently manned ships. The crews that were put on board had little training, and that only in roadsteads, where the French ships spent far more time than at sea. Going aloft in anything like a stiff breeze and a seaway they found to be a very different thing. Gunnery training was largely neglected in harbour and, again, fighting a gun in battle, especially with any sea running, was a distinctly different matter. The British crews were at sea, were hardened and familiarized by it, and were trained by experienced officers. In the difference of personnel, and especially in the training and experience given by sea-time and real gunnery practice, lay the chief explanation of Britain's success and France's humiliation.

By the peace treaty of Vienna the French fleet was reduced by more than a third. Martinique, Guadeloupe, and her African settlements were returned to France, with her Newfoundland fishing rights and her Indian trading posts. The rest of her colonies were lost. Here, and with the navy that had lost them, she had to start afresh.

CHAPTER XIII

# The Last Decades of Sail

With the Bourbon monarchy restored (for the second time) after Waterloo, France, defeated and humiliated, could lick her wounds and recuperate her *amour-propre*. Yet this hardly applied to her navy. Here the need for recuperation had come after Trafalgar, San Domingo, and Aix Roads. Napoleon had seen to that as far as ships were concerned and even to a good extent as far as officers were concerned, but he had given the service no chance to recover its spirit because he had stripped it of seamen to bolster his failing fortunes on land. Now the victorious allies took from France a number of the imperial ships. That still left her fifty of the line, with others building, and a number of frigates. Most, however, were promptly laid up. This meant a reduction in the large corps of officers, the more so as places had to be found for those officers, now returned to France, who had been *émigrés*. Here the restored monarchy was in a particular difficulty. In their own eyes these officers were the only loyal men in their service: they had remained true to their king and had suffered exile and poverty for it; now that he was restored they expected reinstatement in a rank proper to their seniority. In the eyes of France as a whole they were deserters of their country and, in some cases, even traitors who had fought against her in the employ of her enemies. The older ones were given a rank at which they could be retired on pension. Others had to be incorporated into the service and room had to be made for them – this at a time when the number of ships had been forcibly reduced by France's conquerors and would have to be further reduced because of the financial stringency that would not allow them to be maintained afloat. No less than four hundred officers were placed on half-pay, which meant, in effect, that they were retired, though they could, and some did, find employment in the slowly reviving mercantile marine.

Nor, outside the Service, was anyone interested in the navy. Few recalled the victories of the American War: to most the navy presented only a long tale of failure and of defeat. There were no colonies worth mention to protect, for France had lost almost all of them: there was no seaborne commerce to speak of: the troops could guard the coasts. Why maintain a navy! So the maritime prefects were dismissed and the old Intendants brought back, with the consequent separation of administration and active service in the run-down ports. In the short period between the Restoration and 1818 there were no less than four transient ministers of marine, administering an utterly inadequate budget. There was virtually no activity beyond a minimum of care and maintenance, and there was just one event of which mention must be made because, though it is of no historical importance, it brought about a tragedy so ghastly as to attract much attention.

As her trading posts on the West African coast had been handed back to France, a small expedition was fitted out in 1816 with the necessary garrison, civilians, and stores to reoccupy them. It consisted of the frigate MÉDUSE(44) and three smaller vessels under the command of the frigate's captain, Duroys de Chaumareix. This officer was a returned *émigré* who had not been to sea for twenty-five years. He got separated from the rest of his ships, and ran MÉDUSE on to the Arguin shoals too firmly for her to be got off. Her passengers were far too many for the ship's boats, so a large raft was constructed which the boats were to tow. The sea getting rough, the tows were cast off and the boats sailed away to get help. This, of course, was shameful: if one boat had gone for help the others should have stood by the raft. Even in abandoning ship discipline must have been bad, for seventeen men were left behind: on the raft, once the boats had gone, it broke down utterly. The men got at the wine, there were brawls, some fell overboard, some killed themselves, many died of exposure and want, and when at last one of the other ships found the raft, only fifteen of the 150 orginally aboard it were still alive. This disgraceful episode did teach a lesson: an ordinance of 1817, reducing and remodelling the officer corps, got rid of unsuitable *émigrés*.

The revival of the fleet, however, starts from the next year, 1818, when Baron Portal, formerly an organizer of privateers, took over the Ministry of Marine. He roundly told the Chambers that they must choose between doing away with the navy entirely and giving it a sufficient appropriation. The latter having been decided, Portal set to work to reorganize the service. The officer corps, approximately 1 200 under the Empire, was now fixed at approximately 850. The national conscription was used to complete crews. As early as 1819 a small division under Jurien de la Gravière (CA) was sent to join a small British squadron under Vice-admiral Freemantle to visit Algiers, Tunis, and Tripoli to make strong representations about the slave trade, which the powers had agreed at the Peace Congress of Vienna to abolish. An Admiralty Council was set up, and by 1822 France had 121 ships of various kinds in the service,[1] and had sent some to show the flag both in the Levant and in the Antilles. France had need of them, for she was about to go to war again.

By the Congress of Vienna a 'concert' of the great powers was established. Britain regarded this as an agreement to maintain peace for at least twenty years, but did not intend to interfere in the internal affairs of any state unless such affairs threatened the peace beyond its borders. The other powers, having in mind the upheaval and wars caused by the French Revolution, interpreted it as entitling them to suppress any liberal or nationalistic movement. In the midst of the struggle against Napoleon, in 1812, the Junto that ruled Spain in the name of Napoleon's prisoner, Ferdinand VII, had given the country a liberal constitution. Ferdinand rewarded their loyalty by abolishing this on his return, and ruling with such autocracy as to provoke a revolt. This he proved unable to subdue,

[1]Forty-six of the line, thirty-four frigates; forty-one smaller and supply vessels.

and eventually he became a prisoner of his own rebels. Therefore, at the Congress of Verona, in 1823, the powers (Britain dissenting, and therefore breaking the Concert) authorized France to invade Spain and re-establish Ferdinand on his throne. An army was sent under the royal duke of Angoulême, and divisions of the fleet were fitted out, with some difficulty, to give the army such support as it needed.

As a naval war this one was not strenuous. Some ships exercised a light blockade of the Biscayan and Catalonian coasts, and the only real naval operation was against Cadiz. Here Baron Hamelin (CA) went in command with his flag in COLOSSE(80). At its maximum the French squadron off Cadiz consisted of three of the line, with ten frigates and some lesser ships and supply vessels. Angot des Rotours (CA) in CENTAURE(80) was second in command, and Hamelin having asked for his own recall owing to a disagreement with Angoulême, Baron Duperré came out to replace him. (It will be noted that the two admirals who successively held the chief command were men who had proved themselves under Napoleon.) Angot des Rotours took three frigates to bombard Algeçiras, which surrendered, and then, with two frigates, bombarded fort Santi Petri, which is at the south end of the Isla de Leon, on which Cadiz stands, and commands one of the channels leading to the arsenal. This fort having yielded, a gunboat attack was made on Cadiz itself. It failed; but the fleet was able to land some troops. Cadiz surrendered, and the rising against Ferdinand collapsed. It had not been much of a war, but the revived navy had done enough to justify its existence. The maritime prefects were restored in 1826 by the then minister, M de Chabrol, so that the Service was back on its old properly organized footing. To use the word of one historian, it might be said to have completed its 'convalescence'. Despite their many difficulties the restoration ministers had done pretty well.

It was therefore in reasonable trim to meet the next demands made on it. These arose from the revolt of the Greeks, seeking their independence from Turkish misrule. The first risings had been a somewhat sporadic business, starting in 1821. They had occasioned a tremendous wave of popular sympathy in Britain and France, whose people saw the rebels as revivers of the glory that had been ancient Greece. (Russians also had sympathy for them as co-religionists of the Orthodox Church.) Subscriptions from the Western powers flowed into the rebel coffers. There were, of course, some honest Greek leaders like the statesman Kapodistrias (who had been a Russian diplomat) and the sailor Miaoulis; but the majority of the patriots were little better than brigands. It is not too great an exaggeration to say that the general course of events was that they combined, more or less, to defeat each of the ill-prepared Turkish armies sent with bad roads and communications against them; then, having repulsed the Turk, they fought among themselves over the embezzlement of the spoils, in the shape of the funds subscribed by Western sympathizers. When, after several years of failure, the Sultan called in the aid of his nominal vassal, but really independent ruler, Mehemet Ali of Egypt, the situation altered; for Mehemet had an efficient army and fleet (built up by French officers), and an efficient general in his stepson, Ibrahim. The conquest of the patriot rebels was now only a matter of time, and of not too great a time at that.

The Western statesmen, who had seen the nature of the revolt in its true light and were not affected by the emotional sympathies of their peoples, were in an awkward position. Because of the extent of those sympathies they could not leave the Greeks to be overthrown, but they did not wish to weaken the Ottoman Empire, as it was their buffer to keep Russia out of the Mediterranean and away from the vitally important strategical position of Constantinople. Russia, too, could not allow the Christian Greeks to be suppressed and massacred, but did not wish to be embroiled with the Western Powers. Britain and France both had squadrons in the Mediterranean and Russia, in 1827, was sending one round from the Baltic to join them. Their admirals were given the almost impossible instructions to force an armistice on the combatants without any fighting.

The Western allies had the following naval force in Greek waters: three British of the line, with three frigates and some brigs under Vice-admiral Sir Edward Codrington (who had served aboard Howe's flagship at the battle of the First of June), and three French of the line (there would have been a fourth but for a collision) with two frigates and two smaller vessels under the comte de Rigny (CA) with his flag in the frigate SIRÈNE (52). They were shortly joined by four of the line and four frigates from Russia under Rear-admiral count de Heyden. Ibrahim had collected the combined Turkish and Egyptian fleets and his army at Navarino, a bay on the west coast of the Morea. He had seven of the line, fifteen frigates, a large number of smaller warships and some fire-ships as well as many transports, and he intended a seaborne expedition against Patras that might well have terminated what was left of Greek resistance. De Rigny informed Ibrahim of the agreement of the three powers to impose an armistice, threatening an attack on the Patras expedition if it put to sea, and emphasized this by a subsequent visit in company with Codrington. Ibrahim correctly replied that he must await orders from the Sultan and Mehemet, but engaged to keep his fleet in the bay pending their receipt. He interpreted this armistice as applying only to ships, however, and so his troops began a wholesale devastation of the Morea. To stop this the admirals decided to take their ships into Navarino Bay itself, with the obvious threat of destroying the Moslem fleet unless the armistice was applied to land as well as to sea.

The politico-naval situation was already such as to make an explosion probable, and to take the fleet alongside that of the Mohamedans was to hold the match very close indeed to the gunpowder. The allies may have hoped that the Turks and Egyptians would be overawed, but they certainly expected battle; for de Rigny warned the 'axed' French officers serving with the Moslems to leave their ships or take the consequences, and Codrington, who was in overall command, made the most careful dispositions. The fleet was to enter in two columns, the British leading one with the French to follow, and the Russians forming the other: there was to be no firing unless they were first fired upon. The dispositions were not observed, the Russians following in the tail of the French; this was perhaps partly owing to lack of wind and because the smoke from the guns soon lay so thick as almost to obscure visibility. For, of course, the allies were fired upon. The Moslem fleet was moored round the shores of the bay in a sort of horseshoe formation in three lines – the ships of the line and frigates in the first rank, the smaller warships behind them, and the transports inshore. For a while there was an ominous hush; then the *Dartmouth* frigate, told off to watch the fire-ships, sent a boat to require their withdrawal to a less threatening position. Fire was opened on this, and a messenger, sent by boat to tell the ship opposite which SIRÈNE had brought up that the French would not fire unless fired upon, was killed by a shot from a gun-port. The firing promptly became general, and at first the SIRÈNE and the British liners had the brunt of it, for the other French ships could not find their allotted stations in the bank of smoke that quickly covered the centre, and the Russians got into action only late. SCIPION(74) was momentarily in danger from a fire-ship, and the Egyptian that the SIRÈNE was engaging blew up with such a blast as to bring down the Frenchman's mizzen mast. Nevertheless the result was never seriously in doubt. Despite the greater number of Moslem vessels, the gun-power of the allies was much superior, to say nothing of their better gunnery training. One Ottoman frigate struck to ARMIDE, otherwise the Moslem ships fought it out till they were sunk or till, their condition being hopeless, their captains set them ablaze. After two and a half hours' firing all that remained were a few transports and small ships which their captains destroyed the next day. The allies had not escaped unscathed. De Rigny had to shift his flag to TRIDENT, for the other two of the line, as well as SIRÈNE, were so damaged that they were sent back to France. A number of the British-ships had also to return home. The total French casualties were 187 killed and wounded, those of the Russians were much the same, and those of the British were heavier; but the victory was complete.

This one-sided and rather clumsy slogging match in a calm, which demanded no skill beyond that of fighting the guns and a little work with anchors and cables to shift moorings, was the last major fleet action in the days of sail. The Navarino success was by no means welcomed by the statesmen of the victorious countries (the Duke of Wellington called it 'an untoward event'), for it committed them to the awkward diplomatic business of agreeing on the creation of an independent Greece; but it was this naval victory that made Greek independence certain.

Though this was the last fleet action under sail, there were other naval events in the three decades that were to pass before propulsion by steam alone became general. For the French navy these were largely engagements with forts and shore batteries, plus a certain amount of what are now called amphibious operations. One such was the reoccupation of some of the former French establishments in Madagascar in 1829. For this a sixty-gun frigate, TERPSICHORE, was sent, with some smaller vessels, under Captain Gourbeyre. The warlike Hovas were first bombarded from the sea, then driven inland by a few troops and some landing parties of seamen.

Much more important was the great Algiers expedition of 1830. Relations between France and the Dey had become so bad, over such matters as payments due from Revolutionary days and the French right of coral fishing, that a somewhat ineffective blockade of the Algerian coast had been maintained by a few frigates and lesser ships since 1827. During it two small ships ran ashore, and more than half of their crews were promptly massacred.[1] Yet the expedition was due mainly to internal French politics. The restoration king, Louis XVIII, had died in 1824 and was succeeded by his brother, Charles X. This prince was an autocrat and a bigot; and his views on sovereignty and religion were much those of the former English king, James II. Elections having returned a liberal majority he refused to appoint liberal ministers, and considered it condescension that he appointed only moderate conservatives. When this ministry inevitably came to grief he declined to 'condescend' further, and called to office Prince Polignac, with a ministry that looked like a reincarnation of the *ancien régime*. Polignac felt that his only hope of changing the adverse political balance in the Chamber was to bring off a coup that would so appeal to French patriotism that he would win an election. A conquest of Algiers was indicated. The French admirals were opposed to the expedition. They considered that the strong sea-front fortifications of Algiers and the nature of its roadstead made impossible a bombardment adequate to cover the landing of a large army, and that, no other roadstead being available, the landing of troops on open beaches with possible bad weather would endanger the ships and perhaps cause the army to be cut off ashore with no supplies. Nevertheless the expedition was ordered.

Baron d'Haussez, minister of marine, prepared it with care and speedy energy. So in May Baron Duperré (VA) with his flag in PROVENCE(74) and a Villaret-Joyeuse as his flag-captain, sailed from Toulon with the following enormous number of ships:

Three of the line and seventeen frigates fully armed, eight others of the line and seven frigates armed *en flûte*, plus smaller vessels to the grand total of one hundred, including seven steamers. To these must be added (though in two waves) no less than 572 hired merchantmen. They transported an army of 30 000 men with its horses, artillery, and supplies, under the comte de Bourmont; for the expedition was far more a military than a naval affair. To keep this vast mass of shipping together was not easy, especially when, as he neared the African coast, the weather turned foul, and Duperré brought the whole lot back to the Balearics. However, by mid June he had got them into the bay of Sidi-Ferruch, west of Algiers, and the sailors had the arduous task of getting the troops and their stores ashore rapidly, for fear of renewed bad weather. This was done, and each transport was packed off back to France as soon as unloaded. On July 3rd, the troops, having beaten the Algerian army at Stawelly, closed in on the town, and a part of the fleet, including its bomb-vessels, entered the roadstead to support them with a bombardment. This was more heavy than effective, but the soldiers took the Emperor fort, which commanded the town, and the Dey surrendered and went into exile. Ducampe de

[1]They would all have been so had not a courageous Arabic-speaking Maltese insisted, despite a knife at his throat, that they were not French but British.

Rosamel (CA) was then sent in Trident(74) with some frigates and bomb-ships to take Bona and to lay down the law to the Bey of Tripoli; a few other ships were left in Algiers, and Duperré took the rest back to Toulon.

They were not flying the same colours as those under which they had sailed. For the good news from Algiers had not swung the election in favour of the Government, whereat Charles X had ordered fresh elections under a different and narrowly conservative franchise. The result had been the revolution of 1830 – a rising in Paris that had driven him from his throne in favour of Louis-Philippe, the head of the younger, or Orléans, branch of the royal family. The new king had accepted the tricolour, the flag of the Revolution and the Empire, and so the white flag with the gold Bourbon lilies was hauled down for the last time. Bourmont had wished to take a number of the troops back to France to restore Charles, but Duperré, with his Napoleonic background, considered himself the servant of the Government *de facto*, and refused to provide transport. The new monarchy created three new posts of full admiral of France, and Duperré received his reward for the successful conduct of the expedition by being promoted to one of them. One wonders if he, who twenty years before had fought bravely when captain of a frigate to try to save the last colony (Mauritius) then left to France, realized that now, at the end of his career, he had helped to lay the foundations of a new and great colonial empire in Africa.

It will have been noticed that on this expedition steamships appear for the first time as a component of the fleet. They were not popular among real naval sailormen; for they were dirty, and their high funnels and clumsy paddle-boxes made them ugly. Moreover these paddle-boxes, besides making them obviously vulnerable, prevented the mounting of a proper broadside. The ideas of the screw, and of a few big guns in place of many small ones, lay in the future. Yet as early as 1822 General Henri Paixhans, who could see further ahead than his contemporaries, had proposed the building of a number of cheap and expendable iron steamers with one or two large guns firing shells which, he claimed, would smash the wooden walls of British supremacy. Though his proposal was rejected, France was more ready than Britain to accept the new form of propulsion and was somewhat ahead of her in this respect at the time of the Algiers expedition. The few guns of her few steamers had been used to cover the landing of the troops, yet for fully two more decades the main use of the steamer was that of a tug: to get the great three-deckers in and out of harbour, or into place in line when the wind failed.

The year after the Algiers expedition (1831) France became embroiled with Portugal. That country, under the reactionary régime of Dom Miguel, was almost in a state of civil war, and the Government, in its persecution of the liberals, had dealt summarily with some French sympathizers there, and ignored French demands with regard to them. A blockade of the Tagus and the Azores by smaller vessels having had no effect beyond the capture of two minor warships, de Rigny, then minister of marine, sent six of the line, with three frigates and some lesser ships. The commander was Baron Roussin (CA), with his flag in SUFFREN(90)[1], an able officer who started the campaign in a proper seamanlike way by tacking his flagship out of the inner road at Brest in the teeth of a westerly wind. His orders were to force the passage of the Tagus up to Lisbon, and to bombard that city if French demands were rejected. The task was not easy, for the currents and shoals of the Tagus are tricky, and there were strong fortifications. The forts were passed, however, both sides exchanging fire (but the politically divided Portuguese were possibly half-hearted), and the French appeared off Lisbon, where the Portuguese fleet was drawn up. This consisted of one seventy-four, three frigates, and four lesser ships; and it struck directly the leading Frenchman opened fire. Only if the French had suffered heavily in passing the forts would it have had any chance against such odds, and the damage to the French had been negligible. Roussin had done admirably; and the severe and somewhat haughty French demands were granted.

This was followed, the next year, by the Ancona incident. Austrian troops had been sent to help to suppress a revolt in the Papal States; but France was not disposed to allow Austria to extend her already great influence in Italy. So SUFFREN with two frigates and some troops was sent to Ancona, under Captain Gallois. The orders were that they should not land till papal approval had been received. But Austrian troops were close to the port, so Gallois decided to break his orders and land at once. He occupied the place peacefully and Austrian ships that were in the harbour departed. For this piece of initiative he and the commander of the troops were recalled in something like disgrace; but French soldiers remained in the town for six years!

Naval ships also covered with their fire the landing of troops at Bougie in Algeria, to retake it from the Arabs, who had seized it. And in 1834 baron de Mackau (CA), commanding in the Antilles, forced his way into Cartagena (Colombia) with two frigates and three smaller ships to demand, and receive, redress for an insult to the French consul there. Other interventions in the affairs of the former Spanish-American empire were to be less fortunate, as we shall see, but for the moment we should turn aside to note one or two administrative matters.

The Restoration had done away with the Napoleonic training schools for officers, and had established a new one at Angoulême. A naval school so far inland, especially at a time when people had no interest in a navy, had not been a success. So in 1830 Angoulême was closed and the naval school was established in a ship at Brest, where it prospered. The *Inscription Maritime* persisted but underwent a great change in 1835. For then was set up the *Levée Permanente*, which was a conscription out of the

[1]SUFFREN, completed 1829–31, was the last ship built under the direction of Sané, the veteran naval constructor of Louis XVI's days, and she was notable for having no tumble-home. Apart from this and the carrying of a greater spread of sail, improvements to ships after 1815 were in the material used and in armament (as by the introduction Paixhan's shell-firing gun-howitzer). For by 1800 the sailing warship had virtually reached its zenith. ETATS DE BOURGOGNE, designed by Sané in 1786, was still active at sea in 1842 – under her new name of OCÉAN – and was not 'put down' until 1855!

men listed on the Inscription. This meant that there was henceforth a regular turnover of men, and those who had received training were available for calling up even if they went into shore jobs, as many did (eg by 1844, when the Inscription numbered 44 000 all told, there were a further 9 000 who could have been brought back under the conscription law). Now, as has already been said, the idea that the professional French sailor is a poor seaman is wrong; but there is a good deal of difference between a good merchant seaman and even a merely competent naval one. Nor is this only a matter of good seamanship in the quick handling of sheets and gear, and of fleet manoeuvring: it is even more a matter of gunnery. The faulty gunnery of conscripted seamen had been shown up by the Algerian expedition. So in 1837 two gunnery schools were established, in Brest and Toulon, and the idea of a training squadron was revived. This was the Africa squadron of six of the line plus an adequate number of frigates and corvettes. Its commanding officer was Lalande (CA), who as a junior officer in the Napoleonic wars had fought in the spirited little affair of the Sables d'Olonne (see p 271). Immensely keen, forceful and able, during the next two years he brought his small fleet to a really high level of efficiency. The French navy now had forty of the line, carrying from eighty to 120 guns, fifty frigates of from forty to sixty guns, and over 200 lesser and supply ships. It was, in the words of René Jouan, the apogee of the sailing fleet; for among the lesser craft there were already a number of steamers, the ships of the future. Indeed, it was to be only six years before the first screw-driven ship (a dispatch boat) appeared, and eight years before the first screw-driven frigate, POMONE – commissioned in 1845. Here again France was ahead of the British, who did not plump for the screw until 1848.[1]

Meanwhile, in the year with which we have been most concerned, 1837, France had got herself involved both with Mexico and in the Plate river. The republics that had succeeded Spanish dominion in Central and South America were so loosely governed and in such disorder that it was most difficult to deal with them, although, in a fighting sense, they were not formidable. In neither case were the reparations for injuries to French subjects worth the naval trouble they caused. It was at first thought that the appearance of one or two warships would be sufficient to bring the Mexicans to reason. Far from it. So in the spring of 1838 Captain Bazoche in the HERMINIE (60) proclaimed a blockade. He had so few ships that its maintenance was farcical. More and more small warships were sent, but without effect. The only thing that might bring Mexico to its senses was an attack on its main port of Vera Cruz. This was strongly fortified and Bazoche dared not attempt it. His ships' companies were worn out, and decimated by yellow fever. In despair he asked for his recall, and Baudin (CA) was sent out in the early autumn in NÉRÉIDE (52).

With the ships that he brought, and those already there and still serviceable, he had four frigates, five corvettes or brigs, two bomb-vessels, and

[1]Though France more than once took the lead over Britain in matters of steam, she did not long retain it; for the British shipbuilding programme with each development was always the larger.

two 160 hp steamers. One of the corvettes, CRÉOLE, was commanded by the young royal Prince de Joinville. Having spent a month negotiating to no purpose, Baudin formally notified the Mexican governor and commander at Vera Cruz, Rincon, that he would bombard. The town, itself fortified, was covered by the really powerful island fortress of San Juan de Ulloa, which was difficult to approach. Baudin had made his dispositions carefully. Three of the frigates – two towed into place by the steamers, the third under sail – and the bomb-vessels were to do the bulk of the work. Starting in the early afternoon, the bombardment lasted until the evening, when Baudin got his attacking ships away to a safe anchorage for the night. Some lucky shots had blown up two powder magazines in the fort and next morning, as Baudin was moving his ships back to renew the bombardment, Rincon capitulated on terms.

The Mexican Government rejected the capitulation, however, declared war on France (which seems superfluous after a bombardment following some months of blockade), and General Santa Anna came with more troops to occupy the town; for only the island fort of San Juan had been put into French hands. Baudin was ready for this refusal, and had made careful plans to attack the town while Santa Anna was still unprepared. Three landing parties were put ashore at six in the morning, one at each end of the walls and one in the centre. The Mexicans were taken by surprise and little resistance was encountered. Though Santa Anna escaped, another general was taken and the guns were spiked and thrown from the walls. Because he had no troops with him Baudin made no attempt to hold the town, and, indeed, got his men back to their ships quickly as the weather looked threatening. The French casualties, both in the bombardment of San Juan and the seizure of the town, were less than a hundred. Baudin, who was promoted to vice-admiral, was rightly pleased with the good discipline of his men, who had limited their destruction to strictly military objectives; and long-lasting negotiations for peace were opened.

In the Plate river things did not go so well. The Argentine was then a loose confederation of states or provinces – so loose that there was little governmental authority and civil war was endemic. The trade in cattle and horses and their by-products had attracted a number of Frenchmen. These were claiming reparation for damages suffered as a result of civil war, and were objecting to a law of Rosas, the dictator of Buenos Aires, which made them liable for service in the militia. The objection to service was justified: many of the claims were not, being grossly exaggerated. It was thought that diplomacy, backed by a little force, could achieve a settlement. But Rosas was much too subtle for the various diplomats and officers who tried their hands with him, and on the station Leblanc (CA) had only a frigate and five smaller warships. He had to limit his blockade to Buenos Aires harbour and even so was much handicapped by the lack of any base. To obtain one, he supported an anti-Rosas revolt in Monte Video. The Uruguayans proved worthless allies, good at taking subsidies and at little else. When hostilities, hardly to be called war, had dragged on helplessly for two years Dupotet (CA) with another frigate, a corvette, and nine little brigs replaced Leblanc – and did no better. The island of Martin Garcia, commanding the debouches of the Parana and Uruguay rivers, was seized, and some ships were sent up the Parana to fire on batteries and to support an army acting against Rosas. They only covered its retreat! The whole affair had become a bad mess and muddle, of service only to those on both sides who were making profits from it; and Baron de Mackau (VA) arrived in September 1840 to get France out of it on the best terms that he could. He did so, not very satisfactorily, but perforce; for France was in trouble in Europe.

This was due to our old acquaintance Mehemet Ali of Egypt, who was at successful war with his nominal overlord, the Sultan, and seemed about to overthrow the Ottoman Empire. France, who had much influence in Egypt, quietly backed Mehemet, thinking that the different interests of the other powers would keep them so divided that they would take no action. It was a miscalculation. The powers quickly came to an agreement, without consulting France, and she had to stand aside while her protégé was ordered (under force) to restrict himself to Egypt. It was a great humiliation and there was much bellicose talk; but the truth was that France could not make what would be a naval war. Lalande and his efficient little eastern Mediterranean fleet might even have gained an initial success, but, though France had other ships with which to support him, she had nothing like so many of them as had Britain, nor had she sufficient trained crews with which to man them. So the fighting Lalande was recalled, lest he provoke an explosion, and France swallowed an unpleasant pill. It had, however, a fortunate result for the navy, for it was at least partly because of it that Baron de Mackau, when minister of marine, was able to pass his 'law of 193 millions' in 1846. This was intended to provide an adequate budget to maintain an adequately large navy adequately manned by trained men. (It sufficed for the moment, but had to be increased in a few years because of the necessity to provide more and better steam-driven ships.) France had at last realized that she must become a real naval power.

All that she had been able to do by way of a balance to the Mehemet humiliation was to send the Prince de Joinville to St Helena in 1840 to bring back Napoleon's remains to be entombed in the Invalides. But she also became almost aggressively active in the South Pacific. This largely unknown area had interested her ever since Bougainville had reached Tahiti and had discovered the Louisiades and the Solomons; and the ill-fated La Pérouse and d'Entrecasteaux had both sailed for that region before the start of the Revolutionary wars. The subsequent voyages of such people as Dumont d'Urville in the 1830s, and of Duperrey, renewed the interest. French traders made for some of the islands and, in this case, the flag followed trade. By 1840 France had two or three small ships of war there, and a frigate also went out under Dupetit-Thouars (CA). In 1842 this admiral occupied the Marquesas and established a protectorate over Tahiti and the other Society Islands. Captain Laplace of the frigate ARTEMISE, who spent two months there getting his ship off a reef and repairing her, had built up a French party, and the admiral somewhat high-handedly levied tribute. This brought him into conflict with the queen, Pomaré, and in 1843 he landed five hundred men to haul down her flag. This meant a collision with a determined missionary, Pritchard, who was the British vice-consul there. Though British missionaries had preceded the French to the islands, the British Government had raised no objection to the French occupations because they wished to avoid colonial commitments; but they were not prepared to overlook an insult to the flag and, to her people's annoyance, demanded and obtained an apology from France in 1844. The French Protectorate was recognized, however, and was later extended.

In Africa, the Arabs had naturally opposed the French as they spread their occupation from Algiers inland, and had received support, and a refuge, from the ruler of Morocco. Negotiations having failed, the Prince de Joinville, with his flag in SUFFREN(90), took two others of the line, some smaller ships, and, be it noted, six steamers, to bombard first Tangier, then the Sultan of Morocco's own private (and pet) seaside possession of Mogador. In both cases the bombardment was carried out mainly by the old sailers of the line, which were towed into position by the steamers; but a steam frigate assisted at Mogador. The success of the expedition was marred when, on the return, that frigate, GROENLAND, ran ashore and was wrecked.

This had been in 1844. The next year saw further trouble with the Argentine. We have seen that Mackau had had to be content with such terms as he could get from Rosas, and even these had not been observed. Now Rosas closed the navigation of the Parana to foreigners, and enforced this by setting up powerful batteries at Obligado, a short way up the river, and grouping there some armed vessels, including fire-ships, behind a boom. This was too much, and it affected Britain as much as France. Both countries had some ships, mostly small, on the station and they took joint action. The senior officers were Captain Tréhouart and Captain Hotham; and they took up to Obligado five small French ships (one a steamer) and six small British ones (of which two were under steam). The sailing ships attacked in two columns, each of mixed nationality, Tréhouart leading one in the locally armed SAINT-MARTIN brig, while Hotham remained in reserve with the three steamers, using them where they were needed. The boom was broken, the ships behind it sunk, including the fire-ships, and the batteries were silenced: only the SAINT-MARTIN was seriously damaged. This little battle has some small importance in naval tactics, for it was the first in which use was made of the superior mobility of steam, Hotham sending in the steamers, and calling them off again to throw their weight elsewhere, in a way that could never have been done with sail.

In 1851 the people of Sallee, in Morocco, reverted to the piratical activities of their former rovers, and seized and plundered a French brig. So Dubourdieu (CA) was sent with the HENRI IV and four steamers, two of them frigates, to bombard the town. The ship of the line and the frigates did so, and destroyed the town's fortifications, though all three ships suffered some damage. While there had been nothing even approaching a fleet action in the roughly twenty-five years since Navarino, the French navy had seen a good deal of active service in one way and another, and in widely spread parts of the globe.

In 1848 France had undergone yet another revolution. The eighteen-year reign of Louis-Philippe had been humdrum at home and inglorious abroad; and when the revolutionaries took over a demonstration in Paris and turned it into a rising of the wretchedly underpaid working class, the monarchy collapsed, for there was nobody interested in supporting it. Yet France as a whole was unwilling to follow the revolutionary lead of Paris, and the republic that took its place was of a conservative type. The man elected president was Prince Louis Napoléon, nephew of the great emperor, and heir to the Bonapartist claims. At the end of 1851 this prince carried out a coup-d'état whereby, in 1852, he established the Second Empire, reigning as Napoléon III.[1] Such effect as these upheavals had upon the navy was to the good. For the success of a Napoleonic empire would depend upon its prestige, and it was to the interest of that prestige to make the navy a strong force. It was well that this was so, for early in 1854 came the Crimean War.

Ostensibly this, the first international struggle since 1815, arose from a dispute between the Greek Church monks (favoured by Russia) and the Roman monks (supported by France) over the right to care for the Holy Places in Jerusalem, and Thiers called it a war to give some monks the key of a grotto. Of course the real cause of so great a conflict lay far deeper, in the increasing suspicion in Britain and France of Russia's designs upon the Ottoman Empire in general and upon the great strategic position of Constantinople in particular. This was confirmed by the hectoring demands of a special Russian envoy to the Porte for a Russian protectorate of all Greek Church subjects of the Sultan. When this was refused the Russians occupied the provinces under Turkish suzerainty north of the Danube; and this meant war. The British and French Mediterranean fleets had already been ordered to the Aegean and now were sent through the Dardanelles, first to cover Constantinople and later to police the Black Sea, while large armies and further ships, including the effective part of the Brest fleet, were sent out from home. The French fleet which entered the Black Sea was, like the British fleet that accompanied it, very much a mixture. Baron Ferdinand Hamelin (VA) who commanded, with Comte Bouet-Willaumez (CA) as his chief of staff, had his flag in the VILLE DE PARIS(112), a three-decker under sail. Of his fifteen of the line, eleven were the old sailing three-deckers. One, the NAPOLÉON(90), was a powerful screw steamer with also a full rig of sail: she was designed by a constructor named Dupuy de Lôme, and could steam over thirteen knots. The others, including the MONTEBELLO(120), flagship of Bruat (VA), the second in command, were sailers with auxiliary steam; though screw-driven, their engine-power was low; in ideal conditions they could just make nine knots, but they laboured heavily and slowly in a seaway or against a current. He had five steam frigates, and also nine steam corvettes and despatch vessels whose duties in action were to be to tow the three-deckers into position. Some were paddle wheelers. Purely from the standpoint of those interested in naval tactics it is a pity that neither in the Black Sea nor in the Baltic

[1] 'Napoléon II', who never reigned, was the great emperor's only son, who died in 1832.

did Russia feel her fleets to be strong enough to face an action, for it would have been interesting to see how this mixture of ships, of different speeds and with different means of propulsion, would have been used as a combined force. As it was, the war was one of armies; and the services of the combined fleets were limited, if that be the word, to bombardments and to providing a mobile base for the troops, transporting their provisions and munitions, their reinforcements, and, above all, their sick – for, as is well known, the medical arrangements were deplorable, and cholera and scurvy were rife. It was an inglorious service, but an essential one, and one that was desperately hard-worked and testing, for the Black Sea did not belie the name that its weather had earned it.

In April 1854, as Odessa had fired on a flag of truce, that great port was bombarded. The steam frigates, five British and three French, were used. Most of the shipping in the harbour was sunk, and the fortifications damaged: the only serious injury to the allies was to the French VAUBAN, which was temporarily driven out of action. The armies had been landed on the Black Sea coast of Bulgaria, to aid the Turks in defending their Danubian frontier. When the Turks proved unexpectedly able to do this for themselves it was decided to invade the Crimean peninsula to destroy the great Russian arsenal of Sevastopol, on which any operations round the Black Sea (including much of the Turkish frontier) largely depended. Strategically the decision seemed sensible: practically it was unwise, for the armies were not prepared or equipped for attack upon a powerful and well-garrisoned fortress and, by the time the decision was reached, there was only a short campaigning season left before the coming of the dreaded Russian winter.

However, the fleets took the armies to Eupatoria, well to the north of Sevastopol, in September, and covered their advance along the coast, using their guns to help the right wing of the army when it forced the passage of the river Alma. As Sevastopol was approached the Russian ships were seen coming out of the harbour. Yet this was not to fight, but so that they might be sunk across its entry. Their guns had been removed to strengthen the batteries, and their men were used to reinforce the already large garrison. The generals decided that the place was not assailable from the north and moved their forces round to the south of it, the French fleet basing itself near them at Kamiesh, and the British at Balaclava, both places being open roadsteads rather than proper harbours. A large number of seamen and a hundred guns were landed under Rigault de Genouilly (CA) to form thirteen batteries, additional to those mounted by the troops; and throughout the eleven months of operations against Sevastopol, they did notably well. The great assault was delivered in October, and it failed. The fleet's share was to bombard the fortifications of the water front. The French were in two lines, the outer line firing through gaps left in the inner line for the purpose. Shallows prevented the ships from getting close enough and the fortifications were strong. Only the battery called the Quarantaine suffered seriously, while the CHARLEMAGNE was hit in the engine room and other ships (including NAPOLÉON) were damaged. French casualties numbered two hundred, and the operation had been a waste of men, powder, and shot. Land operations having also failed, the allies settled down to a siege – which was not really a siege at all. For the Russians had the greater gun-power and, though the allies had the more men, their armies were not sufficient completely to surround the fortress. The fleets settled down to the wearying task of maintaining these armies, while being themselves in inadequate bases. A gale in November drove the veteran HENRI IV and a frigate ashore, and at Christmas Hamelin returned to France, handing over the command to Bruat (VA) with Captain Jurien de la Gravière as chief of staff.

In the Baltic, too, there had been no real success. The British had sent Admiral Napier there with thirteen of the line and twelve lesser vessels, all under steam. France, anxious to have a hand in the operations, had sent the screw-driven auxiliary steam AUSTERLITZ(80), and followed it with seven of the line and some lesser vessels under Parseval Deschenes (VA) with his flag in INFLEXIBLE(90). These were sailing ships, taken out of reserve, with reservist crews. It had been hoped that the fleets would take the great fort of Kronstadt, commanding the approaches to St Petersburg (now Leningrad), but a careful reconnaissance showed this to be impossible, and they had to be content with a skilfully made bombardment of Bomarsund, the fort of the Aland Islands, and with the capture of those isles. As they could not be retained, for the Russians could retake them across the ice when winter came, the operation served little purpose, and the fleets returned as the Baltic began to freeze over.

In the Pacific, Febvrier-Despointes (CA) in the frigate FORTE(60) had begun the year by occupying New Caledonia (where he took a somewhat high-handed line with British missionaries working there), and then sailed north to join British frigates in an attack on Petropavlovsk, the chief Russian settlement in Kamchatka. The allies took some ships, but were driven off from the town with loss, though the Russians later evacuated it.

In 1855 Pénaud (CA), with his flag in TOURVILLE(90) and two other auxiliary steam ships of the line plus some lesser craft, joined Admiral Dundas in the Baltic, and they carried out a successful bombardment and destruction of Sveaborg, the fortress commanding the approach to Helsinki; but the only object that they and the previous fleets in the Baltic had achieved was that of keeping Russian forces in the north in case troops should be landed. In the Crimea the exhausting maintenance of the armies and service of shore batteries continued, varied by an occasional bombardment of the water front by a ship or so, to keep the Russians occupying the shore batteries, and by an expedition, jointly with the British, to seize the Kertch peninsula and clear the sea of Azov. This was completely successful and, as Sevastopol was being supplied from this area, it hastened the fall of the fortress, which took place in September. A final operation was a joint bombardment of the fortress of Kinburn which commanded the exits of the Bug and Dnieper. For this the French used the MONTEBELLO(120), JEAN BART(80), ULM(70), WAGRAM(70), and some new ironclad, steam-driven, floating batteries[1] – the forerunners of the later monitors, and, indeed, in their armour, of the later battleships. These, whose

[1] DÉVASTATION, LAVE, and TONNANTE

construction had been proposed some years earlier, fully justified themselves; and the fall of the fortress cut the communications of the Russian armies on the Black Sea.

Both sides were now weary of the war. The allies at great cost, and with much mismanagement, had achieved some sort of objective; the Czar whose aggressive statesmanship had caused the war had died, and Russia was gravely strained in finance. Early in 1856 arrangements were made for a Peace Congress in Paris, and Bruat, promoted to admiral, sailed with the fleet for home, though he himself died on the voyage. By the terms of peace the Black Sea was neutralized, and that sufficed to protect the decaying Ottoman Empire till Russia broke the treaty while France was engaged in the war of 1870 with Prussia. The Congress of Paris ranged over a number of other matters besides peace terms, and these included two that concerned naval affairs. It was agreed that privateering and letters of marque should cease, and that, while neutral ships should be safe except for contraband of war, a blockade should be proclaimed only if it could be made truly effective. These decisions took from France not only any idea of a Napoleonic 'Continental System', but also the temptation to turn to her beloved *guerre de course*. She could now concentrate on using her seamen and her navy for their proper purpose. As will have been seen from the numbers and nature of the ships used in the clumsy Crimean War, the government of the Second Empire was already making that navy both formidable and modern. Though some of the great three-deckers (with their tugs and auxiliary steam) appeared in the Adriatic during the Franco-Austrian war in North Italy in 1859–60, and though some purely sailing warships are found for a while in colonial waters, the day of the sailing warship was over.

CHAPTER XIV

# The Naval Revolution: Ironclads and Policies – The Last Imperial Wars and Colonial Expansion[1]

The years following the Crimean War were to produce the first great revolution in naval warfare since the introduction of cannon, a revolution that continued almost till the end of the century. It had been heralded by the appearance of the screw-driven steamer and of the ships built of iron. The competent performance of the little iron-clad, screw driven French gunboats in the bombardment of Kinburn ensured that it should continue. Before the end of the century it was to produce the great armoured battleships with their large, rifled, and shell-firing breech-loaders, their torpedo armament, their attendant heavy and light cruisers, and their flotillas of torpedo craft. Change was continual and rapid, as experience followed experiment. In every naval country committees sat, and commissions were appointed, to plan the nature and size of future fleets, only to find that before the ships to which they had committed themselves were completed these were being rendered out of date by others, already on the drawing-board and embodying some new achievement in engineering or construction, some new device in armament. As the century aged the pressure was intensified by the colonial race, already slowly started by such activities as those of Dupetit-Thouars and Febvrier-Despointes in the Pacific, and growing in speedy competition as the powers increasingly hurried to colonize, or at least to establish spheres of influence in, all lands where Europeans (or their descendants) were not already in possession.

Although she had no good steam coal, and her iron deposits were inconveniently distant from her ports where the ships must be built, France more than once took the lead in the naval race; and always, despite her alliance with Britain in the Crimean War, her naval service was impregnated with the most lively anglophobia. Her *revues* and service journals redounded with articles, often almost openly bearing the stamp of officialdom, on how the British could and should be met and beaten at sea; and it is not until the rise of Tirpitz and his big navy programme for Germany, followed by the Entente Cordiale of 1904, that this anglophobia disappears – or more truly, perhaps, lies only latent. Whenever France took the lead Britain, with her tradition of the navy as her main force, was quick to follow, usually in greater quantity, though not always in superior quality. Laid down by France in 1858 and completed in the next year was LA GLOIRE, an armoured frigate developed from the lessons taught by the Kinburn gun-boats. Designed by the famous Dupuy de Lôme, she was

[1]For map of Indo-China and Courbet's campaign see p 184.

armoured from stem to stern, could steam at thirteen knots, and carried fourteen rifled and breech-loading guns of 24 and 16 centimetres. Britain replied by the end of the next year, 1860, with the *Warrior*, faster and larger but more vulnerable and less manageable. The appearance of such vessels might not of itself have doomed the wooden ship: this was done by the rifled gun, firing shell instead of shot. Rifling had been slowly developed from the eighteenth century: shell had so far been used mainly in bomb-vessels, as being too likely to start fires in large wooden ones. The Crimean War saw a rapid development of the rifled field gun by the French, and a quick following suit by the English: rifled shell-firing cannon therefore appeared on the ships.

At first steam had been a mere auxiliary to sail but now sail became the auxiliary – till it disappeared altogether.[1] Much of the conception of naval strategy was altered because ships became dependent on coaling stations for their mobility. The specialized seaman – the engineer, the signal-man, the coxswain or quarter-master – was now superimposed on the old able-bodied hand, and 'schools' were created at naval depôts to train him. The specialist officer was a natural corollary of the rapid technical developments. France even toyed with the idea of a submarine in 1863. This was LE PLONGEUR, designed by Captain Bourgois. She was 140 feet long, of 410 tons displacement, and was driven by compressed air; but she lived up to her name so excessively and so erratically that the idea was dropped. Yet, on the whole, in 1860 France was equal to Britain in old ships, and ahead of her in new ones and in her building programme, though, despite the introduction of long-service terms for her sailors, she still lacked an adequate reserve of competent seamen.

Meanwhile, anglophobia or no anglophobia, the navies of the two powers had again been acting together, this time in the Far East. The potential trading opportunities for Europeans in China were great, but the Chinese Empire was in a state that seemed to be approaching disintegration. So there was need to make arrangements to protect that trade, the more so as agreements so far made with the High Celestial Empire were more subject to breach than to observance. The Imperial Commissioner at Canton having seized a Chinese vessel sailing under British colours, the British high-handedly demanded a humiliating apology and, this being refused, felt obliged to take action. France resolved to join with them in order to obtain a firmer agreement with the Chinese than that which had been so often broken. A special commissioner was sent out, as was Rigault de Genouilly (CA) in the frigate NÉMÉSIS(50) in 1857. He joined the British admiral Seymour off Canton, with this ship, four corvettes, an aviso, and four gunboats. All but the frigate and one corvette were steamers. The forts were bombarded and some British troops and a French landing party (led by the admiral in person) carried the town. The government of the High Celestial Empire, however, made no move, so it was decided in 1858 to leave a few ships and men to hold the Canton forts and maintain a blockade there while the rest moved north to the Pe-Chi-Li gulf to

[1]A famous report by Baron Hamelin in 1857 had stressed that the navy must become an 'all-steam' force.

approach Pekin itself. This was a daring move with only small forces, even granted that the Chinese were notoriously inefficient. The mouth of the Pei-Ho river, up which lies Tien-Tsin through which Pekin could be reached, was strongly fortified, with a powerful weight of artillery. Nevertheless an attack was made by six gunboats, four French and two English. This was directed by the two admirals, flying their flags together in a small British auxiliary steamer. Though the Chinese stood to their guns bravely they handled them poorly. The forts and batteries were silenced and occupied, and the allied ships went up to Tien-Tsin. At this the Imperial Government gave way and signed a treaty opening six ports to international trade under European control.

This done, Rigault de Genouilly, now promoted to vice-admiral, took some of his ships south to Cochin-China, as he had already been ordered to do. The French had had connections with this region ever since the 1780s; but these connections were tenuous, being limited to the active work of missionaries and the occasional visit of a naval ship. A change of ruler in Annam had been followed by a persecution of the missionaries and their converts, and they had appealed to France for help. Rigault de Genouilly picked up five hundred troops that had been sent out, and some Spanish Filipino soldiers, and made for and captured Tourane bay, some forty miles from the capital, Hué. This had no effect, so in 1859 he took some of his ships and men up the Donnai river to Saigon, which he seized hoping that the occupation of this great food trading centre would bring the Annamites to see reason. They made only threatening moves, however, and as France could at first send no help, being engaged in her war with Austria, the French admiral was in the difficult position of implying a threat when his ships needed overhaul after long campaigning and his already small numbers were reduced by illness and by the fighting.

The war with Austria entailed no naval operations beyond the dispatch of a fleet into the Adriatic, where it took no action worth mention because of the unexpectedly early signing of peace. France was now free to turn her attention to the Far East, where, in addition to the awkward situation at Saigon, the Chinese had promptly broken the recent treaty, rearmed their forts, and repulsed the few allied ships (mostly British) that had remained off the Pei-Ho. So a large Anglo-French force was sent out. The French ships were commanded by Charner (VA) with his flag in the steam frigate IMPÉRATRICE-EUGÉNIE, at the head of fifteen naval sailing vessels (ten with auxiliary steam) plus some gunboats. The expedition was mainly military, but the allied ships covered the landing of the troops, bombarded forts, and sent their marines ashore to give valuable support to the soldiery. Pekin was taken, and a sterner and more permanent treaty was signed.

Charner then brought a number of his ships south to Indo-China. Here, Tourane had been evacuated and the small French force had concentrated on holding Saigon and keeping open its communications. This it had stoutly succeeded in doing, though all but surrounded by the Annamites in their fortified lines. The arrival of Charner with his marines and some of the troops that had been in the expedition to Pekin enabled the French to take the offensive. The Annamite lines were stormed, their army was put to rout, and its magazines were seized. Though the smaller ships had supported the troops by fire from the river, the operation had been mainly a land one; but it was commanded by naval officers, and largely carried out by naval personnel. This was true of the operations that followed, though the little gunboats performed prodigies in getting past the various obstructions in the rivers to shell forts. By this campaigning the occupation was spread west to the main mouth of the great Mekong river and also well to the east of Saigon. Charner went home at the end of 1861, leaving Bonard (CA) to complete his work; and in 1862 a treaty was signed ceding to France the three southern provinces of Cochin-China. The fighting qualities of the Annamites, and their armament, had been poor; but their numbers had been huge, and the difficulties of the country and of maintaining good health had been great. The conquest of this colony by so few had been a fine achievement, and it had been mainly a naval one.

France now became involved in what is called Louis Napoléon's Mexican 'adventure' or 'imbroglio' – according to taste. The chaotic conditions of endemic civil war in Mexico caused loss to European traders, and no satisfactory arrangements could be made with President Juarez for indemnity to them. Therefore, in 1861, Britain, France, and Spain agreed to dispatch a joint force which should occupy Vera Cruz, and perhaps Tampico, till satisfaction was obtained. The French ships (one of the line, five frigates, and some smaller vessels, with transports) were commanded by Jurien de la Gravière (CA). A convention was signed with Juarez, and the British and Spaniards withdrew. France, however, repudiated Jurien's signature and sent out large forces. There was a vocal clerical party in Mexico, opposed to Juarez and wrongly supposed in Europe to represent far more of the Mexican people than was the case. European clericals were pressing the Habsburg Archduke Maximilian to accept the crown of Mexico (which he agreed, after hesitation, to do) and France resolved to support him. This unwise step was taken to please the powerful French clericals, who had been angered by the grave way in which Louis Napoléon's war against Austria in Italy had weakened the Papal position.

For the French navy the resultant Mexican war was as ungrateful and unsatisfactory a business as that of the Crimea. There were the same transporting of troops, wearisome maintenance of them with supplies and other services, tiresome and difficult blockade of the coast, and lack of any sea fighting which might bring credit. There was even the same strain of heavy losses from illness – the killer in this case being not the awesome Russian winter but the deadly fevers of a highly insalubrious shore. Tampico was taken at the expense of a wrecked gunboat; then it was abandoned. Later it was taken again and held by a naval garrison, whose men slowly died off, from fever, till the survivors were withdrawn towards the end of the war. The army reached Mexico City, and the Archduke arrived and was crowned Emperor; but it soon became clear that his native supporters were too few, and that he must depend on French forces to keep his hold on his throne. As most of the French fleet had been recalled once Maximilian was enthroned, the support of the army and the maintenance of a hopeless blockade fell to a mere handful of vessels (two frigates, two corvettes, and some gunboats) under the command of Captain Cloué, whose inexhaustible energy did wonders with them. There were continual petty actions, in the form of landing little parties of men to support the tiny garrisons of coastal places when these were beset by republican guerrillas: there were continual chasings of Mexican corsairs sailing from American ports. On the Pacific coast Bouet (CA) appeared in 1864 with the steam frigate PALLAS and some lesser vessels, and bombarded Acapulco and some other places, but could make only the most temporary occupation of them because of lack of troops.

The war dragged on and on; but by 1866 the American Civil War had ended, the United States was now free to attend to the affairs of the American continent, and France was bluntly told that, under the Monroe Doctrine, the presence of her troops in Mexico could not be tolerated. The Civil War had left the United States with a formidable army, and unless France was prepared to fight a major war thousands of miles from the home country she had no choice but to withdraw. To show a defiant strength she sent out a number of her ironclads early in 1867, but they came only to bring the army away. The unlucky Maximilian, refusing to desert his partisans, was overwhelmed by the republicans, and shot by a firing party. The whole affair is an inglorious chapter in French history: the navy comes out of it with some credit.

Then, in 1870, France fell into an unscrupulous diplomatic trap set by Bismarck, and went to war with Prussia and the German states. The fleet steamed north to blockade the German ports; but, as Prussia had but few warships and the ports were too far up rivers for bombardment, the only purpose served was to keep a number of Prussian troops away from the main battlefields to guard the coast. The sole sea fighting of the war was that between two small ships off Cuba, and it ended when they came into territorial waters.

On land, all went ill. One of France's two armies was pinned in Metz, and the other, making a dangerous flank march to free it, was surrounded and forced to surrender at Sedan, the Emperor Napoléon III being taken prisoner. At the news, the Republic was proclaimed in Paris, the great city prepared to stand a siege, and a *levée en masse* was made to drive the invader from the country as in the early days of the Revolution. It was a fond hope, for the efficient enemy of 1870 was very different from the disunited allies of 1793. The war had become a desperate struggle on land in which the fleet could play no part; but its men could do so, and nobly, and over 57 000 of them served ashore. It was the naval brigade of 12 000 men, hurried into the capital with its gunners, that shared the fighting during the siege with the one army corps that had got away from Sedan; for the civilian population, formed into a National Guard that elected its own officers, strutted much and, for the most part, fought little. They were naval brigades, too, who, with some troops from the colonies brought over safely because of the navy's presence, provided the solid fighting nucleus of the armies of untrained men that were formed in the north and on the Loire in a desperate attempt to relieve Paris and restore the situation. All was in vain; and France had to make a humiliating peace. But her navy was not to blame. It had done all that could be expected of it, and once again it was the happening ashore that prevented it from being put to its proper usage.

Napoléon III had proclaimed that the Empire stood for peace. When it fell he had fought wars against each of the great powers of Europe with one exception – the power that many of his people and almost all his seamen considered the real enemy. Yet it was so that France, plus a naval ally, might be equal to Britain at sea that he had built up a navy of thirty-four ironclads,[1] with over 300 other ships and an establishment of 28 000 men, plus over 40 000 in the reserve. History can indeed turn aside down odd paths. At least during the Second Empire the navy had been strong and had been kept active, even if upon rather thankless tasks.

After peace was signed with Germany in 1871 the general feeling in France was not favourable to a naval service, despite the stout fighting that her sailors had done ashore. She had had to surrender the two provinces of Alsace and Lorraine and her thoughts were focused upon revenge to recover them. Of what use was a fleet for this, which must be achieved by victory on land? She had also to pay a heavy indemnity. Why build expensive ships? In the recent war they had had to remain useless in harbour while their men had come ashore to fight. Even the minister of marine, Admiral Pothuau, had to agree that his service should be sacrificed for the benefit of the army: that the main object of a navy was only to defend one's coasts. So the naval programme for 1872 reduced the fleet drastically to little more than two hundred vessels of all kinds. Yet apart from the consequent reduction in personnel this cut was not so severe as it would seem, for the ships that were scrapped were largely the outdated and less useful. More serious in the long run was the unsettled nature of the government, especially in the first thirty years of this, the third French Republic. France was riven by bitter feuds between Royalists and Republicans, Clericals and Anti-Clericals, and was disgraced by not a few governmental scandals. The French were far too logical to accept a two-party system whereby everyone was born either a little liberal or else a little conservative. There were several parties. As a result governments were a series of unstable coalitions. It was rare for one to hold together for over two years and common for one to last only a few months. In the thirty years from 1871 to 1901 there were no less than thirty ministers of marine. A consistent naval policy in circumstances of such constant change was all but impossible. Nevertheless it was not always change of minister that brought about the abandonment of the naval programmes that were laid down from time to time: sometimes this was due to the rapid development of naval architecture, machinery, and weapons.

The naval programme of 1872 envisaged twenty-eight ironclads in two classes, thirty-four cruisers, and over a hundred smaller vessels. The first new battleship, REDOUTABLE, was to be built entirely of iron on a steel fram-

[1]His latest ironclad, OCÉAN, was of 7 334 tons' displacement, had armour 20 cm thick, steamed over 13 knots and carried four 27-cm guns, with ten of lesser calibre.

ing (the first use of steel at sea), and not, like former ships, of armour over a wooden hull, while the main weapon of most of the smaller craft would probably be that ominous new missile, the torpedo. The torpedo-boat was still only in the near future: the destroyer was not thought of.

In fact, France was to develop a huge colonial empire, and this was to be quite largely due to young naval officers leading handfuls of men. First prominent among these was that gallant young lieutenant Francis Garnier. He had played a part in steadily extending French influence in Indo-China during the 1860s and, having explored the Mekong enough to find that it furnished no route to China but that this was provided by the Song-Koi river flowing through Hanoi, he seized the citadel of this place in 1873 though it had a garrison of 7 000 and he had merely 175 seamen. He was killed in an ambush by the Chinese mercenary company, the Black Flags, the next year, and France withdrew from Hanoi for the time being, subject to an Annamite recognition of her protectorate.

It was not with colonial expansion or ventures in view, however, that in 1878 M Lamy made a notable speech on the navy in the budget debate. Britain with her island situation and overseas empire was a special case for a large navy. Leaving her aside, he said, France had the largest naval budget of any country, but it had not given her the ships. The money was going too much into the creation and maintenance of dockyards and arsenals. They had an absurdly large personnel, and the expenses of colonial troops were borne by the 'Marine'. All this was another way of saying that the navy was in danger of becoming a land organization which happened to have some shipping. France, he urged, should have a fleet capable of meeting those of any two powers other than Britain. In fact, purely where ironclads were concerned, France could reasonably rival Britain at this date, with twelve laid down. The CAIMAN had 19½-inch armour, made 14½ knots and carried a 42-cm gun in a barbette forward, another aft, and two 10-cm guns on each broadside.[1] As Bismarck was now busy building his system of alliances to make a war of revenge impossible for France by isolating her, Lamy probably had in mind the Austrian fleet, plus that of any other Continental power whom Bismarck might coax into his net, or plus that of Germany, though that was as yet small – for her big navy policy lay in the future. Finally, Lamy questioned whether battleships were worth while. They were so expensive to build and with this new torpedo weapon available they seemed very vulnerable.

This report led to more than a decade of furious haggling as to what should be the nature of the fleet; for it brought to the fore what was known as the 'Jeune École' of naval thought, led by a flag-officer named Aube. There was a prolonged pamphleteering war. The Jeune École argued that, as one could build sixty torpedo boats for the price of one battleship, it was best to have coast defence vessels and a number of fast, small, and cheap craft that could attack the enemy's trade: battleships were only pomp and show. This was really a reversion to that old red herring, so much to the taste of Frenchmen, the *guerre de course*. Privateering

[1]Moreover, the guns were breech-loaders, whereas Britain, alarmed by breech-loading accidents, was some time getting rid of the muzzle-loader.

being now forbidden by international convention, the whole navy, apart from coastal defenders, was to go a-commerce-raiding in little ships, hard to spot and moving fast. Against this the more traditional argued for the powerful battle fleet that by its victory would have command of the sea. The conflict of ideas led to confusion of policy and to the slowing up of construction.

It made little difference to the young officers who were forging ahead oversea. They were encouraged by the strong and able Jules Ferry, who held office for two spells of two years in the first half of the 1880s, and who saw that France must extend her overseas possessions into a colonial empire if she was to acquire the wealth and resources to meet the more heavily populated Germany. (This policy was not popular and twice brought about his fall from office, but it stood France in good stead.) So, in 1880, a young naval officer, Savorgnan de Brazza, peacefully acquired the French Congo by a treaty with its native ruler. In the same year Dupetit-Thouars (CA), nephew of the officer who had occupied Tahiti, led a punitive expedition to avenge a murder in the Marquesas Islands, and by their occupation consolidated French control of the islands in the south-eastern Pacific – though an attempt to anticipate the British protectorate of the Cook Islands failed. Much more important was the occupation of the 'Regency' of Tunisia in 1881. Trouble on the Algerian frontier provided the excuse, but the real reason was to forestall Italy. (The move had under-the-counter support from Bismarck, who successfully hoped thereby to bring the angry Italians into a somewhat unnatural Triple Alliance with Germany and Austria.) Admiral Cloué was then minister of marine and sent the Mediterranean home fleet (Garnault, VA) and east Mediterranean squadron (Conrad, CA) to escort the thirty thousand troops. Garnault bombarded Tabarca which was then easily occupied, Bizerta surrendered to Conrad without a blow, and the Bey accepted a French protectorate. The whole bloodless affair had taken less than a month and the troops were mostly recalled. This was premature. The tribes in the south of the Regency rose in furious revolt and the nine ironclads of the joint squadrons had hurriedly to return.[1] Sfax was shelled and carried by landing parties from the ships, plus a few soldiers, and Gabes was also taken, by the marines. The navy having captured these ports for them the troops then returned to finish the business. In Madagascar the natives launched an attack on the French settlements, so ships bombarded Majunga and Tamatave and landed parties to take Vohemar, on the north-east corner of the island. That settled matters for the moment, though there was to be further trouble, later, before the occupation was complete.

Much greater difficulty was experienced in dealing with Indo-China. Here the Annamite ruler, Tu-Duc, was a consistent breaker of treaties; but it was difficult to get the French Chamber to vote supplies for action in Tonkin, and Admiral Jauréguiberry, who was minister of marine, was able to

[1]They were COLBERT, TRIDENT, LA GALISSONIÈRE, FRIEDLAND, MARENGO, SURVEILLANTE, REVANCHE, ALMA, and REINE BLANCHE. Their tonnage varied from 8 000 to 3 900, and their guns varied from small calibres to 10.8 inch.

send out only a few men. With this small force Henri Rivière (CF) occupied Nam-Dinh on the Red (Song-Koi) river in 1883. But this brave young officer, like Francis Garnier before him, was ambushed and killed by the Chinese Black Flag mercenaries. Then, when it was too late, France was really stirred. A proper expedition was sent out in 1884 and, Tu-Duc conveniently dying, peace was imposed on the new emperor.

Now China, who was known to have been behind the Annamite resistance, and still had troops in Tonkin, was presented with an ultimatum. This demanded withdrawal of all troops from Tonkin, recognition of French rule, and an indemnity. Courbet (CA) was ordered to steam with the ships of the Tonkin expedition to enforce it by threatening the Chinese naval base of Foo-Chow and occupying the coal-mining port of Kelung in Formosa.

It was no easy assignment. For Foo-Chow lay up the Min river and could be reached only by passing through two strongly fortified narrows. However, Courbet duly steamed thither to impose his threat, while detaching Lespès (CA) with the ironclad LA GALISSONIÈRE, the cruiser VILLARS, and a gunboat to deal with Kelung. This they failed to do. For when, after a bombardment, the landing parties were put ashore, they could make no headway and had to be re-embarked. Courbet was therefore obliged to attack Foo-Chow and could not use his two armoured cruisers because their draught was too great for the river. With his flag moved to the VOLTA, he had five unarmoured cruisers, three gunboats, and two small torpedo craft with which to deal with a powerful arsenal and a Chinese fleet of eleven ships, plus junks and fireships.[1] With a retreat only through two lots of fortified narrows if his ships were damaged, his position was distinctly precarious. Helped by the inefficiency of the Chinese, he handled the situation admirably. He sent his little torpedo boats in to loose their weapons at close quarters against the two cruisers that were the enemy's most powerful vessels. His own cruisers and gunboats shelled the remainder, and the Chinese fleet was quickly knocked out of existence with heavy loss of life. There followed a two-day bombardment of the arsenal, on which Courbet could make no conquering impression. He therefore steamed down river for the open sea, destroying the forts in the narrows *en route* by clever warping of his ships into position in a three-day battle with them. Then, following his orders, he stood away for Kelung, to avenge the repulse there. He occupied the port, and a neighbouring harbour, and his flagship's picket boats sank a frigate by a daring night expedition up a river; but his attempts to seize the coal mines inland were repulsed by sheer weight of numbers. He was able, however, to cut off the seaborne rice trade, on which northern China depended, and this brought the Imperial Government to seek peace. Though France disliked the Far Eastern policy of its government, it greeted Courbet's achievement with somewhat greater enthusiasm than was merited by a success that, in truth, had been only partial. This does not alter the fact that Courbet had shown himself to be a resourceful, brave, and thoroughly able commander in a tricky

[1]The Chinese ships were much inferior, the strongest being only iron-plated over wood.

situation; and it is sad that he died of cholera before he could return to France.

The rest of the stupendous expansion of the French colonial empire in the last quarter of the century was the work of her soldiers and administrators; but it was work based on foundations which the navy had laid. The navy had still to be called upon on occasion, as when in 1893 demonstrations were made up the Mekong and off Bangkok to impose upon the Siamese a frontier with Indo-China distinctly to the French advantage – greatly to the annoyance of Britain, who wished to preserve Siamese integrity.

At home the Jeune École had come into the ascendant, and in 1886 their leader, Admiral Aube, became minister of marine. He established a naval base at Bizerta, which was all to the good, and embarked on a building programme of cruisers and torpedo-boats. For the latter there was good excuse, as the 1880s were the decade of torpedo development. He also boosted the cause of the submarine, which had been abandoned since the eccentric behaviour of the PLONGEUR. Gustave Zédé, an engineer and constructional officer, had taken up the idea and had produced the thirty-ton, electrically driven, GYMNOTE. This seemed a successful basis for further experiment, and Aube could boast that once again France was in the lead in naval developments. In fact his mind was running along a very old track. For he regarded Britain as the enemy and his programme of cruisers and torpedo craft, and his support of experiments with the submarine, had in view an attack upon British seaborne commerce. (It is interesting to speculate whether, given effective submarines in a war with the United Kingdom, France would have made the unscrupulous use of them that Germany was later to make: it was then thought that the main function of the submarine would be reconnaissance.) Aube had been in office only a year and a half when the ministry fell and, departing, he left behind him a mess: a naval confusion instead of a naval programme.

Now the European scene was altering. In 1890 the aggressive young Emperor William II, who had come to the throne of Germany, forced the resignation of Bismarck; and the system of alliances which he had formed to keep France isolated fell with him. For his successor could not attempt the diplomatic juggling which had kept Russia and Austria in the same net. France saw a chance to escape from her isolation. Her mixture of a fleet paid a formal visit to Kronstadt in 1891, and an astonished world saw the Czar, autocrat of autocrats, stand to attention while the Marseillaise, hymn of revolution, was played. A return visit was followed by a positive Franco-Russian alliance in 1893, and, with Germany later embarking on a naval programme, it was time for France to put her naval affairs in order. So in 1896, Lockroy, the naval minister, created the École Supérieure de Marine, and under the guidance of this advisory body naval experiments and ideas were steadied, and France began to make up the leeway occasioned by confused policies. By 1900 the idea of the *guerre de course* had been abandoned. France decided on building powerful ships instead of many small ones, and at last, in 1901, a definitive class of battleship, the six PATRIES of 15 000 tons, was laid down. Their main armament was the four 12-inch guns that were the standard practice for capital ships of the day, and they could steam over eighteen knots. The ultimate programme, to be achieved by 1907, was for 28 battleships, 20 armoured cruisers, 52 destroyers, and a mass of smaller vessels, with an establishment of 3 000 officers and 112 000 men. Among the small ships were to be 38 submarines. For experiments with these craft had continued. The GUSTAVE ZÉDÉ, a larger GYMNOTE, was launched in 1893 and proved good, save that, relying entirely on batteries, its endurance was very short. Then in 1898–9 an officer named Laubeuf had produced the NARVAL which could make eleven knots surfaced, steam-driven, and stay submerged, electrically driven, at five knots for fourteen hours. This, with its double hull, was a small version of the modern submarine – save for its steam drive, which meant a twelve-minute gap between deciding to dive and being able to do so. When Laubeuf followed this with the AIGRETTE, diesel driven, the modern submarine had arrived, and France had the lead in these craft. It was this that had caused the large number of submarines to be put into the 1901 programme, but in fact, with a change of ministry in 1902, the submarine programme was slowed.

The United Kingdom had hitherto maintained a policy of 'splendid isolation' to avoid entangling alliances. But Kaiser Wilhelm II had flamboyantly proclaimed his country's big naval programme which, since Germany had no colonial need for a large navy, could only mean an intention to seize colonies from Britain or France; and Britain had also realized, from the enmity shown to her during the Boer War (1899–1902), that isolation could be dangerous. So the two Channel powers, though at loggerheads in several parts of the world, resolved their differences by a Cordial Understanding, which was next door to a positive alliance. Russia was joined to this in 1906 and Europe was then divided into that state of two armed camps – Triple Alliance balanced by Triple Entente – which was inevitably to lead, after the crises of Algeçiras (1905), Bosnian Annexation (1908), and Agadir (1911), to the cataclysm of 1914.

With Britain no longer the enemy a certain lassitude crept over French naval developments. A new programme was announced in 1906, however. The strength was to be raised to thirty-four battleships, eighteen each of heavy and light cruisers, with a large number of torpedo craft and submarines. The six Danton class battleships were laid down. Their main armament of four 12-inch guns was supported by a secondary armament of twelve 9.4-inch, and they steamed not less than nineteen knots.

Since British commerce was no longer to be attacked France even gave up the lead that she had obtained in submarines, and her numbers of them actually dropped slightly in each year up to 1914. Those which she possessed appeared to be good by their performance in special manoeuvres for them in 1909, though later, in war conditions, they were to show rather too many technical faults. In short, the Entente Cordiale having been reached, France tended to leave naval matters to Britain. But in 1907–8 there came a further naval revolution for the British had produced the *Dreadnought*, a battleship whose combination of speed and heavy armament made her so superior to all existing battleships as to render them obsolescent. This brought a revival of interest in French naval building, and in 1910 Admiral Boué de Lapeyrère, then minister of marine, carried a programme for building sixteen *Dreadnoughts* or super-*Dreadnoughts* by 1919. (Only three of them were available by 1914.) This meant that France could not afford to maintain her building rate of destroyers, and in overall strength the French had dropped from second to fifth place among the navies of the world. Indeed, back in 1898 France had had to suffer the humiliation of the Fashoda Incident. A brave little body of colonial troops had marched almost right across Africa to occupy Fashoda, in the south of the Sudan, which Anglo-Egyptian forces were reconquering from the Mahdists. Britain demanded their withdrawal; and France had to accede. For if her refusal had led to war, her naval weakness would have left her colonial empire utterly exposed.

During the pre-war years of the twentieth century explosives developed more rapidly than the understanding of how to care for them. Most navies suffered accidents from explosions on board. The French, using a nitrocellulose smokeless powder which deteriorated more rapidly than they realized, had more than the average share of these; and two of them were very serious. In 1907 the old battleship IÉNA, built in 1898 and carrying the pre-*Dreadnought* standard four 12-inch guns, had a series of explosions, rapidly following one another, while in dry-dock at Toulon. Each explosion was more serious than the last and flung fragments that caused loss of life some distance away in the port, as well as of those on board. The LIBERTÉ had to shoot in the dock gates to flood the flaming ship, whose admiral was the last to leave by climbing over an already fire-swept stern gallery. The LIBERTÉ herself suffered much the same disaster in 1911. Completed in 1905, she was lying in Toulon harbour when there was a succession of explosions, culminating in that of the main magazine. This literally rent her asunder, and huge fragments were hurled against two other battleships, doing them serious damage. In both these cases deteriorated ammunition was the cause. The number of things that had gone wrong in the navy had brought an attack on its administration by Delcassé, and the alarm caused by the destruction of LIBERTÉ enabled him, when he became minister, to see that the '*Dreadnought*' programme of 1910 was carried out with vigour.

Considering the period covered by this chapter as a whole, it is clear that the years from 1878 to 1900 were years of great confusion for the navy. To use the apt words of M Le Masson, there was 'an anarchy of ideas'. The frequent changes of minister, the violent journalistic polemics of the Jeune École, and the unreadiness of the Assembly to vote the necessary credits all contributed. (When France at last got some uniformity of design with the Charlemagne class of battleship in 1894–5 the Assembly insisted on reducing the original tonnage, with unhappy results.) By 1900, however, it was clear that the danger had shifted from Britain to the Triple Alliance. There was no question of an attack on its commerce, but rather of meeting an attack by big ships on French ports. Even so, Lanessan's eight-year programme of 1900, while providing for battleships and battle cruisers, retained the little torpedo boats of the Jeune École and made inadequate provision for destroyers and none at all for light cruisers. Still, from about 1900, though progress was slow at first, things steadily improved, and confusion increasingly disappeared. Nevertheless there remained some shortage of ships able to stand Atlantic weather at high speeds, and the effective range of French guns was somewhat lower than that in other navies.

The Entente Cordiale had involved no specified fleet commitments on either side; but after the Agadir crisis of 1911 Britain and France entered on 'naval conversations'. The gist of the result of these was that as Britain was able to keep ahead of Germany in the naval race, she would look after the North Sea, the Channel, and the Atlantic coasts, while France concentrated in the Mediterranean.

This was the position when, in the summer of 1914, the Austrian Archduke Franz Ferdinand was killed by Serbian assassins. The excessive nature of the demands made of Serbia by Austria-Hungary, the support of Serbia by Russia and of Austria by Germany, plus the terms of alliance between France and Russia, led with grim inevitability to the long-threatened general European war at the start of August. For a couple of days there was lively alarm in France lest Britain, having signed no definite alliance, should not honour the 'Entente'; but the violation of Belgian neutrality to carry out Germany's military plan put that out of doubt and, with other countries being later drawn in, much of the world went to war.

CHAPTER XV

# Two World Wars

When war was declared in 1914 the French navy had four *Dreadnoughts*, COURBET class, with twelve 12-inch guns and a speed of at least twenty-one knots, and there were super-*Dreadnoughts* building. It had no battle cruisers. There were eleven other battleships (DANTONS, DÉMOCRATIES, and RÉPUBLIQUES), all built in the twentieth century with the usual pre-*Dreadnought* four 12-inch as their main armament. To them could be added nine older battleships, built at various dates between 1891 and 1899: these included CHARLEMAGNE (1895), BOUVET (1896), and SUFFREN (1899). Armoured cruisers numbered fourteen, carrying 7.6- and 6.4-inch guns; but none was designed later than 1905. Their rate of building had been leisurely; two of the fourteen 7.6-inch gunners of 1905 were not completed till 1910, and the AMIRAL AUBE, designed and ordered in 1899, had not been ready till 1904. The more modern of these steamed twenty-three knots, and there were some older cruisers (some unarmoured) of various dates in the 1890s. The destroyer force, including those building, was over eighty on paper; but more than half of these dated from 1907 or earlier, frail vessels of weak armament unsuited for any strenuous work. There were many torpedo boats, but only seventeen of these were ocean-going, the rest being 'coastal'. Submarines numbered over seventy, but fifty-two of these had been designed in 1907 or earlier. There were also nine mine-sweepers. Commanding the main fleet, with his flag in COURBET, was Boué de Lapeyrère (VA), who had been one of Courbet's junior officers in the Far East.

To accommodate the ships there were large and well-equipped naval bases at Brest, Cherbourg, and Toulon, smaller ones at Lorient and Bizerta Rochefort dealt only with submarines), and lightly fortified oversea bases at Ajaccio, Algiers, and Oran.[1]

Formidable though the force seems in numbers, it will be seen that, till its designed and ordered super-*Dreadnoughts* were completed, the capital ships and cruisers were mostly outdated for a modern fleet action, the fleet destroyer force was too small, and the effective submarines were too few. In fact, though the war was to see several naval battles, these were all between British and Germans, and France had no part in them. The fighting of its fleet consisted mainly of the bombardments of the Dardanelles campaign, though the navy did much other busy work in the Mediterranean. Here the main force was always concentrated, though some armoured cruisers and destroyers operated from Brest and Cherbourg, helping to patrol the Western Approaches and the Channel. The main fleet in the

[1]There were also docking facilities at Saigon.

Mediterranean saw to the transport of the troops from Africa which played a vital part in the battle of the Marne. At the start of the war the effective power of the submarine was a matter of speculation, and the general naval opinion was that it would not be great. In September, however, a sharp lesson was taught when three unescorted British armoured cruisers in the North Sea were picked off, one after the other, as sitting targets. The British and Germans assimilated this lesson quickly: the submarine was a dangerous assailant, and to avoid it ships must steam fast on a zig-zag course and be escorted by vessels of light draught, hard to hit, and quick on the helm to turn on a sighted periscope or run down a torpedo track. The French were surprisingly slow to learn that lesson. The battleship JEAN BART was unscreened when hit in the Adriatic in December 1914, the armoured cruiser LÉON GAMBETTA was patrolling at a leisurely seven knots in the very Straits of Otranto when torpedoed in April of the next year, and the veteran battleship SUFFREN was pursuing a placid course at nine knots, without escort, when she was sunk with all hands as late as November 1916.

In the opening months of the war (autumn 1914) there was much escorting of troopships. MONTCALM, the cruiser flagship of the small Pacific squadron, joined with the Australians in this work in Eastern waters, where the Germans had a cruiser squadron at sea under von Spee. This was dissuaded from pushing home an attack on Papeete by the admirable bluffing defence of a junior French naval officer with a handful of men; but one of its detached ships, *Emden*, sank the destroyer, MOUSQUET, which bravely engaged *Emden* when that cruiser attacked Penang. In European waters, some of the Channel destroyers helped more than once to delay the German advance along the Belgian coast by shelling its flank; and Boué de Lapeyrère took the main fleet into the Adriatic. Here it sank a small Austrian cruiser, but had the JEAN BART hit, and lost the submarine, CURIE, which was captured in the nets, trying to get into Pola.

The comparatively unimportant Mediterranean events of the closing months of 1914 were succeeded in 1915 by the great Dardanelles campaign. Strategically the idea of forcing the Dardanelles to attack Constantinople was one that, if successful, might have had great effect. Tactically its execution came far too near to being a hasty improvisation. The possibility of forcing these straits had been long discussed. It was admittedly a difficult and dangerous business: opinions were divided as to whether it could be done and, among those who thought it feasible, were divided again as to whether it could be done by ships alone or whether troop landings would also be necessary. The straits were thirty-six miles long, were nowhere more than a few miles wide, and at Chanak narrowed to less than a mile. They were strongly fortified by a number of batteries and, as they were so narrow, even the old guns with which some of the batteries were armed would be effective. A current up to five knots flowed out of them, which made it easy to float mines down them, and difficult for the several lines of fixed mines to be swept. When, early in the war, the German *Goeben* and *Breslau* had escaped up these straits, some older French battleships, under Guépratte (CA), had joined in blockading at the exit, and had done some heavy shelling of the forts at their mouth. From the sea this appeared to have been successful: in fact it had been only partially effective. Nevertheless it was decided to force the straits and, at first, to do this by ships alone. The decision, and indeed the whole idea, was primarily a British one. The French fell in with it despite misgivings; but there was never anything even approaching adequate inter-staff consultation.

The British proposed to use twelve capital ships. Augagneur, the minister of marine, agreed to supply four old battleships, a flotilla of small destroyers, and a flotilla of sweepers. The French ships were:

SUFFREN(flag): P F A H Guépratte (CA) (4—12 inch, 10—6.4 inch)
BOUVET Rageot de la Touche (2—12 inch, 2—10.8 inch, 8—5.5 inch)
GAULOIS Biard (4—12 inch, 10—5.5 inch)
CHARLEMAGNE La Grésille (4—12 inch, 10—5.5 inch)

and Guépratte regarded the operation as a diversion that could be broken off, though he took it none the less seriously. In February 1915, the outer forts around the entrance were bombarded and silenced with much allied expenditure of shells. So far so good. The bombardments of forts further up the Dardanelles to the Chanak narrows continued for the first two weeks of March, though with more failure than success as far as silencing the batteries was concerned, and with almost complete failure in dealing with the several well-laid minefields; for, with the current against them and being under fire, the sweepers had an impossible task. Nevertheless, in mid March the British Vice-admiral de Robeck, now in overall command, made the main attempt to force the narrows, as ordered from Whitehall. The four old French battleships flanked the British attack, SUFFREN and BOUVET off the Asiatic shore, GAULOIS and CHARLEMAGNE on the Gallipoli side. The failure was complete. GAULOIS had to creep out of range with grave damage forward, SUFFREN was holed, and BOUVET, coming out of action, struck a mine. The explosion of a magazine quickly followed (there was a fault in construction), and she capsized in a mere two minutes with dreadful loss of life. As the British had lost an old battleship and a battle cruiser, and had had another battle cruiser badly damaged, the operation had cost six of the sixteen capital ships engaged, three sunk (one French) and three others (two French) out of action. (GAULOIS had to be beached before she could be patched up sufficiently to make for Malta.) The minefields had been the main difficulty.

It was now decided to make the operation a military one, as it should have been, though with much more thorough planning, from the beginning. The business of the ships was to cover the landings. Guépratte had been sent the old battleship JAURÉGUIBERRY (in which he hoisted his flag), the coastal defender HENRI IV, the cruisers LA TOUCHE-TRÉVILLE and JEANNE D'ARC, the auxillary cruiser SAVOIE, and the British battleship *Prince George*, which was added to his command. His business was to cover a landing of French troops at Kum Kalé, the western tip of the Asiatic shore, where they were to make an extensive feint or diversionary attack. Though the landing was made in some confusion owing to lack of rehearsal, both it and the subsequent withdrawal were admirably covered by naval fire. The Dardanelles campaign thereafter became a military affair, the ships just giving covering or supporting fire as and when asked, until the troops were evacuated in January 1916.

During this time the Turks had twice made attacks on the Suez Canal, which French ships helped the British to repel. On the first occasion the old coastal defender REQUIN (operating in Lake Timsah with her two 10.8-inch guns) and the cruiser D'ENTRECASTEAUX took part. On the second occasion, as most of their ships were committed at Gallipoli, the British asked for French assistance, and the cruisers D'ESTRÉES, JEANNE D'ARC, and d'ENTRECASTEAUX were sent, while the REQUIN and also the old cruisers DESAIX and MONTCALM, on their way home from the Pacific, operated in the canal. On both occasions they were successful. There were also operations off the Syrian coast against the Turks. Finally, as the attitude of King Constantine of Greece was distinctly equivocal, Dartige du Fournet (VA), who had taken over command in the Mediterranean from Lapeyrère, was sent to coerce him in 1916. This was interference with the rights of neutrals, and the French admiral disliked his task intensely. He managed it by firing a few shells from the battleship MIRABEAU at the royal palace in order to extricate some landing parties which had got into trouble. Constantine came more or less to heel, and abdicated a little later; but Dartige du Fournet came in for some reproach for humanely refusing to bombard Athens.

Otherwise, after the Dardanelles campaign had ended, the French navy in the Mediterranean was concerned with the humdrum business of patrol and escort, with submarine attacks on the enemy, and in coping with the enemy's underwater craft. It cannot be said to have handled these problems particularly well, through employing many auxiliary craft. The situation was not easy. Enemy submarines had only to slide out of the Adriatic or the Dardanelles to attack the allies in almost the whole length and width of the Mediterranean, whereas the allies' submarines had the dangerous business of going up the straits into the Sea of Marmora, or up the Adriatic, to get at the foe. Moreover, questions of prestige caused the Mediterranean to be divided into zones, each under a different power, as agreed at a Paris Naval Conference in 1915. France had four of these zones. One was between lines drawn due south from Cartagena and from Sardinia; another lay round Tunisia; a third was within a line drawn from the Straits of Otranto almost to the southern point of Sicily, thence east to the western end of Crete and finally north to the Piraeus; the fourth was the Levant – the eastern end of the Mediterranean. Britain and Italy had other zones. It was a bad arrangement, but the French rejected the idea of a unified command till late in the war. They were sure that enemy submarines lurked in obscure creeks, and therefore handicapped their patrolling and escorting by using some of their not over-large supply of small ships to seek for them. It was an odd obsession for Pola and Constantinople were the obvious bases. Like the British Admiralty, too, the men at the Ministry of Marine disliked the idea of convoy, and it had to be forced on them in the grim situation of 1917. As a result of all this, losses in the Mediterranean were heavier than they need have been. These included three old cruisers, three auxiliary cruisers, a troopship, and the pre-*Dreadnoughts*

GAULOIS and DANTON. In the last named the abandon ship arrangements did not work well and, though she took forty-five minutes to sink, just over eight hundred men went down with her. As against this, when the trooper AMIRAL MAGON was torpedoed and sank quickly, the destroyers ARC and BOMBARDE saved eight hundred of the nine hundred aboard – a fine piece of work.[1]

Nor were French submarines successful, for several showed unsuspected faults of design and build. Twelve of them were lost during the war, seven in Mediterranean or adjacent waters. Two of these were captured, CURIE in the Pola nets[2] and TURQUOISE, who regrettably failed to destroy her papers, by going aground in the sea of Marmora. Four enemy sinkings were purely French, UC38 being depth-charged after it had torpedoed the old cruiser CHÂTEAU-RENAULT (which, with other French ships, had been supporting Allenby's advance in Palestine); UC35 was sent down by the good gunnery of the little patrol vessel AILLY, an Austrian submarine by the destroyer BISSON and UC24 by the submarine CIRCE. Yet the final verdict on the Mediterranean situation must be that it was not well handled. Enemy submarine losses there, and in the Adriatic and Turkish waters, were only eleven, compared with 157 elsewhere. This was not good, even though enemy submarines were fewer in the Mediterranean than in the Atlantic.

When, under the terms of the armistice of November 1918, the German fleet steamed to Scapa Flow, the French navy was rightly represented at this surrender. The war had cost the French service a number of ships and, though these were mostly old and 'expendable', it had also cost her 11 500 officers and men.[3]

Nor, with the German surrender, could the war be considered as quite finished; for there was the problem created by the Russian revolution to deal with. The allies were backing the White Russians, who were trying to overthrow the Bolshevik Government, and they were backing a losing cause. The old cruiser AMIRAL AUBE went to Murmansk, where there were valuable stores to be protected or withdrawn, and both Britain and France sent naval forces to the Black Sea to support the White Russian armies in the south. The French were originally based on Odessa and Sevastopol, but withdrew to sea because of some mutinies among their war-weary men. They were on the spot in 1920, however, when General Denikin's advance collapsed, and with it the White Russian cause; and French ships played a large part in the doleful business of evacuating those who could be got away. It was no fault of theirs that the whole situation had been misjudged.

[1]The destroyer BOUTEFEU hit a mine off Brindisi, going to the aid of the torpedoed *Dartmouth*.
[2]Her captain was a Dupetit-Thouars: a fourth generation in the service.
[3]Four battleships, four armoured and one unarmoured cruisers, sixteen destroyers, sixteen submarines, and some lesser craft and auxiliaries. Six of the destroyers and some of the submarines and lesser craft were lost, not by enemy action but by collision, grounding, etc. A number of the casualties occurred in the Naval Brigade which served on the Western Front under Ronarch (CA), and in the battle of Ypres lost half of its 6 000 men.

The navy emerged from the First World War rather disgruntled. Few in France were interested in it. The tireless and trying work of escorting vital supplies went unheeded. It was the army that had held the trenches and had brought the final victory. The sea battles that made headlines had all been fought by the British: the French had not even suffered a defeat that might have brought a surge of patriotic interest, such as had followed disaster in the days of Choiseul and Napoleon. Also the navy knew in its heart that it had not done really well. The big ships had been good in the Dardanelles, the cruisers had played their part in the Canal Zone and off Syria, as had the destroyers off the Belgian coast;[1] but the patrolling and the submarine work had been disappointing. The Service fell into a state of somewhat resentful apathy. There was anger that it was a British admiral who had settled the armistice terms with Turkey (the Mediterranean being a French area of command), and that the British Admiral Beatty had complete power in dealing with the German fleet. So the old antagonism to Britain appeared. It was in this mood that France attended the naval conferences of the 1920s.

The first of the naval conferences met at Washington in 1921–2 and was an earnest attempt to make a firm beginning to the problem of disarmament. Aristide Briand represented France, with Admiral de Bon as his adviser. A tonnage ratio for capital ships was agreed which gave each of the UK and the USA a proportion of five to three as against Japan, and for herself France accepted a lower tonnage in a ratio to the others of about one and a half. Though this left her existing tonnage to her and she did not then want to build more for reasons of financial stringency, she was offended at this low ratio. That was probably a matter of prestige, for it put her in much the same position as Italy which had fewer colonies; and she turned 'difficult' with regard to other proposals. America and Britain wished to outlaw submarine warfare. France refused to consider this, perhaps a little from pique, certainly more because of the writings of some of her younger school of officers, who had learned the lesson of the recent war and urged how valuable a weapon was the submarine to the weaker of two naval powers. She also refused to accept a restriction on the number of cruisers and destroyers, though she did accept a limitation to their size and armament. (In fact, it was with cruisers that France started her post-war building, laying down three DUGUAY-TROUINS of eight 6.1-inch guns in 1922, and two TOURVILLES of eight 8-inch in 1924.)

The next naval conference was promoted at Geneva by the League of Nations in 1927. It broke down because of disagreement over cruisers between Britain and Japan on the one side and the United States on the other. France had played little part in this beyond showing opposition to British proposals during the preparatory work, and her own compromise proposal, next year, was rejected by the USA. A London Conference in 1930 'failed because of the intransigence of France and Italy' – Mussolini had begun to shout for Corsica and Nice and wanted the means to take them, while France was determined to have the means to defend them. This really ended any question of agreement on naval armaments, and a

[1] Where, on one occasion, a French admiral flew his flag in a British monitor.

mention of the London Treaty at the next conference, in 1935, is almost superfluous. Japan withdrew, and the treaty signed by America, Britain, and France contained so many escape clauses as to be worthless.

In 1926 Georges Leygues had become minister of marine, and though there were the usual changes of government they were largely reshufflings of the same people, so that he remained minister till he died in 1933. This meant some continuity of policy, and by the time of his death that continuity was more or less assured. For, with what proved wise nepotism, he had advanced his naval godson, Darlan, who became a flag-officer in 1929. This vigorous little man made such an impression on his service, and so quickly, that, though he was not its most senior admiral until, in 1938, he created a special fleet-admiral's rank to make himself so, he was its acknowledged leader early in the 1930s. During that decade he lifted it out of the apathy that had beset it, restored its confidence and pride in itself, and brought back a unity and enthusiasm that had long been absent. Moreover, between the wars the navy acquired what it had long lacked – a naval staff. This kept clear of party politics, prepared for the future so that the Service knew what was expected of it, and gathered its officers and men round it so that they became almost a separate caste in the country. Admiral Durand-Viel, Chief of Naval Staff from 1932 to 1937, did much to prepare the way for the unifying of the Service that Darlan completed.

To the existing *École de guerre navale* was added the *École des hautes études navales* working on the lines of thought of a clever strategical theorist, Admiral Castex. They provided the equivalent of a staff college to round off the revived enthusiasm with an intelligent efficiency. Naval limitations having gone by the board in the failure of the conferences, a building programme was embarked on, from 1932 onwards, to give France six capital ships, six new cruisers, and a number more destroyers and submarines. This was, first, to meet the threat imposed by Germany's cleverly designed pocket battleships, and, later, in 1935, to meet the further threat of bigger battleships imposed by the Italian and German renunciation of naval treaty agreements. For France was rightly anxious to ensure her communications with her empire.

With the weak diplomatic shifts that failed to arrest Nazi and Fascist aggression in the next few years we are not here concerned; but when the humiliating collapse of Anglo-French diplomacy took place at Munich in 1938 and showed that war could not long be delayed, that programme was reasonably on its way, even though it was never to be completed. Moreover, the strained situation at the time of Munich had enabled mobilization measures to be tested. The navy was ready.

When Hitler invaded Poland, and Britain and France declared war on Germany in September 1939, the French fleet was a formidable force. There were two first-class capital ships, DUNKERQUE and STRASBOURG, with eight 13-inch guns as their main armaments, two others, JEAN BART and RICHELIEU (eight 15-inch), nearing completion, three battleships built in 1913–15 and since modernized (PROVENCE, BRETAGNE, and LORRAINE with eight or ten 13.4-inch guns), and two older – COURBET and PARIS – with twelve-inch main armament. There were nineteen cruisers, including one minelayer. Six of these, the LA GALISSONIÈRE class built in the 1930s and one of 10 000 tons, were armoured, carried nine 6-inch guns, and steamed over thirty-five knots. Of the sixty-six destroyers, mostly built in the 1930s, thirty-two sometimes described as 'supers' were all but light cruisers, as they carried five, six, or, in two cases, eight 5.5-inch guns and could steam even a little more than the thirty-six or thirty-eight knots for which they were designed. There were thirty-five sloops ('avisos'), some minesweepers and patrol vessels, and there were from seventy to eighty submarines. These included the giant SURCOUF, an extraordinary vessel which carried two 8-inch guns and a small aeroplane; all the submarines were fit for service. There was one aircraft carrier, BÉARN, a battleship converted for thirty-six to forty planes, and a seaplane tender to her, the COMMANDANT TESTE. There were also, of course, auxiliary ships – converted merchantmen and trawlers.

To man all these there were 91 000 regulars and 61 000 reservists. At the head of this Service, and most completely in command of it, was Fleet-admiral François Xavier Darlan.[1] Born at Nérac in 1881, he had shown capacity as a gunnery lieutenant in the first war, and, as has been said, he had also the good fortune to have political pull through Georges Leygues. Though he was every bit as much a political sailor as any British borough-holding admiral of the eighteenth century, there is no doubting his capacity in his Service or his devotion to it. He had made the navy very much a *corps d'élite*; but he had not yet made good some of its weaknesses. One of these was naval aviation, and another was shortage of anti-aircraft armament aboard ships. France was also much behind Britain in anti-submarine devices and equipment. Moreover, though the Service was keen and perhaps more devoted and better disciplined than ever before, it is permissible to wonder whether all faults had been eradicated. In the fighting, especially that which took place after the 1940 armistice, it was mostly the naval-manned land batteries that were effective. Against this it must be remembered that on all fighting occasions the ships, whether surface or submarine, were operating under difficulty, usually against odds, and that, after the armistice, shortage of oil fuel drove France back into her position of sailing-ship days – only exercise in harbour could be given. The sailors certainly achieved heroic marvels in getting their ships away from the grip of the Germans when the army collapsed.

Before this happened were the months of what became known as 'the phoney war' which saw the annihilation of Poland and Russia's dastardly attack on Finland, but when nothing much happened in the West apart from German submarine activity in the Atlantic. This meant much convoying, in which France joined vigorously with the British: the destroyer SIMOUN sank U54 and VALMY played a part in the sinking of U55. Then, in April 1940, the quiescent war in the West suddenly exploded with the Nazi attack on Norway. French troops were sent to join the British in

[1]Darlan had caused this rank to be created, finding that, as vice-admiral (the highest rank in the French service), formal protocol placed him junior to full admirals of minor countries at official gatherings. He also rightly wished to rank equally with high commands in the army.

the attempt to recover Trondheim and they were supported by six auxilliary cruisers, notably well handled by Jean Cadart (CA), and by the mine-laying cruiser EMILE BERTIN with a number of destroyers and small craft under Derrien (CA). The operation at Trondheim failed, and the evacuation of the troops from Namsos under air attack was a desperate business. BERTIN was damaged and the large destroyer BISON sunk by bombing; and when the ships reached Greenock, from which they were operating, another large destroyer, MAILLÉ-BRÉZÉ, blew up owing to a faulty torpedo.

Germany now unleashed her main assault in the West by the invasion of Holland and Belgium and, though little could be done to help the former, allied armies moved at once into Belgium. Unfortunately a grave error of judgment by the French High Command had left the Ardennes area weakly held, since it was considered impracticable for tank warfare, and it was here that the main German blow was struck. It had complete success: German tanks were soon racing across northern France, almost unopposed, until they reached the coast near Boulogne. The allied forces were split in two and the armies in Belgium were cut off.[1]

If they were to be maintained, or in any way extricated, the one remaining port, Dunkirk, must be held. There was little between it and the German onrush from the south along the coast. But the navy made a brave if short defence of Boulogne against tanks and aircraft, using the fire of destroyers close inshore to hold back the German armour. In this ORAGE and CHACAL were sunk by bombing. Then the British made a last-ditch stand in Calais. Thanks to these two actions, Abrial (VA) who commanded at Dunkirk was able to make some sort of southern defensive line along the little canalized river Aa, and it was the navy's 155-mm (6-inch) guns that stopped the German tanks at Gravelines on May 24th. The guns were partly manned by the crew of the destroyer L'ADROIT, which had been sunk by bombing off Dunkirk. There were also in the fighting line the crews of other small naval craft, and of the large destroyer JAGUAR, which had been sent to the bottom. The port had then been under air attack for seven days, and it soon became clear that to hold Dunkirk would be impossible: it could not be supplied under bombing, and the British, somewhat to French resentment, were already gathering small craft to evacuate their troops. France could only follow suit.

The 'Miracle of Dunkirk' – the evacuation of nearly 340 000 men (129 000 of them French) under fire and by channels subject to air attack – perhaps the most astounding emergency operation ever carried out, though it was mainly a British affair was by no means without French participation. France mustered over a quarter of the number of ships mustered by her ally, and these took off over 48 000 men, nearly all to England, though about 4 000 went to Cherbourg and Le Havre. Her naval losses in the operation were two large and five other destroyers, and thirty armed trawlers: to these must be added more than twenty commandeered mer-

[1]While any troops were still fighting in Walcheren, S Beveland, and on the coast north of Ostend, the guns of French and British small craft gave them admirable support, and these craft then carried out successful evacuations of them from Ostend and Zeebrugge.

chant and private vessels. She had used nine destroyers, six torpedo boats (whose quickness on the helm made them most successful), four sloops, and a number of naval motor launches, and she had blocked the harbour when the last boats left on the night of June 3rd.[1]

Hitler and his tanks gave France no breather, and on June 5th opened their attack on the line that the veteran General Weygand, recalled to command, was trying to form. His task was hopeless. The army was a beaten force: the government left Paris and went via Tours to Bordeaux: all was confusion; and all was collapse. Well before the end of the month Paul Reynaud and his government had resigned in favour of the aged Marshal Pétain, and an armistice was being sought. It was concluded on June 24th.

In these desperate last days the navy had not been idle. Italy had declared shameless war on the beaten France, for Mussolini wanted a claim to grasp some of the spoils. Four of the eight-inch cruisers, under Duplat (CA), promptly went from Toulon to bombard oil installations in the gulf of Genoa, with no more serious reply than a single hit on one of the destroyer escort; and a sloop rammed an Italian submarine off Cape Palos. Off the Channel coast, mine-layers braved the bombing to lay their 'eggs' so that the German coastal advance could have no shipping support, and off the Cotentin the old battleship COURBET shelled the German advance on the road from Carentan till everything in Cherbourg had left or was destroyed. For, of course, the great need was to get the ships away before they fell into German hands, and to wreck the port installations. The absence of any clear orders in the ruinous confusion made this no easy task and threw great responsibility on to the flag-officers in command.

In Brest, de Laborde, a fiery but popular martinet, had been told that troops were retiring on Brittany where they would occupy a line across the approaches to that province to hold it for the allies. Then he got a telephone message that there was no line and that German tanks were entering the province and would reach Brest in about ten hours! In the great harbour and its roadstead were over 150 ships, merchantmen and men-of-war, the latter including the fine new battleship RICHELIEU, just completed. Laborde had a mere ten hours in which to get all these to sea or scuttle them, and to destroy the port installations. He got all the seventy odd warships away, except for four submarines and the destroyer CYCLONE, badly damaged at Dunkirk, and these were scuttled. The merchantmen were also got away, and much damage was done to docks before the Germans arrived.[2] The small craft in Lorient got clear or sank themselves and, greatest triumph of all, the unfinished JEAN BART, sister ship of the RICHELIEU, was got out of St Nazaire. The dockyard scaffolding was still aboard, her engines were not complete, she had neither guns nor fuel, and was in a basin sealed off by an earth dyke. Her captain, Ronar'ch,

[1]Platon (CA), who had done admirably at Walcheren, then, under Abrial, kept things going in Dunkirk, despite the bombing. Later he became a loyal supporter of Pétain and Laval, and was shot out of hand by the maquis as a collaborator.

[2]SURCOUF, with her machinery dismantled to receive new diesel cylinders, put back her old ones and scraped out to sea with them at the last minute. The gold reserve of about a thousand tons had gone two days before.

FLEET-ADMIRAL DARLAN

BATTLESHIP 'LORRAINE' In 1937 Launched 1913, extensively modernized 1935. Saw much service in later part of Second World War

THE 'SURCOUF' The 'Giant' submarine

had realized the position just in time to get sufficient of her engines and boilers completed for her to steam, to get fuel and some anti-aircraft guns aboard, and to get the dyke sufficiently dredged for her to be nosed out by tugs on the last of the high tide. Her engines had not even been turned and steam was still being raised. She had not a properly swung compass, and her main engines were only just able to take over from the tugs when the first of the German planes arrived as she followed de Laborde, with his flag in the destroyer HARDI, out to sea. There can be glorious achievements even in disaster.

At Bordeaux, where the shipping was mostly merchant manned and many of the men, with wives and children in mind, would not obey orders, things went less well – though the destroyer LANSQUENET, under fire from German batteries, got out of the Gironde. The escaped naval ships made, some for British ports, some for French African harbours, some for Toulon. When it is realized that no plans whatever had been made in case of such a military collapse, that the collapse was so sudden, the confusion so great, and the general atmosphere so despairing, one cannot but feel that the French navy, at least, had done well.

The government which signed the armistice on June 24th was that of the veteran Marshal Pétain. Of absolute personal probity and with a deservedly great reputation from the first war, the aged marshal had promptly put aside any question of getting all men possible across to Africa and continuing the war from the empire. His duty, he felt, was to Metropolitan France, and to save its people as far as possible from the horrors and miseries of occupation, horrors and miseries that would certainly be intensified in France itself if her forces continued fighting from overseas. General de Gaulle, who had been a minister in the Reynaud government, had got away from Bordeaux to England at the eleventh hour, and from there had made his broadcast calling on all patriotic Frenchmen to rally to his standard and continue the war. But to the ordinary man in his misery, the words of a French soldier, still in hospital in Britain after Dunkirk, to the writer, 'C'est un homme inconnu', applied; and to officers and officials, especially the more senior, Pétain was the legitimate head of Government, and was to be trusted and obeyed. So, though de Gaulle immediately received a certain amount of support, the numbers of his Free French were for some time only small. Darlan, as he saw collapse rapidly approaching, drafted instructions to the fleet to go to African or British ports and so, at that stage, was clearly prepared to continue the war from the empire. But the orders to sail were never sent, for Pétain made the admiral his minister of marine, and the admiral, all-powerful where the navy was concerned, accepted the marshal's view that further resistance meant only misery for the French people. As an astute man he probably realized, too, that the intact French fleet would be an important bargaining pawn in making armistice terms, and that it would stand between an enemy occupying much of France and the occupation by him of any part of the French Empire.

Yet if the future of her fleet was important to France, it was equally so to Britain. From fighting a grim war alongside a powerful ally, the United

Kingdom now found herself alone, threatened with invasion by a much superior army supported by a hitherto highly successful air force, and the enemy now held the whole of the European Atlantic coastline from the North Cape to the Spanish frontier. Her very existence depended upon her superior sea power, and with Italy now in the war and Japan ominous, not merely her Mediterranean position but her actual naval superiority was threatened should the French fleet fall into Axis hands and be joined to those of Italy and Germany. By article No 8 of the humiliating armistice signed at Compiègne the Axis had solemnly declared that it had no intention of using or claiming the French fleet, which was to be immoblized except for such ships as were needed for the colonies. But the words of Hitler and Mussolini were not to be trusted for a moment. During the fortnight in which she was collapsing Churchill had paid visits to France and had twice been assured by Darlan that the fleet would never be allowed to pass into German hands; and Darlan had repeated that assurance to Admiral Sir Dudley Pound at Bordeaux on the eve of the armistice. Yet was Darlan to be trusted? He was so notably a political sailor, had never shown himself to be anything of an anglophil, and was now minister in a government, some of whose members were fulminating against their recent ally. ('Il faut déraciner les Anglais!') In any case he might be suddenly removed from power before he could give any orders to his ships. When France had asked to be released from her engagement to fight with Britain to the end, Churchill had stipulated that the fleet should either join the British or be scuttled; but the Council of Ministers who had accepted this was no longer the Council of Ministers that governed France. Churchill and his Cabinet decided that to put trust in any of these engagements was too great a gamble: the French fleet must be secured, or rendered harmless.

The dispositions of the French ships after the armistice were as follows: In British ports, mostly at Plymouth and Portsmouth, were the old battleships COURBET and PARIS, two large and eight other destroyers, six submarines (including the giant SURCOUF), some sloops and small craft. With the British fleet in Alexandria were the battleship LORRAINE, four cruisers, three destroyers, and three submarines. At Toulon were only four cruisers, thirteen destroyers, and a few submarines, for the bulk of the ships were at the naval base of Mers-el-Kebir just west of Oran. Here were the great DUNKERQUE and STRASBOURG with the older battleships BRETAGNE and PROVENCE, six large and five other destroyers, the seaplane tender COMMANDANT TESTE, and five submarines. At Algiers were six cruisers, three destroyers, and some submarines. The powerful RICHELIEU was at Dakar, and the powerless JEAN BART (her guns had not yet been smuggled to her) was at Casablanca. The carrier, BÉARN, was at Martinique. The French considered that the removal of all but a few unimportant ships to oversea ports placed them out of reach of the Axis. In view of the strength of the Italian fleet, Hitler's well-known treachery, and the weakness of the French Government, Britain could by no means be so sure.

To seize the unsuspecting ships in British ports was easy. They were suddenly, if somewhat treacherously, boarded and their crews bundled ashore: only on SURCOUF, where two British officers and a marine sergeant and one Frenchman were killed, was there any fighting.

Admiral Somerville was ordered to sail from Gibraltar for Oran with Force 'H', which consisted of the battle cruiser *Hood*, the battleships *Resolution* and *Valiant*, the carrier *Ark Royal*, two cruisers, and eleven destroyers, and on July 2nd 1940 he was given the text of a letter to Gensoul, the admiral commanding at Mers-el-Kebir. It was, in fact, an ultimatum. Opening with the statements that the French Government had declared on June 18th that the fleet would either join the British or be scuttled and that the words of Germany and Italy as given in the armistice were not to be trusted, it gave Gensoul three choices: he might bring his ships to join Britain in the fight against the Axis; he might send them with reduced crews to a British port under a solemn promise of repatriation of the crews and ultimate restoration of the ships; he might steam with reduced crews to Martinique under British supervision for the ships to be demilitarized or possibly put under the care of the United States. Otherwise he was required to sink his ships within six hours, failing which force would be used. Admiral Cunningham was sent similar orders with regard to the ships at Alexandria. Both admirals were appalled, but were told that orders were to be obeyed.[1] So on July 3rd, at about ten in the morning, Captain Holland, who had been naval attaché in Paris, entered Mers-el-Kebir harbour to deliver the letter and to attempt to negotiate with Gensoul. That admiral had been told of the assembly of ships off the port, had ordered his own ships to raise steam and reduce moorings, and refused to see Holland, exchanging messages of negotiation by way of his flag-lieutenant. The terms were, of course, such as no proud nation, especially one smarting under defeat, could accept – save that further negotiation over the conditions of sailing to the West Indies, combined with the withdrawal of any supervision, might, barely possibly, have avoided battle. But Gensoul, who was in touch with his government, seems not to have communicated this part of the proposals, or at least not to have stressed it, for the Pétain government spent little or no time discussing it. The acceptance of any of the conditions would have been a breach of the armistice by France, and, for his part, Somerville, with his ships lying exposed to air or submarine attack, could not delay indefinitely; and, besides, he had his positive orders. The situation was one in which it was tragically impossible to avoid an action that came close to being fratricide. Somerville having signalled the hour at which he would open fire, Gensoul consented to see Captain Holland at some time between three and four in the afternoon and told him that the promise not to let the ships fall into Axis hands absolutely held good, but that, if attacked now, he would fight. The sorrowful Holland left the harbour at 4.35, just on five o'clock Somerville opened fire, and PROVENCE replied at once.

The battle, if it can be called such, was brief. Gensoul had cleared for

[1]According to Kenneth Young (*Churchill and Beaverbrook*, published by Eyre & Spottiswoode), the French-loving Churchill was 'horrified' at giving the order – which he did at the urging of Sir Dudley Pound and Lord Beaverbrook. Churchill's memoirs do not mention this.

action, shortened moorings, and had ordered his ships to steam from harbour in line ahead on STRASBOURG, the destroyers to act separately. The British had got the range and their gunnery was good. BRETAGNE was heavily hit at once and blazed furiously, some magazines exploded, and in ten minutes she had capsized with the dreadful loss of over 950 men. DUNKERQUE, whose mooring position obliged her to get under way only slowly, was heavily hit by at least three shells and, with both forward and after engine rooms damaged, could only stagger to an anchorage in the shallows. PROVENCE got off some salvoes which fell short and was then hit in two turrets. She caught fire, had to flood magazines, and was obliged to beach. The large destroyer MOGADOR was in the entrance to the harbour engaging a British destroyer when she was struck by a shell that exploded her depth charges, and came to anchor as a barely floating wreck. STRASBOURG and three destroyers (which worked up to forty knots) got out of the harbour: the British then turned under smoke, and STRASBOURG avoided attacks by *Ark Royal*'s planes and got clear away to Toulon. An attempted submarine attack on the British got nowhere. The remaining destroyers and some small craft reached Algiers with the loss of a sloop. Three days later *Ark Royal* sent in her planes to attack the damaged DUNKERQUE. Alongside that ship was a small craft containing depth charges. She was hit and her explosion made a huge hole in the flagship and brought the total French losses to all but 1 300 men. Such was the 'battle' of Mers-el-Kebir.

At Dakar the aircraft carrier *Hermes* sent in planes which damaged RICHELIEU with an aerial torpedo; but at Algiers the cruisers had been ordered to steam for Toulon directly it was heard that the British were off Oran, and at Alexandria things went better. Admiral Cunningham, with the French ships safely among his own, was in an easier position than Somerville, and had no intention of obeying his orders if he could avoid it. Godfroy (VA) was equally unwilling to obey the orders from Vichy, which were quite hysterical and to the effect that he should fight and at least sink something: it would have meant a useless and heavy loss of life. Lengthy negotiations, with all keeping their heads, ended with Godfroy agreeing to reduce his crews (for repatriation) and his stocks of fuel and ammunition in return for a British undertaking to leave the French ships alone and to keep the remains of their crews supplied.

Could the affair, the fifteen-minute 'battle' or sinking, at Mers-el-Kebir have been avoided? Speaking with the advantage of hindsight the answer is 'Probably, yes'. There is some confusion over how much the British Admiralty knew of Darlan's signalled instructions to his fleets. Churchill did not know the stringent orders Darlan had written to his admirals, signed with a private code name, to ensure that ships should be sunk rather than ever fall into Axis hands. Had he seen these he might have called off the operation, or at least have removed the condition of British supervision if they went to the Antilles that the French found so insulting. That removal might just possibly have led to a French withdrawal to Martinique. Darlan almost certainly meant to keep his word never to yield the fleet, and, a little before his death, after the Toulon scuttling, wrote to Churchill that he had done this. Whether the Axis would have allowed him to do so had not the incident made the navy so anti-British is another matter. At the time, however, the whole disaster built itself up to its catastrophe with the inevitability of a Greek tragedy. In Britain it was genuinely regretted and there was much feeling of shame. Yet the shame was that of having had to hit a man when he was down, and the regret was largely that the French had not accepted the, to British minds, reasonable suggestion of going to the Caribbean: the necessity for the action was generally accepted. In France as a whole, though the Press fulminated and there was deep anger, that anger did not last. A dignified pronouncement by Pétain that there would be no attempt at retaliation turned the minds of the public back to the plight of the nation as a whole. But in the navy itself all the latent anglophobia was brought seething to the surface: the British were regarded as being almost as much the enemy as the Germans, and future adjustments when, with America's entry into the war, France began again to take the field, were made difficult. At least it had for France one small advantage – Hitler no longer demanded that the ships should be immobilized.

The ships of war seized in British ports were, in many cases, old and in poor condition and were little used. The 115 merchantmen were mostly valuable, in view of the heavy sinkings by German submarines. The newer of the small naval craft were put to good use, so were some of the submarines, among which the mine-layer RUBIS did excellent work. The two large modern destroyers, TRIOMPHANT and LÉOPARD, were later sent to the Pacific, to Noumea, where they subsequently worked with the Americans. (LÉOPARD returned to Europe in 1943 and ran aground off Cyrenaica.) There was a considerable manning problem. Hard-pressed in the situation created by the fall of France, and with enemy agents found among the many refugees who arrived, Britain was not over-considerate with the crews of the seized ships and kept them in conditions not very different from those of prisoners of war. They resented the Gaullist propaganda to which they were subjected and naturally wished to be home, to stand by their families. So the majority plumped for repatriation. Even so, over three thousand joined de Gaulle, and the numbers steadily increased. To command was Vice-admiral Muselier, really on the retired list, who had made his way to England when France fell. Both he and de Gaulle were 'difficult' men, and it was a pity that they did not get on well. In sum, while it was good to keep the ships that came to Britain out of German hands, and though its men, manning British corvettes, had some good success against submarines in patrol and convoy work, the Free French marine otherwise played no great part in the war at sea. Sixty of its vessels, including merchantmen, were lost, and its naval brigades did valiantly among the land forces.

After Mers-el-Kebir the 'legitimate' French fleet was mostly at Toulon; but seven destroyers and some submarines were kept at Oran, a light cruiser, nine destroyers, and a dozen submarines were with the immobile JEAN BART at Casablanca, and RICHELIEU with some smaller vessels and some submarines was at Dakar. There were, too, detachments, mostly of small craft, at Diego Suarez, Beirut, and Saigon. Their tasks were to keep France fed by getting supplies into the Mediterranean ports, and to prevent the Germans from seizing the empire. The officers had nearly all taken a personal oath of loyalty to Pétain, and so regarded de Gaulle as a traitor soldier: they distrusted the politics of those around him, and saw the Free French as doing only harm to the unity of the country and inviting retaliation by the Germans. The navy would live up to its motto, 'Valeur et discipline'. It had already shown valour in disaster: it would show discipline by obeying the legal government.

Nevertheless, in the late summer of 1940 the southern part of the African Empire, Congo, and huge adjacent territories went over to de Gaulle – more, perhaps, for economic than for purely patriotic reasons. The Free French therefore thought that they should improve this, their new central African position, by an occupation of Dakar, which, they believed, would readily come over if sufficient force were provided to overcome the resistance of Vichy loyalists. They would provide the landing forces and the agents to win over waverers, but Britain would have to provide the transports, with some powerful ships to deal with the RICHELIEU and the shore batteries if these should fight. In other words, it was to be a nominally Free French expedition which hoped that Free French forces would be sufficient; but British ships and troops were to be with them in case they needed help. Though it had been difficult to keep this expedition secret, it was purely coincidence that the Vichy government had sent the cruisers GLOIRE, MONTCALM, and GEORGES LEYGUES, with three large destroyers, under Bourragué (CA), to Dakar. His orders were to retake Pointe Noire, the port of the Congo. A delay in passing a signal at the British Admiralty gave them a good start and though GLOIRE, which had engine trouble, was overtaken by the cruiser *Australia* and obliged to turn into Casablanca, the rest reached their destination. The presence of British ships prevented them from carrying out their orders with regard to Pointe Noire, but they served greatly to stiffen the resistance at Dakar when de Gaulle arrived. The port was on the alert, for it was known that some sort of expedition was somewhere about. So when, on the foggy morning of September 23rd, de Gaulle sent in the Free French sloop SAVORGNAN DE BRAZZA and two other small vessels, they were fired upon. His agents, landing on the airfield or infiltrating into the town, were seized, an attempt by the SAVORGNAN DE BRAZZA to land some men a little outside the town was repulsed, and Admiral Sir John Cunningham's big ships had to take over. Lacroix (VA) who had been flown out to take over from Bourragué, had the RICHELIEU (which could not steam), two cruisers, three destroyers, three submarines, and some powerful coastal batteries with which to oppose two battleships (*Barham* and *Resolution*), the carrier *Ark Royal*, the cruisers *Cumberland* and *Australia* and a number of small craft. There were three days' fighting, during which RICHELIEU and the shore batteries were well handled, while the planes from *Ark Royal* were less well used than normally and suffered losses. On the first day the large destroyer AUDACIEUX was hit by a salvo, caught fire, and was wrecked with loss; and the submarine PERSÉE, attempting an attack, was sunk. As against this, the shore batteries knocked the *Cumberland* out of action with a hit in the engine-room; and two destroyers were also hit. The second day saw the surrender of the submarine Ajax, which was forced to the surface; but on the third day the remaining submarine, Bévéziers, made a successful torpedo attack which did serious damage to *Resolution*; and, *Barham* also receiving hits, Sir John Cunningham called off the action. Though it cost the loss of two submarines, with some damage to Richelieu and other ships, the repulse of this powerful expedition in a damaged state was an unquestioned victory for the French navy. It did much to raise the morale of the long-suffering French, and was only slightly offset when, two months later, the Free French made their position in Equatorial Africa secure by the capture of Libreville, with the loss to the navy of a sloop and a submarine.

In January of the next year the navy was to have another victory, on the other side of the world. After the defeat of France her position in Indo-China was obviously weak, and it was impossible to avoid making airfield and other concessions to Japan. But that the Siamese should seize Cambodia, as they were attempting to do, was another matter. Admiral Decoux' forces were only the cruiser Lamotte-Picquet – Captain Bérenger, eight 155-mm (6-inch) guns – four sloops, of which two were old, and some river gunboats. Bérenger took this force to sea to seek out the Siamese fleet. He came upon it in the Koh Chang archipelago, just off the south-eastern tip of Siam. It consisted of two coastal battleships, two powerful gunboats, and some torpedo-boats, the whole being superior in gun power to the French. Bérenger attacked in the mists at dawn, damaged both battleships (so that one later capsized), and sank three torpedo boats. The victory saved Cambodia from Siam, and brought some respect from the Japanese, though, of course, it did nothing to stop their demands to 'be granted the use of' further airfields.

In February 1941 Darlan became Vice-President of the Council of Minsters, ie Prime Minister under Pétain as Head of the State. He was furiously angry with the British, who seemed by the Mers-el-Kebir action to have put no faith in his word, and it is perhaps for that reason that he allowed German planes to have the use of airfields in Syria when they attempted to seize Iraq. The British thwarted that attempt, and riposted by sending troops, including Free French, to occupy Syria. Darlan ordered General Dentz, the French commander, to fight; for only by so doing could he prevent the Axis from making good a claim (which they were continually pressing) to use the French north African ports. The naval forces in Beirut consisted only of two large destroyers, Valmy and Guépard (Captain Gervais de Lafond), three submarines, and a sloop. The British were using three cruisers and some destroyers to support their land forces. Against such weight Gervais de Lafond could do little, but he handled his destroyers cleverly, keeping them close under the coast where they were a poor target, caused some harassment to troops using the coast road, and badly damaged the destroyer *Janus*. An attempt to attack the British by submarine brought only the sinking of the Souffleur; and the destroyer Chevalier Paul, trying to dodge through with ammunition, was torpedoed by a British plane before a capitulation was signed at Acre.

By the end of the year the Japanese had launched their surprise attack on Pearl Harbor which brought America into the war and so gave new hope to France. Indirectly it led, however, to a worsening, if that were possible, of Anglo-French naval relations. For the British seized Madagascar with an attack on the naval base at Diego Suarez. This was allegedly, and probably truly, just to forestall the Japanese, whose capture of Singapore and the destruction of Britain's Eastern squadron had given them sea power in the Indian Ocean. The French felt it to be unnecessary (as it possibly was) and merely a cheap success to offset the British disaster in Malaya. The hopeless French resistance cost the loss of a sloop – D'ENTRECASTEAUX, which put up a notably brave fight for two days – three submarines, and an auxiliary (converted merchantman) cruiser. America's entry also led to an odd imbroglio when Muselier (VA), sailing in one of several British corvettes now manned by some of the increasing number of Free French sailors, and having the SURCOUF with him, suddenly seized the little French fishing islands of St Pierre and Miquelon in the Gulf of St Lawrence.[1] As these islands, together with the French West Indies, had been taken into the protection of the United States, the Americans were much annoyed, and de Gaulle, whose relations with his incalculable admiral had never been good, sacrificed him by putting him again on the retired list.[2]

More important was that, in February 1942, the Germans, tired of being thwarted by Darlan by one pretext, delaying action, or another from obtaining any use of the French North African ports, had forced his dismissal from the Premiership. He remained commander-in-chief of all France's armed forces, however, and Deputy Head of State to Pétain. Hitler had turned eastward against Russia in the previous year, and Britain was moving to recover the grip upon Libya which Rommel's Afrika Corps and the Italians had taken from her. Above all America was now in the war. Such was the general situation when Operation Torch was projected.

This operation was the allied invasion of French North Africa in November 1942. The large bodies of troops there would have been a formidable army but for the fact that their equipment and ammunition had been getting out of date and deteriorating with every month since the armistice. The naval forces, however, were slight in ships since Mers-el-Kebir, though there were many naval men manning coastal batteries at the ports. The few ships were mainly at Casablanca and Dakar. At the former was the great JEAN BART, still not ready for sea. Some of her 15-inch guns had been smuggled out to her by heroic efforts, and had been mounted in her fore turret, so that moored in a favourable position, as she was, she was a powerful defensive proposition. With her was the 'second light squadron': ten destroyers led by the old light cruiser, PRIMAUGUET. There were also eleven submarines, of which four were out of fighting trim, and a number of sloops and 'avisos' – small craft with crews of about sixty and light armament. At Dakar were the great, but damaged, modern battleship RICHELIEU, three cruisers (GEORGES LEYGUES, GLOIRE, and MONTCALM),

[1]SURCOUF was sunk with all hands, later in the year, by being rammed in the Caribbean by a Yankee freighter.

[2]But he turned up, once more on the active service list, in Algiers in 1943.

two destroyers, and some submarines. In the Algerian ports was even less – at Oran were one large and three other destroyers, four submarines, and some sloops and sweepers, at Bizerta a few submarines and small craft, and at Algiers itself only two submarines and two 'avisos'.

The political position was highly complicated.

The navy, always inclined to be anglophobe, had become and had remained pronouncedly so since Mers-el-Kebir and Dakar; and that feeling was extended to soldiers and officials. Moreover, among the majority, at least of officers and officials, there was an almost equal antipathy to de Gaulle and his Free French, who were based on England. Because of the oath of personal loyalty which officers had taken to Pétain, who, with his ministers, was, after all, the legally rightful government of what was left of France, and because the Free French general was therefore considered a traitor, any liberation of French Africa had to be, at least nominally and mainly, an American and not a British affair; and any French support given to it would have positively to exclude Gaullist co-operation. This suited the Americans well enough, for they had never shared the British trust in the somewhat difficult de Gaulle and had always believed that they could somehow work with the Vichy government, while most of the French in North Africa had always believed that America would some day free them to resume war against the Germans and would somehow manage to do so without their having to break with the marshal.

After the armistice Weygand had been in command in North Africa. He had strenuously resisted Axis demands to use French North African ports and had established some links with the USA. At the end of 1941 Axis pressure obliged Pétain to recall him. America had dealt openly with him, but after his recall arrangements for freeing French Africa became a matter of conspiracy, and it fell to Murphy, the American Consul-General in Algiers, to organize this. At the head of it were generals Béthouart and Mast. The senior naval officers were considered unapproachable so that the leading naval conspirator, Barjot, later to be an admiral, was then no more than a dismissed CF.[1] To ease French susceptibilities, the linch-pin of the conspiracy was to be General Giraud, who had added to his reputation as a soldier by a daring escape from what the Germans had considered an unbreakable war prison. He was to slip out of France in a small craft and transfer at sea to a British submarine. This duly transferred him, not without adventure, to a seaplane which landed him in Algeria. Here he thought he was to command all allied armies, though America intended that his command should be limited to the French, who would not be in a condition seriously to take the field for some time. Meanwhile Murphy arranged a meeting, at a little place called Cherchell, of the main conspirators and five Americans led by General Clark, who were smuggled ashore – also not without considerable adventure. Though the great invading force was already at sea the Americans did not trust the conspirators with this information. In view of the loyalty of the main French commanders to Vichy, and the uncertain attitude of Pétain's government, this is understandable;

[1]Captaine de Frégatte: equivalent to a RN commander. Barjot had been removed from the active list for working on behalf of the Allies in Dakar.

but it was to lead to tragic loss of life on both sides, and particularly in the ill-fated French navy. If the conspirators had been trusted they could have approached their objectives gradually to arrange that the landing should be unopposed, with precautions for such eventualities as a failure in timing. As it was, when at last given the date, they had to act hurriedly, for the invasion was almost upon them. On top of that, the American forces arrived later than expected, and in the absence of any adequate communications the conspirators moved too soon. To cap it all, Admiral Darlan, commander-in-chief of all French armed forces and Pétain's trusted deputy, arrived unexpectedly in Algiers, where his son was gravely ill. The result of the breakdown of the conspiracy, and of Darlan's presence to give orders to resist (as at first he did), whereas other senior commanders might have hesitated, was serious fighting at Oran and Casablanca between Americans who were coming to free the French and French who did not want to oppose them, and, at Bizerta, the establishment of an Axis bridgehead in Tunisia that was to hold up the Allies for months, at great cost in men and material.

Darlan may well have found through his Intelligence service that landing forces were on their way. He may have intended to change sides and order all French forces to join the Allies, but he ordered resistance when the attack on Algiers started. The fighting there was limited to the sinking of the British destroyer *Broke*, which forced the boom at twenty-six knots, and some firing on landing parties; and Darlan soon surrendered the city while leaving his generals and admirals outside it to continue resistance. It seems that while the little admiral preserved a firm outward front his mind was constantly shifting behind it, and that his game was to make enough show of fighting both to save Metropolitan France from German vengeance and to put himself into a good bargaining position with the Americans. He possibly underrated the gallantry, devotion to duty, and obedience to orders of his own service that were to cost it so unnecessarily dear; and in any case he was taken by surprise for, though he had expected the Americans to come, he thought that the attempt would be made at Dakar in the spring of the following year (1943).

The invasion forces were in two parts. One, partly American but British as to its naval escort, was to come from England via Gibraltar and to make for the Algerian ports. The other, entirely American, was to come direct from the States to Morocco. The conspirators, reckoning that the navy would resist, had advised landings at Rabat and Magazan where there were no seamen. Alternatively, if the Americans rejected this advice (as they did, for plans had already been settled), they advised landings each side of Casablanca and Port Lyautey, the naval air base to its north-east, from which negotiations could be opened to avoid fighting. This advice partly coincided with the planning. On approach to the coast, the huge invasion force, most skilfully assembled and brought across the Atlantic, divided into three parts. The northern wing, under Rear-admiral Kelly, with the old battleship *Texas*, two light cruisers, two escort carriers (one of these being merely a transport for planes), and seven destroyers, took its numerous troop transports direct against Port Lyautey. The central force under Admiral Hewitt, commanding in chief, with General Patton in command of the troops, had the heavy cruiser *Augusta* (flag), a light cruiser, the aircraft carrier *Ranger*, an escort carrier, and ten destroyers, and made for Fedala, just north-east of Casablanca. The southern wing, under Rear-admiral Davison, with the old battleship *New York*, a light cruiser, an escort carrier, and ten destroyers, took its transports south-west to Safi. There was also a covering force under Rear-admiral Giffen, with his flag in the new battleship *Massachusetts*, two heavy cruisers, and four destroyers.

In command of all forces at Casablanca was Michelier (VA), a calm and determined officer: under him, Gervais de Lafond (CA) led the light forces. Michelier had some troops and naval gunners who manned the coastal batteries, but pathetically few ships (see p 325) to meet so overwhelming an attack. Moreover, the supply of oil, and indeed of all fuels, had long been desperately short. The shipping for the South of France (which the British allowed through the straits of Gibraltar with little inconvenience[1]) could be given only the tiny avisos as escorts; it was not possible to maintain regular patrols out at sea, and the light craft could be allowed only an occasional brief exercise for one or two at a time, to keep their crews in trim, and on return to harbour shut off all fires and lay as dead as at a refit when the dockyard mateys have taken over the engine room. The great JEAN BART could not go to sea, but was moved to a quay in an advantageous position where her four huge guns could bear over a wide sector. When Michelier got news of General Béthouart's plotting, the conspirators having moved too early, he considered that he had to meet a Gaullist plot in the army, and prepared to defend Casablanca against an attack on the landward side. Nevertheless he brought such ships as were fit for sea to three hours' notice for raising steam and weighing. Even so, the attack from the sea was a complete surprise.

The Americans arrived at their stations some miles offshore before dawn on November 8th, and proceeded, with some clumsiness and loss of life, to get their troops into the landing craft. There was no attempt at negotiation. True that German submarines might be radioed to the scene when the news of the landings reached Europe; true that Michelier might order out his own submarines, and perhaps his destroyers, to attack the transports. Yet the escort was powerful, it would be hours before any U-boat could arrive, and Michelier might have hesitated to resist when he had been told the size of the force against him. Surely a few hours might have been spent in negotiation to try to save life among their own men and among those to whom they were supposed to have come as friends and allies, the more so as it was not then known that the conspiracy to obtain an unopposed landing had failed. With Admiral Hewitt ordering, on the

[1]A occasional ship carrying possible contraband, eg graphite, was seized, and on one occasion the seized ship was retaken by the Casablanca forces; the great majority were allowed to take their supplies to Marseilles and Toulon. The so-called Vichy navy did much convoying of supplies and fuel that kept France from starvation and utter breakdown. Carrying contraband in convoys was avoided as much as possible. For that, fast motor launches were used.

way across, that they should strike and strike hard, and with Admiral Giffen signalling to his men that they need not mind killing the French for, if they resisted, it would show that they preferred Hitlerian slavery to liberty, it is hard to avoid the conclusion that the Americans were determined to have a fight and to destroy. This was to cost the French navy alone, at Casablanca only, 462 killed with 642 wounded, to say nothing of the loss of shipping that would have been valuable. It was a sad piece of aggressive stupidity.

At Port Lyautey (Mehdia) the planes were sent in to attack the naval air base. Though some of the thirty available French naval aircraft were caught on the ground, others became airborne and did some damage to the landing parties. These Devoitines and Glenn-Martins were no match, however, for the American Wildcats and were crushed; but the American air losses were quite heavy. For the French, warned by the conspirators having moved a day too early, were on the alert. The American ships opened a heavy bombardment on the naval batteries at 7 am, and eventually put them out of action. Nevertheless Port Lyautey continued a fierce resistance, and its last defenders had still to be overthrown when the cease fire was ordered on the tenth.

At Safi there were few guns, resistance was quickly overcome, and the landed tank forces could move towards Casablanca.

Here the attack was made by the covering force and the planes from the carriers. It was just after eight when the planes streamed in and the great sixteen-inch guns of *Massachusetts*, and those of the heavy cruisers *Wichita* and *Tuscaloosa*, let loose a deluge of bombs and shells on the port and on the naval battery at Cape El-Hank just to its west.[1] As the little submarine base was a death-trap for any ships therein under bombing, even if only a Gaullist *putsch* was to be feared, five of the submarines had been got to sea while it was still dark. Two others were just clearing the entrance to the basin as the planes arrived, and got away. The three remaining in the basin were sunk, and the destruction of ordinary shipping in the port was great, and included serious damage to three destroyers moored just outside the submarine basin with engine trouble. The shelling passed over the other destroyers moored just inside the main outer jetty, however, and did little or no damage to the naval battery at El Hank, or to the JEAN BART. That ship hastened to reply, and though sighting was awkward in the smoke, her heavy guns drove the Americans further to sea, until her turret was jammed by an unlucky fragment of an American projectile. The El Hank battery, though bombed and heavily shelled, maintained a fire that kept the Americans at a distance, and continued to do so for the three days of the fighting – a courageous and determined achievement. Meanwhile Hewitt in *Augusta* and his squadron were covering the landings a few miles away at Fedala. The batteries here were weak and ill-protected, and though they caused one destroyer to creep out of action with a shot in her engine-room they were soon reduced.

[1]Admiral Hewitt solemnly declared that the French were the first to fire: it may have been so; there is no doubt about the ready promptness and weight of the American reply.

## MERS-EL-KEBIR 1940

'BRETAGNE' is hit

'BRETAGNE' capsizes on fire

PLATE XVI

# THE TOULON 'SABORDAGE' 1942

'STRASBOURG' scuttled into the mud

Sloops wrecked in the old harbour

So, just after 9.0 am, the Second Light Squadron (seven destroyers) unmoored to make a desperate attack on the Fedala landings. As PRIMAUGUET, which had been having a partial refit, was not ready for sea despite tremendous efforts, Gervais de Lafond (CA) flew his flag in the destroyer MILAN. The odds against them were stupendous. Not even the sacrificial bravery of La Jonquière defending his convoy surpassed that of this 'Charge of the Light Brigade', this last sally of the French navy against ships at sea.

The destroyers left under heavy fire, and the drama was heightened by the dark-robed presence at the outermost end of the quay of the chaplain of the PRIMAUGUET, regardless of the shells bursting around him, his arm raised in blessing on the ships and men steaming at eighteen knots to almost certain destruction. Gervais de Lafond kept close to the coast, to make his ships a difficult target, tactics he had used off Syria. Here the odds were much greater, and though his appearance caused high alarm among the American transports, the *Augusta, Brooklyn*, and their destroyers made speed to cover them. Not only was fire heavy from the enemy ships, but the carrier *Ranger* flew its aircraft to attack the French, and they had no air cover and little anti-aircraft armament. Upper-deck casualties mounted, wireless and means of signalling were shot away, and, though they engaged the enemy destroyers, the heavier gunfire of the cruisers forced them to turn away, under smoke, making for PRIMAUGUET which, by prodigious effort, had struggled to sea to give its pathetically inadequate support. They turned to make a further attack, but now came under the fire of the *Massachusetts* and the heavy cruisers of the covering force. FOUGUEUX was sunk, still fighting her guns: MILAN was hard hit and drifted ashore. The admiral then moved his flag to PRIMAUGUET which was hard hit in its turn and forced to make for anchorage, and the squadron lost cohesion. One section, BOULONNAIS and BRESTOIS, turned to sea to make a top speed torpedo attack on the American cruisers, but BOULONNAIS was blown out of the water by a full salvo from one of these, and BRESTOIS, gravely damaged, could only creep back to port, where she sank later in the day, as did FRONDEUR. ALBATROS, terribly battered, dropped anchor inshore, and only ALCYON survived to join the avisos in picking up survivors. These were hurriedly formed into a land force and sent next day to help the defence against the troops advancing from Fedala. The battery at El Hank continued to fire, but the brave sortie had failed.

As for the submarines which got to sea, only two made a torpedo attack on the cruisers, and that in vain. Two were never seen again, sunk by bombing, two got to Dakar, one went ashore, damaged, and only one returned to base. The last, TONNANT, after extraordinary adventures, scuttled herself after putting her crew into safety. Admiral Hewitt found, to his horror, that the great *Massachusetts* had used no less than three-quarters of her sixteen-inch ammunition, and this mostly on a few destroyers and (harmlessly) on the El Hank battery.[1]

[1]After the cease-fire, when he received the American leaders, Admiral Michelier was standing between two huge dud American shells, retrieved – and rapidly polished – from this battery.

The fighting was now limited to land – almost, but not quite. For on the 10th, when the attack on the naval battery and men that formed the eastern defence of the post was growing intense, the surviving avisos crept along the coast to help with a flanking fire from their light armament. The American ships opened on them, at which JEAN BART, her turret having been restored to working order, brought her heavy guns into effective covering action. Thereupon a furious Hewitt ordered his planes to 'sink that yellow-painted ship'. They bombed her heavily but, though much damaged aft, she was not sunk, despite the fliers' report that there was 'no more JEAN BART'. (The report was made 'jubilantly', a strange attitude to take for attacking a few brave and hopelessly outnumbered men of a country one wanted to make one's ally!) This happened five hours after Darlan (between 10 and 11 am) had ordered a cease-fire; and the reason that the fighting at Casablanca had continued was mainly that General Eisenhower's communications had broken down. Michelier heard Darlan's cease-fire, but also an order from Pétain for continued resistance (at the demand of Laval who was about to see Hitler); but when Admiral Auphan, Naval Minister at Vichy, added to the 'continue resistance' the words 'if still possible', Michelier could, and did, cease fighting with honour.

Looking back at the Casablanca battle it might seem, at first sight, that the French had been caught napping, and that they had shown their old fault of wasting their bravery by their inefficiency – JEAN BART's gunnery had not been used to shell the transports off Fedala which were within her range, and the submarine attack had been only partial and a failure. This would be unjust. The complicated political position made surprise and confusion almost unavoidable, and the shortage of material, especially of fuel, had made proper wartime training impossible. Nothing can obliterate the heroism shown against odds, nor can one avoid the reflection that the battle might have been avoided, or at least conducted with less ferocity – though, even so, it must be remembered that loss to civilian lives and property was carefully avoided.

The tragic battle has a tragic footnote. When the attack on the submarine base was made, one submarine, CONQUÉRANT, was in dock. By superhuman endeavour she was got to sea, though in no condition to do more than creep slowly to Dakar. On her way she was bombed and sunk by an over-eager American aircraft on the third day after the cease-fire. One of the other submarines, SIDI-FERRUCH, was sunk with all hands by bombing after the cease-fire had been announced.

Events at Oran were less sad only because fewer ships were involved. The approaching invaders ran into a small French convoy, which signalled the presence of a big force: so the port was alerted. The ships there were the large destroyer EPERVIER (just leaving dock), three other destroyers, four submarines, and some lesser craft. When two small British sloops rushed the harbour with landing parties they were smashed to bits by the concentrated fire of the destroyers. They and some sloops then left to reconnoitre a report of landings further away and ran into the main escort. The sloop SURPRISE was sunk, the destroyers TRAMONTANE and TORNADE driven ashore in a sinking condition, and an attack by the submarines failed, with the loss of two (a third getting to Toulon). EPERVIER, trying later to break through to sea, was set on fire and had to beach. The other ships were scuttled.

There remained Bizerta. Here the Allies made a grave mistake in not taking the risk of sending a token landing force to see what would happen. Admiral Derrien, in command there, only wanted someone or something to give him a lead. As it was, confused by the contradictory reports and signals coming by radio, he did nothing while Axis troop-carrying planes were landing, until it was too late. Then, at last ordered to resist, he was powerless and had to surrender with all his ships. They were old submarines, and some small craft which the Axis used. Far more important than their loss was that, given the word to fight, he might have stopped the Axis landings and the subsequent Axis occupation of Tunisia.

These almost unnecessary fights and losses were bad enough; but worse was to happen at Toulon.

It did not take the Americans long to realize that the 'king-pin' on whom they had relied, Giraud, was going to be of little use to them. Darlan, the little admiral in the tinted glasses, was the man whom the French in North Africa were going to follow. His order to cease fire on the morning of November 10th was obeyed and, fortified by a private signal from Pétain that despite public broadcasts to resist the marshal left things to him with confidence, he was rapidly climbing back into the saddle and was beginning to negotiate with the Americans on something approaching equal terms. German tanks, though avoiding the town of Vichy, were now overrunning the free zone of France, and on November 11th Darlan sent to Laborde, in command at Toulon, to bring the fleet to Africa if it was in danger of capture. That rigid old martinet dismissed the signal with a crude and contemptuous monosyllable, and did not obey. True that the signal was more of a request or invitation than a positive order, that it conflicted with Pétain's public utterances and might, therefore, have been a fake; but three much more powerful reasons caused de Laborde's disobedience. His personal loyalty to Pétain was profound; he trusted the Germans to observe their undertaking not to occupy Toulon or seize the fleet; and his anglophobia went too deep for him even to consider any association with the British. The first of these reasons does him credit; but the second was almost absurd and, indeed, though the Germans halted the advance of their tanks outside the Toulon area, this was only to wait until they had established their air power in a way that would make the escape of the fleet impossible. As to the third reason, it was surely a strange patriotism that would sacrifice a whole fleet rather than bring it to fight alongside a nation which, however much detested, was battling to free one's country from its main enemy.

He had, however, made thorough arrangements for scuttling; and it was soon too late, because of the presence of German aircraft and the lack of any French air cover, to do anything else. Even so his orders were left almost beyond the eleventh hour for execution. The Germans had been ready for their seizure operation since 1940. It was in the early hours of November 27th that their tanks suddenly bashed their way through the fortified gates into the Toulon dockyard, and had they not then lost their way and got entangled in the mazes of that large establishment, they would have been able to grab at least some of the ships. As it was, the order to scuttle went out just in time. There was then a scene not unlike that of the evacuation in 1793. Gunfire as the men arming the gates tried to stop the invaders, flames as some ships were set afire, explosions as turrets and parts of engines were wrecked, bombing and shells as the planes and tanks tried to hit a few submarines that were struggling to get to sea, a mass of ships slowly but irrevocably sinking to the bottom of basins or of the harbour – and, amid it all, old Admiral de Laborde, on the bridge of the sinking STRASBOURG, furiously if somewhat absurdly demanding of the Germans on the quays why they were breaking a solemnly signed agreement!

Three battleships, STRASBOURG, DUNKERQUE, and the old PROVENCE, seven cruisers, thirty-two destroyers, sixteen submarines, and nineteen other craft – seventy-seven in all – shared this vast suicide and 'lay on the bottom in one chaotic heap', their upper works protruding from the water and listing this way and that, with guns and machinery mostly in ruins. Some destroyers and smaller ships, inefficiently scuttled because they had only small maintenance crews, were later raised and towed to Italy; but the Axis got little use of them.[1] Five submarines alone got away: one was so damaged by bombing that she had to be sunk, one went to internment in Spain, three only reached Algeria – the remains of the Toulon fleet. The navy had kept its word. It had not been allowed to fall into the hands of the Germans.

Yet, though the sacrifice can be dramatized and was honourable, was it not also rather stupid? That the fleet should have been kept from any sort of union with the British, even at the risk of a Mers-el-Kebir, so that it could stand between the Axis and the French Empire, is understandable. But when that empire, under the navy's own chief, had gone over to the Allies and hostile tanks and planes were already making for positions where the fleet would be left with no choice between seizure and felo-de-se, surely Pétain and his navy minister, who had kept one code secret from the Germans, should have used that code to order the fleet to join the Allies in Algiers. It was already clear that to liberate France huge troop landings would have to be made, requiring huge cover from air and sea. How valuable those seventy-seven ships would have been.

France still had some naval forces, mostly scattered round the empire. Even though JEAN BART had been so damaged at Casablanca that her repair had to be left to the end of the war, there remained the capital ships RICHELIEU and LORRAINE, a carrier, nine cruisers, eleven destroyers, nineteen submarines, and nearly a hundred small ships. All needed fitting with such new equipment as radar and AA guns. To these were added some ships given or lent by the British and American navies and, later, some from the surrendered Italian fleet, and some taken from the Germans. It formed a somewhat motley *marine*. Only the ships in Indo-China, left to their own devices in maintaining such independence as they could of the

[1]Because most of the ships were in the inner harbour at Toulon, which is shallow, the Italians, working feverishly, managed to raise thirty hulls; but by then their own surrender was imminent.

French position vis-à-vis the Japanese, could be written off. There remained also six air squadrons and a naval brigade; and to man the ships were some forty thousand officers and men.

It was a bitterly divided service, however. The Gaullists were perpetually trying to entice men from their ships or places of duty into what had been the Free French vessels (to the intense anger of Muselier, who was now in Algiers as head of naval police), and most of the others long considered themselves as serving their country in accord with the orders given by Darlan as deputy for the now powerless Pétain. Moreover, the service was soon to lose the man who was working to unify it. For, having proclaimed himself in charge of French affairs in Africa in the name of the marshal on November 13th, Darlan was assassinated on Christmas Eve.[1]

He was given an honourable funeral, and it was attended by the leading officers of the Allies. Otherwise he was, and has been, somewhat execrated. This is unjust. Certainly he had played politics, but he was also the able and personable professional who had lifted his Service out of apathy into vigorous keenness. Certainly he had been pronouncedly anglophobe after the collapse of France and the tragedy of Mers-el-Kebir and had given the Axis some actual assistance; but he had had a most difficult hand to play, and had succeeded in keeping the enemy out of the French Empire. His path had certainly been tortuous, but the devious courses that he steered were undoubtedly taken for the salvation of France. He was not a great man, perhaps not even a great admiral, yet he has some claim to be the most notable French naval officer since Suffren – though cast in a very different mould and working in very different circumstances.

Now that the marshal was obviously little more than a prisoner, Darlan's death left the French who followed him without a real leader, so that inevitably, however slowly and unwillingly, they tended to accept de Gaulle. Giraud attempted to fill the admiral's place, and in June 1943 a joint committee of liberation was formed under him and de Gaulle, from which Giraud was shouldered into retirement a few months later. This body promptly removed Michelier, a poor reward for the expeditious way in which that admiral had got most of the remaining ships to America for refit and for modernization in equipment such as anti-aircraft weapons.[2] It was not until August 1943 that the political divisions were overcome, and the navy became a truly united one.

The activities of this revived remnant of the navy before D-day saw the allied invasion of Europe can be briefly summarized. RICHELIEU was sent at the end of 1943 to join the fleet that was forming in the Far East,

[1]His young assassin is said to have been the tool of the extreme right, ie an anti-Nazi royalist; the precise truth may never be known.

[2]Robert, the flag-officer commanding at Martinique, had even worse treatment. Between the Americans, the rival French factions, and the threat of a communist rising to make civil war on the island, he had been in a most difficult position. He had managed to preserve order, his ships, the French gold, and French independence; and he had rejected frantic orders from Laval at Vichy to scuttle the ships and sink the gold. He handed over his charge to the new committee and returned to France. His reward, later, was to be disgraced and given a sentence of hard labour – which was never carried out.

and had some share in the defeat of the Japanese. The rest had much patrolling and convoying, in which they were successful in seizing some blockade runners and some Axis vessels in the Mediterranean. In particular the ships were used for the rapid transport of troops; and the liberation, or recapture, of Corsica was entirely a French operation, carried out by forces disembarked from cruisers and destroyers. The navy gave assistance in the bombardments that covered the Allied landings in Italy, and the submarines, especially CASABIANCA, did well against shipping. Two were lost, one, alas, by being bombed in error. The larger destroyers twice cut up enemy convoys.

The navy also played a part, though only a small part, in the D-day landings in Normandy in 1944. The cruisers GEORGES LEYGUES and MONTCALM under Jaujard (CA) and the destroyer LA COMBATTANTE[1] shared in the bombardment to cover the landing, while a number of smaller vessels formed part of the escort for the vast expedition. In the Provence landings, though the British Admiral A B Cunningham was in overall naval command, and the American Hewitt in local charge, the French had a much larger share. Jaujard's cruisers (GEORGE LEYGUES, GLOIRE, and MONTCALM), those of Auboyneau (CA) (EMILE BERTIN and DUGUAY-TROUIN), three large destroyers, and the old battleship LORRAINE all took part in the covering bombardments, and the escorts for the convoys were largely French destroyers and sloops. Toulon was retaken on August 28th; but it needed a fortnight's mine-sweeping and unblocking before, on September 13th, a French squadron once more returned proudly to its roadstead and harbour. The Germans still held grimly to some other French ports, and this necessitated bombarding, LORRAINE being largely used, and some stiff fighting which caused unnecessary casualties; the naval manned tank regiment, too, was with the troops closing up on to the Rhine; but otherwise the navy's fighting was finished and it could devote itself to the arduous and dangerous tasks of sweeping mines and clearing harbours. Its war, dreadful through so much tragic disaster, was over.

There remained two major things to be cleared up – the position in Indo-China, and the resettlement of the navy in France.

The first was dealt with satisfactorily, though only temporarily so. Admiral Decoux had had a hard time. He could not prevent the Japanese using his territories as a base, but he had retained French control. His great problem was to keep his people fed. He had laid up his cruiser, LAMOTTE-PICQUET, in 1942, and she was later sunk by American bombing in error, as was some of his shipping; but he managed to get in sufficient supplies. In 1945, however, the Japanese, facing defeat, sent an ultimatum demanding surrender of the colony. The French fought; but the remaining gunboats and sloops were overwhelmed in the rivers. A little later, with the collapse of Japan, all was chaos, and the Americans, with their curious prejudice against 'colonizers', would give no help. Fortunately the British did so and, with the arrival also of several French warships, a measure of order was

[1]Also, in August, this destroyer played the star rôle in the complete destruction, off Fécamp, of a German coastal convoy with its escort of motor gunboats (*Schnellboote*).

restored. Much of the territories remained divided between nationalist rebels, pirates, and the Chinese. But a powerful force of warships and troops was sent, the nationalist leader Ho Chi Minh elected to join the new French Union, the Chinese were 'persuaded' to withdraw, and French authority was restored. Later the fragile understanding with Ho Chi Minh broke down, and all went wrong; but that is outside the scope of this history.

The resettlement in France presented no great difficulty as to what existing ships were to be retained in commission, and what ships of the Toulon fleet were to be raised and restored. But otherwise it went grievously. De Gaulle's prominent supporters contained some political extremists of the left, such as Thorez, who was a naval deserter in 1939, and Tillon who, when a petty officer, had been rightly sentenced to imprisonment for an attempt at deliberate treason aboard his ship. Such men were not likely to have much respect for admirals who had obeyed the orders of Darlan, and something not far short of a clean sweep was made of flag-officers. They were put on the beach without pay or pension, and a number were brought to trial. Nor were these trials legal, for in only one or two cases were the officers granted the court martial that was their right, the others being tried by courts whose bias can be guessed from the fact that at least one admiral went into hiding rather than attend for trial at a hint from the Minister of Justice himself. A few, like de Laborde, were condemned to death, others to imprisonment (from which one or two, like the veteran Esteva, were released only to die at home on breakdown of their health), most were sentenced to 'national disgrace'. All this was for no other reason than that they had obeyed the orders of their commander-in-chief as naval officers should do. The Resistance and the Free French had done bravely indeed, in the face of great difficulties and much discouragement. They deserve all praise; but it was a pity that some of their number descended to such petty and spiteful vengeance. The sentences of death were not carried out and, in a while, good sense prevailed: the prisoners were released and pensions were restored. Yet the persecution made a sad ending to a sad story.

CHAPTER XVI

# Conclusion

With the end of the Second World War came the end of the capital ship. Maritime nations might retain her for a while, and even complete any already on the stocks; but she was doomed. Starting in the sixteenth century, when improvements in construction and rigging plan made it possible to mount some guns as a broadside, she had grown into the great three-decker with its hundred and more guns and its attendant frigates and despatch vessels. Then came the naval revolution of the mid nineteenth century, with steam, iron building, and rifled cannon to change her into a sailer with auxilary engines, a certain amount of armour, and a few medium weight guns. From this, in a mere fifty years, she had grown into the huge armoured super-*Dreadnought* with its eight or more great guns, its high speed, and its attendant cruisers and destroyers. Even during the First World War there had appeared a dire threat to her supremacy in the unexpected effectiveness of the submarine, and by the end of that war it was clear that there would soon be an even greater threat in the bombing aeroplane. The Second War proved it to be so, and the supremacy of the battleship was over. Her place was to be taken by the aircraft carrier, not herself a fighter but bearing the far-flying bombing planes, and her attendants were destroyers, almost as large as light cruisers, and corvettes, all these armed mainly with anti-aircraft guns, anti-submarine devices, and, later, guided missile weapons. To complete such a navy were submarines of greater power and endurance than ever before. Against these and the bomber the costly capital ship, for all her strength, was too vulnerable. In one sense, though for reasons and in ways of which he had never dreamed, Admiral Aube had been right.

France steered the same course as other maritime powers. By the early 1960s her fleet had as its main vessels the two 'carriers' of the CLEMENCEAU class, that bore thirty planes and were armed only with anti-aircraft guns. There were two lesser carriers, two anti-aircraft cruisers, some destroyers (of which eighteen, displacing 3 750 tons when fully loaded, might well have been called light cruisers), some *escorteurs rapides* (frigates), some old corvettes ('avisos'), and, of course, submarines.[1] There were guided missiles; otherwise the armament was primarily, indeed almost entirely, anti-aircraft. It was not a large navy: France had ceased to try to rival any major maritime nation in strength. Its most notable action was that of a withdrawal – in 1959 President De Gaulle, to show his displeasure

[1]Most of the ships mentioned on p 336 had been scrapped by the early 1950s, the remainder by 1959, except for a few that were retained as accommodation hulks until 1969.

with the attitudes of Britain and the United States, recalled the French naval units serving with the Mediterranean command of NATO!

This history has been that of the capital ship as it served France. It has been mainly a sad story, for it has been largely a record of failure. How much of this failure has been due to the Service itself: to the ships and the men who sailed them, to the naval administration?

To the ships and the men who manned them, little. In the days of sail the ships were well designed and powerfully scantled. They were held to be superior to those of the main enemy, England. It has been said that the finest naval unit of those days was a captured French ship manned by a British crew. When the naval revolution brought steam to replace sail, France often took the lead in naval inventions and developments (eg with armour and the submarine), though she did not retain it. As to the men who sailed the ships, the officers, except in the early days and during the Revolutionary period, were mostly professionally able and personally courageous. They also had tradition; for, until the Revolution, one finds family names recurring, as, with different names, one does after 1800.[1] Before the Revolution the haughty independence of the *noblesse* among them was a weakness, for it occasioned some ill-discipline, some poor relations with '*roturier*' officers, and, with some, a certain lack of care for their men, who were inconsiderable plebs. Till the end of Louis XIV's reign, too, some high commands were given to great nobles because of their greatness; and that did not make good their lack of nautical competence. But it was not due to any of these things so much as to lack of peacetime training at sea that French officers occasionally were found wanting.

As to the mariners, where they were seamen by calling they were good; but such were in short supply, the lack of a large merchant service meant that there were few reserves, and it was not till the nineteenth century that there was a regular enrolment of long service ratings. Before that ships often went to sea carrying too large a proportion of men with little or no training. They were brave enough, but they were handicapped in their seamanship and in their gunnery.

As to the administration, the blame lies here more heavily. It does not help a service that the minister in charge of it should consider it a wasteful excrescence, as, for example, did the Pontchartrains and Berryer, nor that he should be a landsman without technical knowledge of ships and navigation as was often the case in the old royal navy. With a shortage of real mariners, a wise administration would have seen to it that the men were kept properly clad and have taken such care for their health as the medical knowledge of the day could provide. It would also have seen that the Inscription Maritime was working according to rule and that that rule was modified to meet changing conditions. Nor would a wise administration have created such widely divergent branches as the *épée* and the *plume* or have allowed their ruinous rivalry to develop.

More, far more than in these things, the blame for failure lies in policies,

[1]An ancestor of Admiral Darlan fought at Trafalgar. Many other examples could be given.

and in the heart of France itself: the attitude of its governments and of its people.

An excess of cleverness and of logic sent French strategy and tactics along the wrong road. Why were wars fought? To defend one's own possessions and to occupy those of the enemy. One cannot occupy a piece of ocean. Therefore the navy was only an adjunct to those who could occupy land, the soldiers. It was rarely indeed that a French fleet put to sea to seek out and sink the enemy: it sailed with a 'mission'. It was taking reinforcements or supplies to a threatened place, it was covering the safe sailing of a convoy or of transports carrying troops to invade or seize a place, or it was to attack a convoy, or to co-operate with troops operating along a coast. The mission was more important than a battle – for if one fought, and was too damaged by one's victory to continue one's voyage, one's mission had failed. So French tactics were to engage from leeward, whereby one could the more readily disengage, and to fire somewhat high to damage the enemy's masts and rigging, so that he could not follow when one slipped away. This might, and sometimes did, obtain the success of achieving the campaign's immediate object; but it did not bring real victory, and a refitted enemy could later reverse the result. Perhaps the crowning instance of the folly of this over-intellectual approach to naval warfare was the solemn approval given to d'Estaing's failure to take a single one of Byron's battered ships at Grenada, because to do so was not the main object of his campaign.

These missions, too, were often designed by a government of landsmen or by planners with little nautical knowledge. The sailors (except, perhaps, for the *plume*) were not consulted. It was sheer mismanagement that sent Tourville to defeat at La Hogue, that gave the inadequate Conflans a task that would have been almost impossible to a highly able officer, that provided no reserves to send to d'Orvilliers in the Channel. Some of the landsmen ministers, such as Sartine and Castries and, above all, Choiseul, did admirably up to a point, but they were vulnerable in the widest fields of naval strategy. Even when, in later days, the minister was often an experienced officer, he sometimes failed: Decrès was defeatist, Aube's energy created only confusion.

Further, the readiness to turn to privateering, the *guerre de course*, when the main fleets failed to achieve success was a mistake. It was a great temptation. War was costly. Why should it not pay for itself, at the enemy's expense and to the relief of one's own exchequer? Such an attitude was too much like that of medieval knights who fought to win ransoms rather than for victory. This it could not bring. It diverted ships, supplies, and, above all, the scarce prime seamen from the main squadrons. True that it made the enemy suffer and brought in the shekels, at least for a while; but its success was ephemeral: it could not settle the issue and in Napoleon's time it even rebounded, boomerang-wise.

Above all, the French navy has suffered from the general policies of French governments and from the internal happenings of the country. Sometimes the one and sometimes the other has caused a grave break in the continuity of its naval service: more so than in other countries. The first tentative efforts to establish a navy, in the sixteenth century, were ended by the religious wars between Catholics and Huguenots, and all had to be begun afresh by Richelieu many years later. The navy that he created was just finding its sea legs when another set of civil wars, those of the Fronde, virtually blotted it out of existence. On the foundations that he had laid his disciple, Colbert, built a splendid force which rather more than held its own against English and Dutch combined; but when the English Succession war began to strain finances it was the navy that suffered. Pontchartrain laid up the fleets and employed the ships in commerce raiding of a privateering type. When that war ended some attention was paid to naval needs. It did not last. After the indecisive disappointment of Malaga a fine fleet was laid up in favour of privateering, and parsimony became the order of the day. This persisted, with only brief and inadequate loosenings of the purse-strings, till the disasters of the Seven Years War. Until Choiseul's time it combined with the rivalry betwixt 'pen' and 'sword' to be the curse of the Serice. France began a naval war well enough – Toulon and Minorca – only to fade right away because, once a fleet had been sent to sea well found and well manned, the arsenals were exhausted thanks to parsimony, and the reserves of men were lacking owing to mishandling of the Inscription Maritime. Then it seemed that France had learned her lesson from the disasters of the mid eighteenth century and, thanks to Choiseul and his successors, a fine navy was built up that obtained their independence for the Americans. In spite of the defeat of the Saintes it might still have triumphed over Britain had not the strain of mishandled finances forced France to make peace. Even so, the naval future of France seemed bright, when there came the wrecking hurricane of the Revolution. The absurd decrees of the Revolutionary governments and the violence of revolutionary mobs drove the officers out of the service, deprived it of its trained seamen gunners, and left its arsenals stripped and empty. This was the third time since its foundation by Richelieu that the navy was brought near to desolation through no fault of its own.

Then the slow recovery in the nineteenth century produced, by the time of Napoleon III, a powerful service, keenly awake to the possibilities of new inventions, and taking a lead in the naval revolution. At this the Empire must needs take the unwise step of going to war with Prussia; and the failure of the army in that war saw that powerful navy left stranded. It was publicly relegated to being very much the junior service, and even so the inventive intelligence of its officers was largely wasted by the frequent changes of policy that came with the frequent changes of government. Not until the 1930s, under the lead of Durand-Viel and of Darlan, did it regain an *esprit de corps*, an efficiency, and an organization to equal those of its last two great periods under sail.

Now came the final tragedy. The collapse of the army before Hitler's tanks and under Goering's bombers once again left the navy stranded. After performing prodigies in retreat, its loyalty to a government that found itself forced to abandon its ally saw some of its ships grimly sunk by that ally, others later destroyed by its American friends, and, last of all, its own pitiful suicide at Toulon.

What the future of naval warfare will be in an age of atomic power, atomic bombs, and guided missiles is not the business of this history. We have followed the naval fortunes of France from the building of the first royal ships that could fire a broadside to the end of the great super-*Dreadnoughts* of the Second World War. This has covered roughly four hundred years; and those years have seen mostly defeats and failures. These have been less the fault of the Service itself than of the way its country has treated it. During this long time France has had little just cause to be ashamed of her navy : the navy may have had some just cause to be ashamed of France.

# The French Navy Since the Second World War

## By Antony Preston

The assortment of ships which formed the French navy after the 1939–1945 War was fortunately not put to the test for some years. Officers and men had to be trained and the tactical lessons learned could not be applied properly until the administrative machine was functioning smoothly again. All navies were faced with the need to cut back expenditure, but France had to bear the additional burden of reconstructing so much that is essential to a fighting service, particularly her dockyards and ordnance factories. All this had to be done against a background of financial crises and political instability.

The French navy had barely entered the era of peace before it was plunged into a war in Indo-China. While this war did not call for heavy ships and did not involve action against an enemy fleet, it placed a great strain on French naval resources, and ultimately involved a considerable number of surface warships of all types.

Chinese Nationalist forces had occupied Indo-China at the close of World War II as far south as the 16th Parallel. In February 1946, after negotiations with the Chinese, a French occupying force of troops in the aircraft carrier BÉARN was moved to Tonkin. When the supporting squadron under Rear-admiral André Jubelin, including the large destroyer TRIOMPHANT, moved upriver to Haiphong it came under heavy fire from the river banks and had to withdraw. The cruiser EMILE BERTIN and other ships bombarded Haiphong on March 6th 1946, and two days later the French troops were able to land as previously agreed. But the retreating Chinese proved to be much more of a problem than had been expected. The aftermath of the Japanese occupation had thrown up highly organized guerrilla units of the Communist-inspired Viet-Minh, who were able to interfere with French traffic using the Red river to Haiphong.

Historically, the bloody war between France and the Viet-Minh broke out with the massacre of French inhabitants in Hanoi on December 19th 1946. Throughout the next eight years the French navy was to play a vital role in keeping the army supplied and controlling the rivers. The rivers were (and still are) the key to Indo-China, for nearly 90 per cent of the usable communications were in the form of rivers and canals. During the monsoon season roads are flooded, and in addition the Viet-Minh guerrillas were able to sever many of the road links. Other wars had seen rivers used to support armies, but for the first time the rivers themselves had become a battleground. The French navy was faced with the problem of devising new tactics and equipment at short notice, at the worst possible moment, when its basic administrative structure was being rebuilt. To cope with the new conditions Vice-admiral Ortoli listed five principal missions:

1 to control the coast in order to provide freedom of access and manoeuvre;

2 to interdict coastal waters to the enemy;

3 to clear mines from ports and waterways;

4 to use naval aviation for patrol, precision bombing, and direct support of land and naval forces;

5 to serve in clearing and patrolling the network of interior waterways which serve as the principal means of access to the life of the country.

The first step for the navy was to establish a river fleet and its supporting base at Saigon. The river squadron was placed under the command of a brilliant Catalan officer, Commander François Jaubert. In a brief space Jaubert requisitioned former Japanese junks and river launches, and soon fourteen LCAs (Landing Craft Assault) and six LCVPs (Landing Craft, Vehicle and Personnel) were brought in by the carrier BÉARN from the British base at Singapore. None of these craft was equipped for a riverine war, and it was necessary to rearm them with whatever weapons came to hand and in whatever fashion local conditions might indicate. Many of the ideas pioneered by the French navy between 1946 and 1954 were subsequently adopted by the United States in its war against the Viet Cong. Many ingenious conversions of assault craft were devised, using tank turrets and light automatic weapons such as 40-mm Bofors and 20-mm Oerlikon anti-aircraft guns. The basic tactical unit was the 'Dinassaut' or Assault River Division, which comprised varying numbers of the converted landing craft. Whenever the 'Dinassaut' formations met their elusive enemy the actions were intense and bloody, as the lists of battle casualties show.

Indo-China saw the rebirth of French naval aviation, as it became more and more necessary to provide support from carrier-based aircraft. In 1946 the French Naval Air Arm was at a low ebb, having only one obsolete aircraft carrier, the old BÉARN. She had been in service since the 1920s, and since 1943 had been useful only as an aircraft transport. The two modern carriers under construction in 1939 had been destroyed in the general débâcle of 1940, and the only modern vessel available was the ARROMANCHES, transferred by Great Britain in August 1946. With the war in Indo-China dragging on the United States made available a light fleet carrier in 1951, the USS *Langley*; she was henceforward known as LAFAYETTE to commemorate an earlier friendship between France and the USA. Two years later her sister *Belleau Wood* was transferred and renamed BOIS BELLEAU.

Turning aside for the moment from purely military matters, it must not be forgotten that the French navy laboured under many handicaps for some years after 1945. Not only had most of the old navy been damaged or sunk, but all the lessons of World War II had to be incorporated into tactical doctrine as soon as possible. Great Britain transferred an aircraft carrier, and at the same time turned over her share of surrendered German tonnage, including four large destroyers and four torpedo-boats. Wisely the decision was made to rely for some time to come on the United States for weapons and aircraft, although a good German medium-calibre gun, the 105-mm, was adapted to French requirements.

The problem of keeping such a varied collection of ships in running order was a tremendous worry to the dockyards at Brest and Toulon, for the French navy now had German, British, and American ships and equipment in service, to say nothing of pre-1940 French equipment for which there were few spares. To make things worse, in 1948 two ex-Italian light cruisers were taken over as reparations. The British and Americans were willing and able to provide spares for most of their ships, but it proved necessary for the Americans to turn over an additional ex-German destroyer to the French, simply in order to provide spares for the other four.

Fortunately there was a considerable amount of good tonnage available, which was either incomplete or capable of modernization and repair. Chief of these units was the battleship JEAN BART, which needed repairs to her war damage and a complete overhaul of her secondary armament and fire control. This was done, and she emerged in 1950 as one of the most impressive capital ships of her time, with extensive command facilities. The cruiser DE GRASSE had been under construction since 1938, and after her launch in 1946 she was redesigned as an advanced anti-aircraft cruiser. Five submarine hulls, including one which had been towed to England in 1940, were completed to a revised design, and thus gave French designers invaluable experience before they pressed ahead with newer designs.

Gradually a new all-French navy was taking shape. In 1952–7 a large series of warships, including a cruiser, two aircraft carriers, seventeen large destroyers, and eighteen frigates, was laid down, as well as fourteen submarines. The ex-German ships had served their turn, and could be discarded, but some of the ex-British and American ships were retained to provide training ships, and, in the case of the aircraft carriers, to bridge the gap until the new French-built carriers were ready. A large number of minesweepers were added, in view of the apprehension felt in NATO circles about the Russian mine threat, and the first of a series of amphibious warfare ships were laid down, including two assault ships comparable to the British *Fearless* and *Intrepid*.

In the Suez Crisis of 1956 French naval forces were as vital to the success of land operations as were the British naval forces. A shroud of secrecy has always masked the movement of French ships at the time, but it is known that the cruiser DE GRASSE and the battleship JEAN BART sailed from Toulon just before news of the crisis became public. The carriers ARROMANCHES and LAFAYETTE were based on Toulon and aircraft from one of these ships attacked an Egyptian destroyer off Alexandria, setting her on fire. The French concentration force in the eastern Mediterranean nearly proved embarrassing, for on October 31st 1956, the night before the Anglo-French ultimatum to President Nasser expired, the Egyptian destroyer *Ibrahim el Awal* ran into a large French task force comprising a cruiser of the MONTCALM class and five large destroyers. The wretched Egyptian destroyer had tried to shell the oil refinery at Haifa, and the flashes of her guns had alerted the French squadron; in some embarrassment the senior French officer had to ask Israeli forces if they would take the surrender of the

*Ibrahim el Awal.* News of the incident, or at least the correct version of it, did not leak out until a year later when the minister of marine decorated various officers and ratings who had distinguished themselves in the Suez operations. Since 1956 the French navy has changed entirely. With its three nuclear-powered ballistic missile submarines it has a fire power far in excess of the battleships and cruisers that once loomed so large in reference books. New missile-armed cruisers and frigates have appeared, and ships are armed and equipped purely from French resources. With the exception of the American 'Crusader' all carrier-borne aircraft are of French origin. The highly successful 100-mm shipboard gun has been sold to the West German navy, and French shipyards have been very successful in obtaining contracts for warships from Israel, Portugal, and South Africa.

The outstanding feature of post-1945 French planning has been the willingness to face reality in the procurement of weapons. Whereas the British consistently tried to produce an all-British weapon system, the French have willingly saved time and money whenever necessary by ordering a foreign version, but without ever losing sight of the ultimate intention to replace it with a French product. Thus the big destroyers of the SURCOUF class were armed with a French gun, but using standard American ammunition. Again, when the time came to rearm some of the same class with guided missiles it was realized that the French missile programme was well behind schedule. To save time and to gain useful experience in the field, four ships were equipped with the American 'Tartar' system; a missile defence for the fleet having been created the delay in developing the 'Masurca' missile did not involve any danger. It is not coincidental that the French navy should have an effective surface-to-surface missile, 'Exocet', before the Royal Navy.

As ever, the design of French warships is a source of admiration abroad, and French features are full of novelty and originality. Although the French navy is no longer part of the NATO naval force, it remains one of the most powerful in Europe, and one which cannot be ignored by a potential enemy.

# Select Bibliography

The number of books and articles on naval matters is enormous.

Probably the best bibliography is that by Professor R G Albion, *Naval and Maritime History*, published at Mystic, Connecticut, which has supplements. Unfortunately it gives only books written in (or translated into) English; and even so it omits some older works which, though out of print, are still on the shelves of many libraries and still have value. These are covered by the bibliography issued by the Historical Association in the 1920s. I do not know of any purely naval bibliography in French, though the Comité Internationale has published one of Histoire Maritime which deals rather with historical geography and exploration than with the navy. However, some of the authorities given below have a good deal of bibliographical matter, eg the three-volume *Puissance navale dans l'histoire* by de Belot, Reussner, and Nicolas (Éditions Maritimes).

Other books in French, or dealing directly with the French navy, are:

De la Roncière, C, *Histoire de la Marine française* (Plon, 1909–32); 6 vols ending with the treaty of Utrecht: contains much matter not strictly naval

Chevalier, E, *Histoire de la Marine française* (Hachette, 1902); 5 rather anglophobic vols ending with 1870

Lacour-Gayet, G, *La Marine militaire sous Louis XIII et Louis XIV*

Lacour-Gayet, G, *La Marine militaire sous Louis XV*

Lacour-Gayet, G, *La Marine militaire sous Louis XVI*

(Champion, 1905–11). These three works contain valuable appendices of biographical notes.

Troude, O, *Batailles navales de la France*; 4 vols, old but indispensable: gives even single ship actions in the manner of William James, qv below, to the end of the Crimean War.

De la Roncière, C, & Clerc-Rampal, A, *Histoire de la Marine française* (Larousse, 1934), & Farrère, C, *Histoire de la Marine française* (Flammarion, 1956). These two books are chiefly useful for their profuse illustrations.

Jal, A, *Abraham Duquesne et la marine de son temps*

Delarbre, J, *Tourville et la marine de son temps;* gives a large number of his letters

De la Gravière, Jurien, Admiral, Guerres maritimes (Charpentier, 1906); 2 vols analysing the main campaigns of the Revolutionary and Napoleonic wars: with an appendix giving much of Villeneuve's correspondence

Desbrière, *Campagne de 1805* (Oxford University Press, in translation by C Eastwick); best read in translation, for the translator has added many valuable notes

LOIR, M, *L'Escadre de l'Amiral Courbet* (Paris, 1894)

GUÉPRATTE, ADMIRAL, *Expédition des Dardanelles*

THOMAZI, A, *La Guerre navale aux Dardanelles* (Payot): *Napoléon et ses marins* (Berger-Levrault, 1960): good on personnel and personalities

DOCTEUR, ADMIRAL, *Darlan: Amiral de la Flotte* (Editions La Couronne)

DE BELOT, *La Marine française pendent la campagne 1939–40* (Paris, 1954)

AUPHAN, P, *La Marine dans l'histoire de France* (Plon, 1955)

AUPHAN, P, & MORDAL, J, *La Marine française dans la Seconde Guerre Mondiale*; published in translation by A C J Sabalot by the American Naval Institute, Annapolis

MORDAL, J, *La Bataille de Casablanca* (Plon, 1952)

VARILLON, P, *Mers-el-Kebir* (Amiot-Dumont, 1949)

VARILLON, P, *Sabordage de la flotte à Toulon* (Amiot-Dumont, 1954)

LE MASSON, H, *Propos maritimes* (Editions Maritimes, 1970); a pleasant and useful collection of essays

LE MASSON, H, *The French Navy* (Macdonald, 1969); deals with the navy in the Second World War and is a sort of pocket JANE'S in two small vols, with a valuable introduction

The *Revue maritime* has for many decades provided a wealth of articles on all kinds of naval topics and events, and there has been much writing on naval problems by theorists such as Admiral Aube.

To the foregoing should be added, by authors writing in English:

JANE'S FIGHTING SHIPS (Sampson Low, Marston & Co); published annually since 1897, gives complete lists of French naval vessels with technical details of their armament and engines

LAUGHTON, J K, *Studies in Naval History* (op) is largely concerned with the French service.

MALLESON, COLONEL, *History of the French in India* (op); old but good.

MAHAN, A T, *Influence of Sea Power on History,* and *Influence of Sea Power on the French Revolution and Empire* (Sampson Low, Marston & Co); famous works, concerned a good deal with the French service

TOMPKINS, P, *The Murder of Admiral Darlan* (Weidenfeld & Nicolson, 1965) gives a good picture of the political confusion of the time.

HECKSTALL SMITH, A, *The Fleet that Faced Both Ways* (Blond, 1963)

There are many books dealing with the British navy, its battles and its commanders, and, since many of Britain's wars have been against France, most of them refer to the French navy to some extent. They will be found in the bibliographies mentioned above. The big standard work is *The Royal Navy,* edited by W L Clowes and published between 1897 and 1903. Its massive tomes can be massively brought up to date by adding the volumes of the official *Naval History of the Great War (1914–18)* by Corbett and Newbolt, and the five volumes of *The War at Sea* by Captain S W Roskill, which cover the Second World War. Many of the older works are still useful, such as W M James's exhaustive six-volume *History of the Royal Navy,* which deals almost entirely with the Revolutionary and Napoleonic wars; for the First World War there are H W Wilson's two-volume *Battleships in Action* and Gibson and Prendergast's *The German Submarine War* (Constable) – though A Moorehead's *Gallipoli* largely replaces the first of these. Though there seems to be no up-to-date biography of Lord Howe, there are such of most other leading admirals who fought the French; eg Carola Oman's *Nelson*, Tunstall's *Admiral Byng*, D Hood's *The Admirals Hood*, and R F Mackay's *Admiral Hawke* (Oxford University Press, 1965). There are good separate accounts of Quiberon, the First of June, and the Nile in Batsford's 'Battles' Series; and Dudley Pope's *England Expects* (Weidenfeld & Nicolson) gives a detailed picture of the confused fighting at Trafalgar. A T Patterson's *The Other Armada* (Manchester University Press) dealing fully with d'Orvilliers and the Invasion Plan, Captain S W Roskill's *Naval Policy between the Wars* (Collins), and J A Williamson's *The English Channel* are valuable. So are *The Most Formidable Thing* (submarines) by W Jameson (Hart Davis, 1965), F L Robertson's *Evolution of Naval Armament* (Storey, 1968), and *The Man of War*, by Macintyre and Bathe (Methuen, 1968). The big standard life of de Ruyter is an old one by Brandt; a shorter one is by P Blok, translated by Renier (Benn, 1933).

# INDEX

## INDEX

Ships and their commanders are not indexed where they are merely listed as units in a fleet or squadron, nor, as a rule, where the action or operation in which they took part was of small importance or their share in it was only a minor one. In indexing proper names the preliminary 'de' is usually, but not invariably, omitted – eg Grancey, Marquis de, but DuQuesne. As in the text the names of French ships are in small capitals, and the ships of other countries in italics.